Astrophysics and Space Science Library

Volume 470

The Astrophysics and Space Science Library is a series of high-level monographs and edited volumes covering a broad range of subjects in Astrophysics, Astronomy, Cosmology, and Space Science. The authors are distinguished specialists with international reputations in their fields of expertise. Each title is carefully supervised and aims to provide an in-depth understanding by offering detailed background and the results of state-of-the-art research. The subjects are placed in the broader context of related disciplines such as Engineering, Computer Science, Environmental Science, and Nuclear and Particle Physics. The ASSL series offers a reliable resource for scientific professional researchers and advanced graduate students.

Philip Judge • James A. Ionson

The Problem of Coronal Heating

A Rosetta Stone for Electrodynamic Coupling in Cosmic Plasmas

Springer

Philip Judge
High Altitude Observatory
Natl Center for Atmospheric Research
Boulder, CO, USA

James A. Ionson
Thayer Center, Suite C
JDC Inc.
Oakland, MD, USA

ISSN 0067-0057 ISSN 2214-7985 (electronic)
Astrophysics and Space Science Library
ISBN 978-3-031-46272-6 ISBN 978-3-031-46273-3 (eBook)
https://doi.org/10.1007/978-3-031-46273-3

Cover illustration: The image of the solar corona on the front cover of the book shows the corona as observed and processed by Shadia Habbal and Miloslav Druckmuller during the total eclipse of 2017 August 21. Red shows emission from ions of iron with 10 electrons removed, green with 13 electrons removed. In the center, the corona as seen in light of iron ions with 11 electrons removed is shown, from NASA's AIA instrument on the SDO spacecraft, obtained shortly before the eclipse image.

This Springer imprint is published by the registered company Springer Nature Switzerland AG
The registered company address is: Gewerbestrasse 11, 6330 Cham, Switzerland

Paper in this product is recyclable.

Preface

This book addresses a subject that is far from new. The first, 1881 edition of C.A. Young's *The Sun* discussed most of the solar and solar-terrestrial phenomena still studied in depth today. The astronomers of Young's era were able to record their results through developments in photography and spectroscopy. Young described the unusual spectrum from the mysterious corona observed during total eclipses. He reported the use of new, powerful spectroscopes invented earlier in the century by Fraunhofer, to identify unique signatures of Earthly and unknown elements in the light from the corona. Young also discussed solar flares, sunspots and their variations, prominences and the chromosphere sandwiched between the visible surface and the corona, and wrote suggestively about possible solar effects on Earth, such as spectacular aurorae and the curious behavior of telegraphs.

Young's writings are especially remarkable when viewed against the background views of at least one nineteenth century philosopher. Regarding stars, Logical Positivist Auguste Comte famously stated in 1835 ("The Positive Philosophy"):

> …nous ne saurons jamais étudier par aucun moyen leur composition chimique ou leur structure minéralogique, et, à plus forte raison, la nature des corps organisés qui vivent à leur surface.
>
> (We would never know how to study by any means their chemical composition, or their mineralogical structure, and furthermore the nature of beings living on their surfaces.)

Happily, after the identification of the new element helium first observed in the 1860s by Janssen and Lockyer, and following spectacular advances in physics at the beginning of the twentieth century, many of Young's observations were soon given solid elementary explanations. Thus, we must justify the addition of another monograph to the extensive literature on the solar atmosphere, 140 years after the publication of Young's book, and after so many modern advances.

Firstly, in spite of so many answers, the Sun still presents us with elementary scientific questions. It is widely known that the Sun can generally be understood as a fluid. Most who have flown in an aircraft or have watched the flow of a stream around obstacles will have a feeling for turbulence, a fundamental "unsolved problem" in physics. Under the visible surface, the Sun exhibits turbulent behavior as fluid motions are required to carry the flux of energy of core nuclear fusion outwards to where it is radiated into space. Fluid turbulence is readily studied on Earth, for example, within our atmosphere. However, the Sun provides us with our best natural laboratory for studying *magnetized plasmas*. Many of the peculiar properties described by Young, such as sunspots, result from the presence of a solar magnetic field. Variable in space and time, magnetic fields interact strongly with solar plasmas (ionized fluids). While plasmas have been routinely studied in laboratories since the 1950s, solar plasmas exhibit phenomena which, because of the large physical scales involved, do not occur in laboratory devices. Physicists say that solar plasmas evolve close to an *ideal limit* where non-linear effects control the coupled dynamics of magnetism and plasma. The detail with which we can observe the Sun makes it a rosetta stone for understanding the behavior of magnetized plasmas across the Universe. Some of the otherwise gravitationally-bound energy is linked to external plasmas through magnetic fields, leading to plasma heating, particle acceleration and large-scale dynamical fireworks. The solar corona, along with coronae of other stars, accretion disks, emission-line regions in active galaxies, galaxy-scale jets of plasma driven from black holes, all are examples of this *electrodynamic coupling*. Institutions worldwide now ask that scientists understand and predict the ejections of plasma, magnetism and high-energy radiation from the solar corona to protect a broad variety of valuable electrically-powered assets such as spacecraft, or even astronauts. The purpose of the present book is to re-examine one unsolved aspect of electrodynamic coupling, namely the heating mechanisms responsible for the corona, eight decades after Bengt Edlén demonstrated that the coronal plasma was far hotter than the surface.

Finally, the authors recognize that the research environment is changing rapidly with the maturation of massive networked computing, the advent of novel space probes and ground-based observatories, and emphasis on teamwork. With the growing influence of machine learning and AI algorithms, all offer hope of new understanding. But with these advances comes a risk. Not only is this environment daunting even for experienced scientists, but there is the threat that research will be driven by available tools, not basic questions, because of societal pressure. Risks of over-specialization and over-publication are real and present. Based firmly in physics understandable to undergraduates, this work aims to help creative young scientists find and solve a problem in natural science. Some of the opinions within the book are based upon decades of experience in the field, and upon detailed knowledge and reasonable interpretation of observational data. The reader may not always agree, but it is hoped that those more subjective opinions will stimulate further, healthy debate.

Bern, Switzerland
Oakland, MD, USA
August 2023

Philip Judge
James A. Ionson

Olivier: “I came to you for answers. Very specific answers”
Gersonides: “So you keep telling me. And I will repeat
the only answer I know. I have none. Not that I haven’t spent
the last forty years looking, but I find answers are as rare as
golden eggs or unicorns. All I can do is help you look for yourself.

—Iain Pears, *The Dream of Scipio*

Acknowledgments

PGJ thanks Prof. Dame Carole Jordan FRS for her generous and patient supervision of a naïve graduate student in the early 1980s, and Holly Gilbert for her gift of freedom to work on this project. This book began many years ago with Dr. Thomas Holzer of the High Altitude Observatory, and we thank him for his wisdom, knowledge and encouragement. Thomas Bogdan and Boon Chye Low, also HAO alumni, served as patient mentors to PGJ over three decades. PGJ is forever grateful to Terri for her patient support, love and companionship through life. The authors are grateful for helpful comments on the manuscript by friend and colleague Jean-Claude Vial.

Contents

List of Figures

List of Tables

Chapter 1
Introduction

1.1 This Book in a Nutshell

During the 1970s, the first surveys of the sky at UV and X-ray wavelengths uncovered a largely unanticipated abundance of sources hot enough to exceed the surface temperatures of most stars. Hot plasmas were discovered almost everywhere where gravity aggregated mass, because of electrodynamic coupling between the central object and its environs via embedded magnetic fields **B**. Of the myriad of objects detected, our best, most detailed view of magnetized plasmas is of the solar corona. It serves as our best laboratory for studying the physics of hot plasmas in astrophysics. It is also the seat of the solar wind and of phenomena such as jets and large mass ejections from the Sun, influencing the Earth, humanity and our electrical infrastructure, and the planets.

After 80 years, is the problem of coronal heating really "solved"? Maybe. Our understanding of the problem evolved dramatically during the 1970s. Data from SKYLAB and other sources revealed that the solar corona, shaped by magnetic fields, arises because of electrodynamic, not mechanical activity. But magnetic fields do no work on particles, and yet the plasmas are hot. The corona therefore arises from the joint action of magnetic and electric fields.[1] While the corona is threaded by large-scale magnetic fields, plasmas cannot easily sustain internal large-scale electric fields. Generally, irreversible heating must occur as the coupled plasma and large scale magnetic fields evolve, through the natural generation of electrical energy (currents driven by electric fields) and kinetic energy of differential bulk flows (through viscous dissipation) on small scales.

[1] Magnetic fields are, to order u^2/c^2, independent of the velocity u of the plasma as measured by an observer. Electric fields transform as $\mathbf{E}' = \mathbf{E} - \mathbf{u} \times \mathbf{B}$, which does no work in either reference frame if $\mathbf{E}$, the electric field in the frame moving with the plasma is zero (Chap. 2).

P. Judge, J. A. Ionson, *The Problem of Coronal Heating*, Astrophysics and Space Science Library 470, https://doi.org/10.1007/978-3-031-46273-3_1

Since SKYLAB we have accumulated a vast and diverse treasury of coronal observations, with greatly improved measurements of evolving vector magnetic fields at the solar surface. Our theories have advanced both analytically and with numerical experimentation. In fact, researchers have been so successful that over twenty heating models remain viable which might sustain the corona. Why then, is this new book needed?

To date, observational studies of heating mechanisms have related the structure and evolution of magnetic fields to coronal EUV and X-ray radiation (e.g. Fisher et al. 1998). They have sought causal links between the injection of energy into the corona on observable scales (>200 km), and the after-effects of its transport and dissipation on scales which can theoretically approach tens of meters where electrons and ions become dynamically independent. We still cannot follow observationally this complete range of scales which are intrinsic to the heating problem. Instead we are obliged to assume that the corona naturally develops the small scales at which dissipation can occur. The coupled transport and dissipation mechanisms are generally blithely ignored. Even when explicitly treated, too often dissipation occurs via a numerical algorithm which converts ordered energy to heat on scales far above physical scales. Consequently, results are contingent upon assumptions, premises, and unavoidable limitations in numerical algorithms. As a result we are left with a large number of unrefuted competing models.

In this book we develop a new strategy for solving the problem. Firstly the problem is explicitly defined, and its scope is by necessity reduced to that of accounting for steady, not explosive behavior. The literature on flares and coronal mass ejections (CMEs) is simply too vast. Steady coronal heating occurs seemingly independently of imminent explosions. But where explosive behavior can shed light on the heating problem, the text draws on lessons to be learned. The problem's challenges are then reviewed, and strengths and weaknesses of methodologies are combined to offer new opportunities for advancement. Magnetohydrodynamics (MHD) forms the theoretical foundation upon which the problem is framed, but kinetic physics is discussed as required in the weakly-collisional plasmas. By emphasizing refutation as a working methodology, various novel, information-rich observations are identified. The observation that active region plasma loops occupy so little of the available volume supports a theoretical proposition (Einaudi et al. 2021) that the corona is a self-regulating part of a larger system. But as a first step, the hydromagnetic lower boundary conditions for the corona must be better determined through thermodynamic and magnetic field measurements near the top of the Sun's chromosphere. Some robust theoretical predictions and unexplained properties of the corona lead us to suggest novel experiments and instruments.

Bayesian methods (Jaynes 1959; Jeffreys 1961) offer ways to assess the viability of multiple theories and identify the most likely models which remain compatible with data. It is argued throughout that much of the data acquired of the solar corona to date are "information-poor", additional data adding little new insight into mechanisms of coronal heating from what was acquired already by about 1980. Newer data may well inform on other important problems, such as the development of instabilities and subsequent coronal dynamics, but not on heating mechanisms per se. Here we define data as "information-rich" as those which strongly influence the Bayesian "posterior" probabilities, i.e. the probability of the hypothesis or model being compatible with the data. One goal of this book is to identify those data which will greatly influence the posterior probabilities of proposed coronal heating mechanisms. Then it is to be hoped that novel instruments, strategically selected to address the accompanying theoretical challenges, will vastly improve our understanding of the corona, fundamental processes such as MHD turbulence, and gravitationally-bound hot plasmas in the cosmos. With new capabilities from the Daniel K. Inouye Solar Telescope (DKIST, Rimmele et al. 2020), spacecraft in low solar orbits, and in AI-related computational science, we may yet find the corona's Achilles' Heels, and finally lay the problem to rest...

1.2 Hot Plasmas and Electrodynamic Coupling

The sky at night remains a source of wonder, when viewed away from the light pollution of cities. It spurred ancient civilizations to keep track of wandering stars, to associate these and the patterns of the "fixed" stars with legends and myths, to ponder at occasional "new" stars, and to use their motions as reliable time-keepers and calendars. Today we scrutinize these nighttime objects in exquisite detail, but always in the same essential way as our ancestors. We can only study those objects that are bright enough to see (an obvious but salient point given the accumulated evidence for universal dark matter). We also can scrutinize objects between us and these bright sources which affect the light transmitted towards Earth. A mundane example is the interstellar medium, a more exotic example might be the bending of light via gravitational lensing.

The raw energy allowing us to see across the cosmos comes from two sources: gravity and nuclear fusion. Gravity is a weak but ever-present attractive force, and nuclear fusion involves the conversion of rest-mass energy mostly into heat in the cores of stars through the strong nuclear force. The inexorable compression of matter by gravity leads to the enormous temperatures and pressures needed to trigger fusion deep inside the compressed objects.

A third force of nature is needed to transport this energy via light and radiation, the electromagnetic force. By far the dominant source of light in the Universe comes from the surfaces of stars as they rid the excess energy of nuclear fusion into space. The weak nuclear force is important in shaping the distribution of heavy elements in the Universe, through the radioactive decay of isotopes and neutron exposure of nuclei during, for example, supernova explosions. Electromagnetism dominates chemistry and biology, enabling us to exist and witness the night sky. It plays an additional role in astrophysics because magnetism, once generated, is difficult to destroy on cosmic scales. The coupling of magnetic fields to the free charges present in ionized fluids and plasmas is a natural consequence of the inexorable effects of gravity. In our Sun, these consequences are of increasing significance to a society dependent upon technology so wholly dependent on a stable electromagnetic environment.

The universal gravitational clumping of matter on scales of galaxy clusters to black holes is generally accompanied by the concentration of magnetic fields, because with even a small fraction of ionized matter, magnetic fields are dragged along with the embedded charges. Galaxies, stars and other objects inevitably host magnetic fields which naturally extend into surrounding space. Figure 1.1 shows two phenomena, both unanticipated from first principles, resulting from the coupling of ionized matter and magnetic fields. This "electrodynamic coupling" within the galactic nucleus of galaxy M87 and the Sun and its corona drives the enormous extra-galactic jet and the coronal mass ejection (CME) respectively.

We speak of electrodynamic, not merely magnetic coupling. Electric fields are not easily sustained in hot plasmas, they tend to be shorted out by the rapid motion of the free electrons within. Yet evidence for plasma heating and accelerated particles is abundant. Both these phenomena, as well as dynamic re-arrangements of plasmas through magnetic "reconnection", sometimes leading to impulsive "flares", require at least the ephemeral action of electric fields, because magnetic fields do no work on elementary particles. It is not yet clear if coronal heating entails these particular processes, or if it occurs even in their absence. This question will be of interest throughout this book.

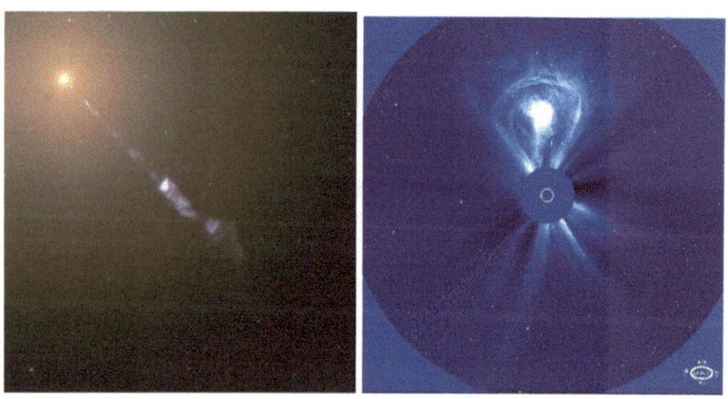

Fig. 1.1 The left panel shows the nucleus, diffuse stellar emission and jet of the active galaxy M87, obtained in visible light with the Hubble Space Telescope. The jet spans about 5000 light years (it is about 20× larger at radio wavelengths), about 5 10^{19} m. The right panel shows a coronal mass ejection observed with the LASCO instrument on the SoHO spacecraft. The CME is the "light bulb" shape on top of the disks at the center, the Sun is under the white circle. The CME is about 20 solar radii in radial extent (1.4 10^{10} m). The rays coming radially from the Sun are called streamers, denser regions of the solar wind emerging from above active regions on the solar disk. The jet and CME are both examples of electrodynamic coupling in astrophysics, both are cause by the transmission of energy via magnetic stresses from the gravitationally-bound object to the very hot plasmas of the jet and CME

Magnetic reconnection (e.g. Biskamp 2003) is encountered almost always when discussing the tenuous plasmas surrounding gravitationally bound objects like the solar corona. It is a dynamical process in which highly conducting plasma can slip across magnetic lines of force, breaking the condition of a fixed topology (Sect. 2.5). The initial ideas of Sweet (1958) and Parker (1957) were conceived to explain the fast rearrangements of magnetized plasma observed during flares. Reconnection is frequently associated with plasma heating, as if heating is inevitable from the process. It may indeed be, but if so it must be accompanied by other processes to convert the flows induced through the changing topology into heat. Later we will visit the problem of the stability of current sheets within systems large enough to be almost ideal, both in large-scale sheets assumed in the Sweet-Parker picture and those developed naturally in MHD turbulent calculations. We will find in Sect. 4.9 that plasmoids form naturally, which have built-in smaller-scale reconnections approaching scales of dissipation (Huang and Bhattacharjee 2010, see Fig. 1.10).

Returning to the solar corona, remarkably, the majority of stars emit X-rays (Schmitt 2013), and as a by-product of modern instruments devoted to finding extra-solar planets, many show evidence of stellar flares (Cliver et al. 2022). Neither X-ray emission or flares would occur in the absence of *electrodynamic coupling*. X-ray emission is one result of electrodynamic heating of plasmas to form a "corona", so-called because of the resemblance of the bright structure surrounding the moon to a crown during total eclipse, Fig. 1.2. Flares result from large-scale reconfigurations of plasma and magnetic field following the slow build-up of magnetic energy within coronae. This book is concerned with understanding why the Sun, as the best studied star, must possess heated plasmas from irreversible energy dissipation, while maintaining sufficient free energy for dynamic flares and coronal mass ejections (Fig. 1.1).

Compact objects such as black holes, neutron stars, white dwarfs and some large diffuse nebulae exhibit some different but physically related effects of electrodynamic coupling. The energy densities associated with neutron stars and black holes spontaneously generate pairs of anti-particles, emitting copious X- and even γ-rays (Blandford and Znajek 1977). Galactic-scale jets of plasma emerge along the axes of rotation of massive black holes in a qualitatively different manifestation of electrodynamic coupling (Fig. 1.1) which results from higher levels of symmetry than are present on our slowly rotating, middle-aged Sun.

This book reviews the electrodynamic coupling between the solar interior and the corona. It is our closest example of the coupling phenomena accessible to detailed observations, a local "Rosetta Stone" signaling our way to a better understanding of the broader astronomical landscape (e.g. Stencel and Ionson 1979; Niedner et al. 1981; Ionson and Kuperus 1984; Kuperus and Ionson 1985).

We will focus on the long-unsolved—or perhaps solved—problem of coronal heating, in relation to the more general problem of dynamics as revealed to us in the solar wind, flares and coronal mass ejections.

Fig. 1.2 The solar corona radiates with the brightness of the full moon, around the limb of the new moon during a total eclipse on 18th May 1901. The data were processed by M. Druckmüller from scans of photographic negatives in the possession of the High Altitude Observatory. The Sun's spot cycle was at a minimum at this period

1.3 An Old Puzzle

The Sun's corona is a near-perfect magnetized plasma. A plasma consists of large numbers of free positive and negative charges (usually ions and electrons), treated using classical mechanics. The dynamical evolution of plasmas presents a special kind of "many body problem", which, because of the huge number of possible particle interactions, has no closed-form solution. The free charges arise usually because an ordinary gas has been heated to the extent that electrons, otherwise tightly bound to the atomic nuclei, are knocked out of bound states by the heat (rapid random motion of constituent gas particles) and sent into free motion. In turn these free electrons can, with sustained heating, knock more electrons off until there are few electrons left bound to any ions. The ions are at least 1836 times heavier than electrons, but contain at most 92 bound electrons, so that the plasma's mass is almost all in the ion nuclei. The lighter electrons move almost instantaneously to short out any electric fields from local net charges in a particular volume. Inside the plasma it remains neutral on average, with no large-scale electric fields sustainable in the rest frame of the plasma. In contrast, magnetic fields permeate plasmas because of the absence in nature of magnetic monopoles, hypothetical magnetic "charges" on which magnetic lines of force might theoretically begin and end (like electrical charges, Parker 2007). Yet the Sun possesses magnetic fields, and so the corona is a magnetized plasma (Parker 2007). Figure 1.2 shows an image of the Sun's corona surrounding the limb of the moon during an eclipse in 1901.

Spectroscopic observations have shown the solar corona to be over 100 times hotter than the visible solar surface. We also know that it also is heated somehow by virtue of possessing the magnetic field, which often shapes the corona into familiar patterns seen in iron filings around a bar magnet. The goal of this book is to examine why the solar corona must be hot enough to be essentially in a fully ionized state, and understand if and why we still face a "heating problem".

The puzzle is that the corona is evidently heated, yet it is also a "nearly-perfect" plasma. "Perfection" here is taken in the hydromagnetic sense of a large magnetic Reynolds number Rm= uL/η in a macroscopic fluid flow of characteristic speed u, length scale L and magnetic diffusivity η (Tables 1.1 and 1.2, these quantities will be made clear in Chap. 2). It means the corona offers almost no resistance to the flow of electric current, like a copper wire. In contrast, the bars of a household resistive heater glow red with heat, because the power station's electromotive forces drive a current through this very poor conductor. But before reaching this resistor, the electrons flow easily along all the conducting wires from the power station. Within the fire's heating element the kinetic energy of the electron flow is converted by friction (particle collisions) into heat at the familiar rate given by Ohm's law: I^2R. Coronal plasma offers such a weak resistance (large conductivity σ) that the corresponding volumetric heating j^2/σ is negligible, unless the plasma finds a way to generate intense electric currents $\mathbf{j}$ associated with unobservably small scales of the permeating magnetic field. Nor is it sufficiently viscous to cause frictional heating by gradients in fluid speeds (it typically has a large Reynolds number R= uL/ν_0, L = physical size of the corona, ν_0 = fluid viscosity), except for particular, magnetic field-aligned compressive fluid motions. This comparison between electric fire and corona fails to acknowledge the fact that the resistance on the one hand is fixed within a wire, but on the other hand is spread throughout a malleable fluid. But still, given such nearly perfect fluid properties, the central puzzle of the corona is simply,

> If the corona possesses so little friction, why then is coronal plasma heated from a few thousand K to over a million K?

We will see that such near-perfection leads naturally, like the literary tragic hero, to its own downfall.

Table 1.1 Physical constants and solar data

Symbol	Physical quantity	Value
e	Electron charge	$1.602\ 10^{-19}$ C
m_e	Electron mass	$9.109\ 10^{-31}$ kg
m_H	Proton or hydrogen mass	$1.673\ 10^{-27}$ kg
k_B	Boltzmann's constant	$1.381\ 10^{-23}$ J K^{-1}
μ_0	Permeability of free space	$4\pi\ 10^{-7}$ Hm^{-1}
ϵ_0	Permittivity of free space	$(36\pi\ 10^9)^{-1}$ Fm^{-1}
c	Speed of light in vacuo = $1/\sqrt{\mu_0\epsilon_0}$	$3\ 10^8$ m s^{-1}
a_0	Bohr radius = $\frac{4\pi\epsilon_0\hbar^2}{e^2 m_e}$	$5.292\ 10^{-11}$ m
R_∞	Rydberg energy = $\frac{e^2}{8\pi\epsilon_0 a_0^2}$	$2.180\ 10^{-18}$ J
G	Gravitational constant	$6.670\ 10^{-11}$ N m^2 kg^{-2}
$R_\odot$	Solar radius	$6.9599\ 10^8$ m
$M_\odot$	Solar mass	$1.989\ 10^{30}$ kg
$L_\odot$	Solar luminosity	$3.826\ 10^{26}$ W
$F_\odot$	Solar bolometric flux density	$6.284\ 10^7$ W m^{-2}
	Spatial size of 1 arcsecond at the Sun	$0.725\ 10^6$ m
$\overline{\mu} = \sum_i \frac{m_i}{m_H} n_i / \sum_i n_i$	Mean mass per solar particle (mostly neutral case)	1.26
	(mostly ionized case)	0.60
$\overline{\mu}_H = \sum_i \frac{m_i}{m_H} n_i / n_H$	Mean mass per solar hydrogen nucleus	1.36
n_{tot}/n_e	Total density ÷ electron density (ionized plasma)	1.92
$\dot{M}$	Mass loss rate in solar wind	2–3 10^{-14} $M_\odot$ year^{-1}

Note: Physical constants are from NIST. Solar data are taken from Allen (1973)

1.4 A *Very* Short History

The atmosphere of a star is defined as those outer layers from which radiation readily escapes. Historically, stars were almost the only source of light observable in the Universe. Analyses of their visible spectra formed the basis for astrophysics. Building on Saha's extension of quantum statistics to include effects of ionization (Saha 1920), Cecilia Payne's detailed analysis of stellar spectra (Payne-Gaposchkin 1925) first revealed the true state of atoms and atomic ions in plasmas, and the chemical composition of elements in the Universe. She set the stage for the study of stellar atmospheres, with the solar atmosphere at stage center, to become of prime importance in humanity's desire to understand the cosmos. Still today, the James Webb Space Telescope will rely on our understanding of stellar atmospheric radiation to shed light on the nature of the very early universe, by measuring properties of the first generations of stars in distant galaxies.

By the latter half of the twentieth century, the evolution of stars and of the chemical elements were broadly understood (Burbidge et al. 1957). Contemporary spectroscopic data were limited to visible and near infrared light, where the continuum and overwhelming majority of spectral lines originate from the *photospheres* of stars like the Sun, those deeper atmospheric layers from which over 99.99% of the stellar luminosity is radiated. But already by the 1920s, it was widely known that the cores of the most opaque spectral lines, for example, Fraunhofer's H and K lines of singly ionized calcium, Ca^{+} emerge from far higher atmospheric layers. It was totally impossible for the light from ions such as Fe^{9+} to be generated by standard photospheric models with temperatures near 6000 K, built upon the well-established principle of radiative energy conservation (Schwarzschild 1906). Earlier measurements during total solar eclipses had revealed brighter emission in the Ca^{+} and other so-called "chromospheric" lines, and a geometric chromospheric thickness far above the predictions of photospheric models (e.g. McCrea 1929a,b; Menzel and Cillié 1937). Narrow-band photographs of the cores of these lines showed structures quite different from the photospheric convective granulation (Hale and Ellerman 1904), instead revealing long-thin structures, later related to a much larger honeycomb-like network pattern identified with "supergranulation" (Simon and Leighton 1964). Early chromospheric work is concisely reviewed in terms of physical ideas by Bray and Loughhead (1974), building on basic insights into the incomplete coupling of radiation and plasma (non-LTE -non local thermodynamic equilibrium) by Thomas (1965).

By 1943 Edlén (1943) had identified mysterious "coronal lines", already reported in spectra obtained in the nineteenth century during total eclipses, as belonging to atomic ions with 8 or more electrons removed. A modern picture revealing this extraordinary truth at Extreme Ultraviolet Wavelengths (EUV) is shown in Fig. 1.3. By seeking extensions to Payne-Gaposhkin's theory of ionization appropriate to lower density plasmas, it was revealed that such highly charged ions are created only when plasma temperatures exceed 1 million K (Woolley and Allen 1948; Miyamoto 1949). In 1948 Tousey and associates used captured V2 rockets to acquire UV spectra and even X-ray images of the Sun (Tousey et al. 1951; Tousey 1953). Finally, the Soviet spacecraft Luna 1 and 2, on emerging from the umbrella of the Earth's magnetosphere, confirmed Parker's 1958 prediction of a hot, steady wind of plasma from the Sun.

Well before these developments, in 1859 British observers Richard Carrington and Richard Hodgson independently witnessed a sudden, small brightening on the solar disk. This phenomenon turned out to be the first report of a solar flare. American Charles Young and others had recognized that this event might be related to brilliant aurorae observed over the following days (Young 1892). Not only this, but it was suspected that subsequent disturbances of Earth's magnetic fields and even interference with the telegraph systems of the day, seemed perhaps to be more than mere coincidence.

Taken together, observations point to the conclusion that the outer regions of solar and stellar atmospheres do not behave according to rules largely applicable to their photospheres immediately below. In a pivotal article with the understated title, "On the Probable Existence of a Magnetic Field in Sun-Spots", Hale (1908) discovered the first fingerprint of magnetic fields in the Sun's atmosphere. Hale invoked the Zeeman effect to detect magnetic fields directed along the line of

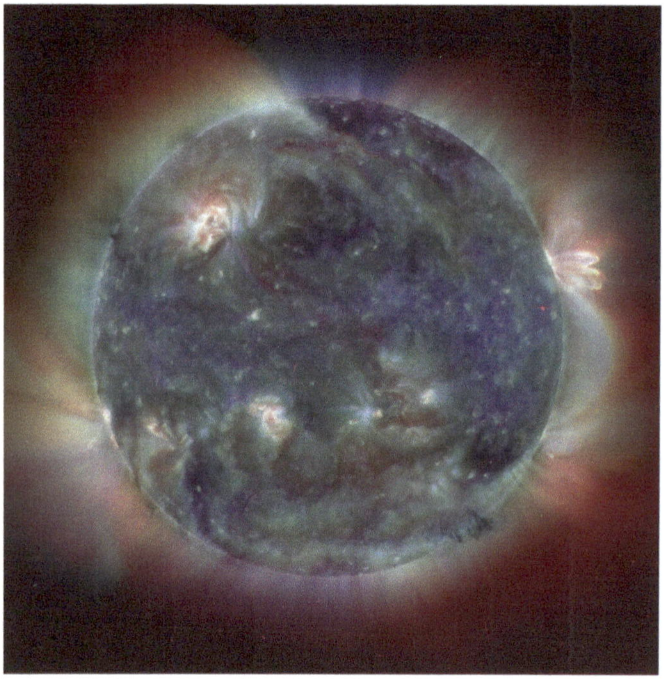

Fig. 1.3 A three-color image of the corona as seen in three near-contemporaneous images from the EIT instrument on the *SOHO* spacecraft, acquired in May 1998. Blue represents Fe^{8+} and Fe^{9+} emission at 17.1 nm, yellow Fe^{11+} emission at 19.5 nm, red Fe^{14+} emission at 28.4 nm (with a small contribution from He^{+} at 30.4 nm). On large scales, the corona is thermally "organized" in the sense that there are huge areas of 17.1 nm emission (blue) physically separated from areas of 19.5 (yellow) and 28.4 (red) nm emission. The active regions and smaller "bright points" show mixed temperature material, when observed at the $\approx$4 Mm EIT angular resolution. Image taken from the *SOHO* web site (http://sohowww.nascom.nasa.gov/)

sight of the observer (the "longitudinal" Zeeman effect, Lites 2000). Curiously, these fingerprints are present in spectra of sunspots shown by Figure 36 of Young (1892), a clear observation of the Zeeman effect before Zeeman's work in the laboratory (Zeeman 1896). We now understand that these magnetic fields emerge from processes deep in the solar interior, most likely because of a hydromagnetic *dynamo* (e.g. Parker 1955; Cattaneo 1997; Ossendrijver 2003; Charbonneau 2020). The Sun's convection zone and atmosphere is the physical system where this magnetism converts small amounts of the benign, steady and life-giving energy into rapidly-varying UV- and X- radiation, and into fast ejections of plasma and magnetic fields (see Figs. 1.7 and 2.4). These effects can cause havoc for our global electrical infrastructure, as first suggested by contemporary reports of magnetic anomalies on Earth following the 1859 Carrington event (e.g. Young 1892; Eddy 2009).

The purpose of the present work is, however, to pare down the vast literature on the higher, more tenuous regions of the Sun's atmosphere, with the aim of revealing the reasons why the Sun must develop such a hot corona. In Chap. 3 we will suggest, on observational grounds, that the heating problem is best studied separately from large dynamic re-arrangements of magnetic structure arising from magnetic reconnection, which leads to flares and coronal mass ejections (CMEs). Dynamic phenomena on observable scales will be discussed further only when needed to address the heating problem. As we will see, this simplification might be regarded as heresy or canon, depending on the individual scientist, such is our profound lack of knowledge. In Sect. 1.7, a working definition of acceptable solutions to the coronal heating problem is given. Of all criteria, the power requirements of the corona are merely a first hurdle, which will be seen to be readily satisfied by many mechanisms already proposed. Power requirements as measured over time are discussed in Appendix B.

We recall that without plasma heating on unresolved scales, the atmosphere would be essentially neutral. Plasma heating is therefore a requirement to permit coupling of ion and electron fluids to the magnetic field, before any large-scale dynamics can occur (Fig. 1.1). In this sense, a satisfactory theory of coronal heating is a pre-requisite for building our understanding of coronal dynamics on large scales, and of consequent terrestrial influences from the Sun.

1.5 Reductio Ad Minimum

The regions from which photons escape from the solar interior into space are collectively called the Sun's atmosphere. Our first task is to identify *typical* physical conditions to establish a basis upon which to build physical understanding of the Sun's atmosphere. If the reader is unfamiliar with basic solar properties and jargon, they might refer to books by Foukal (1990), Stix (2004), Golub and Pasachoff (2009). Those in need of an introduction to plasma physics might consult the excellent pedagogy of Kulsrud (2005). Like Earth's atmosphere, the pressure initially drops exponentially with height, but with a scale height of 0.1–0.15 Mm (1 Mm $\equiv 10^6$ m). The solar radius is 700 Mm, thus the "edge" of the Sun appears sharp to eyes which see light from the solar photosphere. Figure 1.4, intended as *a starting point only*, shows the extension of the photosphere outwards through the chromosphere, before a sudden increase of temperature into the corona. There is no such structure on the Sun, the model is constructed to match some average features in the UV spectrum (see Appendix B). Observations reveal that the atmosphere is dynamic, theory requires dynamical processes to transport energy to be deposited as heat in the chromosphere and corona. Nonetheless, the figure illustrates the steep average drop in the total plasma pressure p_T that must occur across the photosphere and chromosphere, as well as the relatively constant electron pressure p_e above the photosphere. The average pressures and (especially) temperatures become less meaningful as height increases above 1–2 Mm. Their rms variations exceed the means owing to the dynamical structuring by inhomogeneous magnetic fields discussed below. On top of this essential stratification, other properties can be

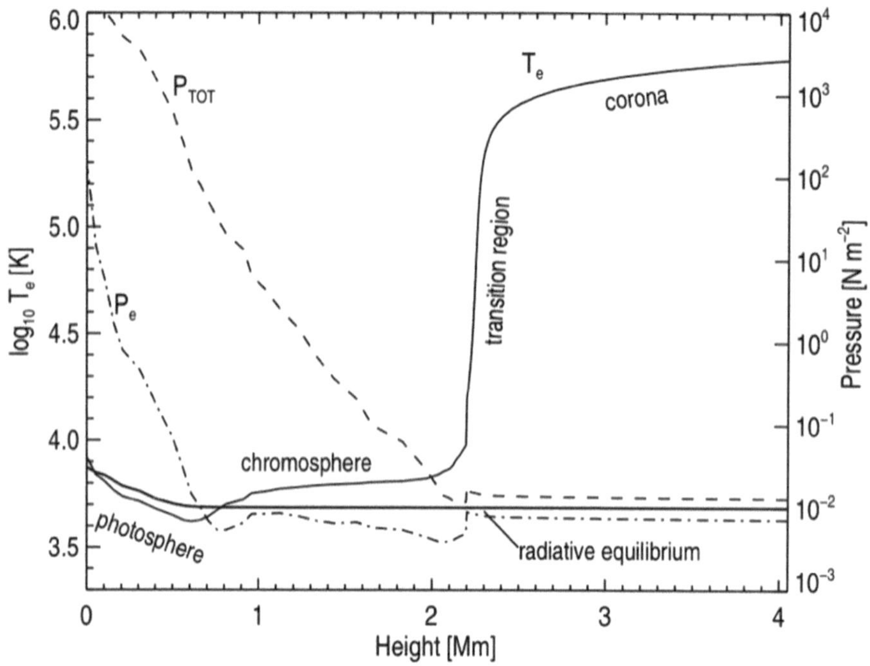

Fig. 1.4 Thermal properties of a 1D model atmosphere which reproduces just *space- and time-averaged* UV, EUV and X-ray intensities of the quiet Sun. No structure exists on the Sun exactly like this (see this model superposed on the distribution of pressures versus height in Fig. 1.6). In particular the small masses of plasma above 10^4 K mean that above heights of $\approx$2 Mm, the atmosphere is more likely to be horizontally inhomogeneous and potentially dynamic, with large variations around the average shown. Data are from Pietarila and Judge (2004), except below 2 Mm, the structure is an atmosphere which fits UV observations from Vernazza et al. (1981). Appendix B describes the two methods used to construct this model. The temperature of a radiatively heated atmosphere approximated by a gray opacity is the solid line labeled "radiative equilibrium". Atmospheric temperatures above this line require *non-thermal heating*

identified based upon repeated observations and/or well-understood physics. These include:

(I) The Sun's central temperature is 15 million K, dropping to below 5000 K at the visible surface (photosphere). Edlén (1943) showed that the temperature then increases to above half a million K in the more tenuous corona (Fig. 1.6). Heat flows from hot to cold, so a source of energy other than thermal radiation, conduction or convection is required to be deposited as heat within the corona and chromosphere. This puzzle constitutes the heating problem.

(II) The photosphere spans continuum optical depths[2] in the 500 nm continuum (mostly H^- opacity) $\tau_{500} \approx 4$ to $\tau_{500} \approx 0.04$ (Allen 1973). With known properties of radiative equilibrium hydro-statically-stratified models, the gas pressure varies from $p \approx 10^4$ to 100 N m^{-2} (Vernazza et al. 1973, and references therein, see Fig. 1.4). Photospheric plasma is generally well-described by local thermodynamic equilibrium (LTE) conditions, where knowledge of electron temperature and density suffices to set the local thermal state of the atmosphere (Mihalas 1978). Just 1 in 100 to 1 in 10^4 photospheric particles are ionized (see the pressures shown in Fig. 1.4).

(III) Coronal plasma pressures of $p \approx 0.02$ Nm^{-2}, 4 orders of magnitude smaller than photospheric pressures, are inferred from the coronal brightness (white light and EUV), brightness measurements of optically thin radio emission (Kundu 1965; Mercier and Chambe 2015) and density sensitive emission line ratios (Mason 1991). Typical electron temperatures T_e lie between 1 and 2 MK (mega-Kelvin, 1 MK $\equiv 10^6$ K), based on the broadened photospheric spectrum scattered from electrons, and decades of analyses using UV, EUV and X-ray spectra primarily sensitive to T_e (the first physically complete calculations were presented by Pottasch 1964; Seaton 1964; Cooper 1966). The corona is almost fully ionized, with less than 1 in 10^6 particles being neutral. Coronal plasma is optically thin to all radiation except at radio wavelengths.

(IV) Without reference to geometry or dynamics, we can identify the observed chromosphere as partially-ionized plasma mediating the large pressure drop (100–0.02 Nm^{-2}, spanning 9 pressure scale heights), and gradual increase in ionization, between upper photosphere and corona. Multiple spectral lines of abundant elements are optically thick in the chromosphere.

(V) Similarly, the change from neutral to fully ionized plasma occurs across "the transition region", in which diverse physical structures contribute to spectral features formed between $\approx 10^{4.3}$ and $10^{5.7}$ K (Mariska 1992, recently re-analyzed by Judge 2021). In the quiet Sun, these structures contain relatively little mass, $\lesssim 5\ 10^{-6}$ kg m^{-2}, even compared with the corona ($5\ 10^{-5}$ kg m^{-2}), and far less than the chromosphere.

Most spectral lines are optically thin across the transition region, exceptions include H Lyman lines, EUV lines of helium, and several resonance lines of abundant elements such as carbon and silicon.

(VI) The lightweight transition region plasmas are subject to forcing from above as well as below, so the variability of emission from them is correspondingly large. *There may perhaps be no meaningful permanent transition region*, instead there is a highly variable transition in ionization state from chromosphere to corona. Again, this might be considered heresy or canon.

[2] Optical depth τ, measured from the observer, is a metric of depth downwards into the atmosphere, measured in photon mean free paths. The probability that photons will emerge into space is $e^{-\tau}$.

(VII) The power emerging from photosphere, chromosphere and corona in identifiable radiation (from infrared to X-rays) is 6.3 10^7, $\approx 10^4$ and $\approx$10–100 W m^{-2} respectively (typical values are listed in Table 1.2). Without non-radiative heating, neither chromosphere nor corona would exist (see the curve for "radiative equilibrium" in Fig. 1.4).

(VIII) The total radiative power observed in transition region lines depends on whether H Lyman-α is included as a chromospheric or transition region line. Lines other than of hydrogen radiate $\approx$ 40 W m^{-2}, Lyman-α $\approx$ 400 W m^{-2}. In either case, transition region lines typically emit as much or more than the quiet corona. Thus the loss of energy from the quiet corona due to heat conduction by electrons to lower temperature plasma is an order of magnitude faster than in-situ coronal radiative losses (Table 1.3, see also Fig. 3.7).

(IX) Spectra of trace elements ("metals" in the astronomical sense, including C,N,O,... Fe etc.) show that the entire atmosphere is "well mixed" (Woolley and Allen 1948), with modest variations reviewed, for example, by Laming (2015). The atmosphere is stirred and mixed by fluid motions caused by sub-surface convection. Thus elemental abundances are not highly fractionated in layers according to atomic mass, unlike in Earth's stratospheric and higher layers. This does not imply that fractionation is uninteresting or unimportant. But we will ignore it in our discussion of coronal heating, because it is primarily of chromospheric origin (Laming 2015).

(X) Doppler velocities from widths and shifts of optically thin lines from the chromosphere and above are generally sub-sonic, the adiabatic speed of sound varying from 6 to over 100 km s^{-1} in the plasmas from 6000 to 2 10^6 K. Widths of lines in coronal plasma, outside of dynamical events, require non-thermal velocities ξ_0 generally between 0 and 30 km s^{-1} in addition to thermal widths (see Appendix A for a definition of ξ_0 and other details).

(XI) The widths of optically thin emission lines, reflecting plasma motions at unresolved scales, almost always exceed line shifts (resolved motions) in the chromosphere and corona. One exception is in falling plasma that has cooled dramatically in the corona ("coronal rain"), a particularly clear example is shown by Schad et al. (2016).

(XII) The Sun and its atmosphere are not demonstrably different from stars of similar age, mass, and metallicity (e.g. Gustafsson 1998; dos Santos et al. 2016; Judge 2018, 2023).

Figure 1.5 shows some of the best images of photosphere and corona obtained to date, on the same scale. They were obtained 8 years apart and are intended to show mainly the different quality of the two types of data. Each pixel might be viewed as its own instance of a 1D atmosphere like that of Fig. 1.4. But clearly these instances must be very diverse to reproduce such different structure across these images. The bright features in the corona are directly related to magnetism, as are the far smaller, narrower, bright and tortuous features embedded in dark regions between

Table 1.2 Typical physical parameters of the corona

Physical quantity	Chromosphere	Corona	Units	α
Thermal properties				
T_e electron temperature	7000	10^6	K	3
n_e electron density ($n_p \approx 0.8n_e$)	10^{17}	$3\ 10^{14}$	m^{-3}	20
p_e electron pressure $n_e k_B T_e$	0.01	$4\ 10^{-3}$	$N\,m^{-2}$	60
p total pressure $nk_B T$	100 – 0.01	$7\ 10^{-3}$	$N\,m^{-2}$	60
ρ mass density	$4.2\ 10^{-6}$–$1.0\ 10^{-10}$	$7\ 10^{-13}$	$kg\,m^{-3}$	20
ξ_0 rms line width FWHM/$2\sqrt{\ln 2}$	2–10	20	$km\,s^{-1}$	1
c_S adiabatic sound speed $\sqrt{\gamma p/\rho}$	6300	$1.5\ 10^5$	$m\,s^{-1}$	1.7
λ_D Debye length $\sqrt{\epsilon_0 k_B T_e/n_e e^2}$	$1.8\ 10^{-5}$	$4\ 10^{-3}$	m	0.4
Plasma parameter $\frac{4\pi}{3}\lambda_D^3 n_e$	$2.5\ 10^3$	$8\ 10^7$		1.2
ω_{pe} electron plasma frequency $\sqrt{k_B T_e/m_e}/\lambda_D$	$1.8\ 10^{10}$	$9.8\ 10^8$	$rad\,s^{-1}$	4
Electron inertial length (skin depth) c/ω_{pe}	0.02	0.3	m	0.25
ω_{pp} proton plasma frequency $\sqrt{k_B T_p/m_p}/\lambda_D$	$4.2\ 10^8$	$2.3\ 10^7$	$rad\,s^{-1}$	4
Ion inertial length c/ω_{pp}	0.72	13	m	0.25
τ_{ee} electron self collision time $1.6\ 10^4\ T_e^{3/2}/n_e$	$1.0\ 10^{-7}$	0.05	s	0.25
τ_{ep} electron momentum changing time	$\approx \tau_{ee}$			
τ_{ep}^{ε} electron-proton energy exchange time	$1.6\ 10^{-4}$	80	s	0.25
τ_{pp} proton self collision time $1.4\ 10^6\ T_p^{3/2}/n_p$	$8.0\ 10^{-6}$	4.7	s	0.25
τ_{pH} proton-hydrogen collision time	$3\ 10^{-8}$–$8\ 10^{-4}$	$1.8\ 10^5$	s	0.67
$\lambda_e = \sqrt{3k_B T_e/2m_e}\ \tau_e$ (electron mfp)	0.024	$1.2\ 10^5$	m	0.4
$\lambda_p = \sqrt{3k_B T_p/2m_p}\ \tau_p$ (proton mfp)	0.034	$2.3\ 10^5$	m	0.4
$\kappa_\parallel$ (heat flux $q_\parallel = \kappa_\parallel T_e^{5/2} \nabla_\parallel T_e$)	0	10^{-11}	$W\,m^{-1}\,K^{-7/2}$	1
η_0 proton viscosity coefficient $9.0\ 10^{-18} T_p^{5/2}$	$3.7\ 10^{-8}$	$9\ 10^{-3}$	$kg\,m^{-1}\,s^{-1}$	16
ν_0 viscous diffusivity η_0/ρ	$9\ 10^{-3}$–370	$1.3\ 10^{10}$	$m^2\,s^{-1}$	0.8
σ_0 electrical conductivity $3.9\ 10^{-4}\ T_e^{3/2}$	230	$3.9\ 10^5$	$S\,m^{-1}$	5.2
η magnetic diffusivity $1/(\mu_0\sigma_0)$	$3.5\ 10^3$	1.8	$m^2\,s^{-1}$	0.2
P_M magnetic Prandtl number ν_0/η	$3\ 10^{-6}$ – $1\ 10^{-1}$	$7\ 10^9$		4
Magnetic properties				
B typical field strength	0.01–10^{-3}	10^{-3}	T	20
β plasma/magnetic pressure $2\mu_0 p/B^2$	3–$2\ 10^{-2}$	$2\ 10^{-2}$		0.15
c_A Alfvén speed $B/\sqrt{\rho\mu_0}$	$4.2\ 10^3$–$9.0\ 10^4$	$1.2\ 10^6$	$m\,s^{-1}$	4.5
ω_e electron gyro frequency	$1.8\ 10^9$–$1.8\ 10^8$	$1.8\ 10^8$	$rad\,s^{-1}$	20
$r_e = \sqrt{3k_B T_e/2m_e}/\omega_e$ electron gyro radius	$2.3\ 10^{-4}$–$2.3\ 10^{-3}$	0.03	m	0.09

(continued)

Table 1.2 (continued)

Physical quantity	Chromosphere	Corona	Units	α
ω_p proton gyro frequency	$9.6\ 10^5$–$9.6\ 10^4$	$9.6\ 10^4$	rad s^{-1}	20
$r_p = \sqrt{3k_B T_p/2m_p}/\omega_p$ proton gyro radius	0.01–0.1	1	m	0.09
$^\ddagger\omega_e\tau_e$	110–11	$7.8\ 10^6$		1
$^\ddagger\omega_p\tau_p$	0.028–0.69	$2.2\ 10^5$		1
"Closed" fields (loops)				
L chromospheric scale height, coronal loop length	$1.2\ 10^5$	$1.5\ 10^8$	m	0.67
F_M energy flux density needed to sustain plasma	$1.4\ 10^4$	$6\ 10^2$	W m^{-2}	15
Avg. heating rate $\varepsilon = F_M/(m = p/g)$	$1.4\ 10^5$	$1.5\ 10^7$	W kg^{-1}	0.25
R Reynolds number $\xi_0 L\rho/\eta_0$	$3\ 10^6$–0.3	200^a $5\ 10^{13a}$		1
S Lundqvist number $c_A L\mu_0\sigma_0$	$1.4\ 10^5$–$3\ 10^6$	$8\ 10^{13}$		15
L/c_S sound crossing time	20	330	s	0.4
L/c_A Alfvén crossing time	10-1	140	s	0.15
Rad. time scale $t_{rad} \approx \frac{3}{2}n_{tot}k_B T_e/[n_e^2 P_{rad}(T_e)]$	90	$3\ 10^4$	s	0.15
Conductive time scale $\approx \frac{3}{2}n_{tot}k_B T_e/(F_M/L)$		830	s	2
"Open" fields				
H_{HSE} pressure scale height	$1.2\ 10^5$	$5\ 10^7$	m	...

Note: $T_p \approx T_e$ is assumed in the table. $^\ddagger\tau_e$ and τ_p are the collisional lifetimes of electrons and protons resulting from all collisions. For the chromosphere, when two values are listed they correspond to heights of 700 and 2100 km of model 'C' of Vernazza et al. (1981), with rough values of B (Fig. 1.6). n_e, n_p and t_{rad} are roughly constant within the chromosphere. ε, the heating rate per particle is 100 times larger in the corona than the chromosphere (Anderson and Athay 1989).

[a]Compressible and incompressible cases. To estimate data for active regions, multiply the coronal values with the factor α, for example, $T_e = 3\ 10^6$ K in active regions.

photospheric granules. The hugely different angular resolutions of the two images would make the identification of connections between photosphere and corona next to impossible. But we can nevertheless add to the above list of properties:

(XIII) Measurements using the Zeeman effect have shown that magnetic fields threading the photosphere (outside of sunspots) are organized by granular motions (Livingston and Harvey 1971; Lin 1995). Magnetic fields are buoyant and so collect, like flotsam and jetsam over a drain, along narrow (< 200 km) regions of rapid convective down-flows (within the dark lanes of Fig. 1.5). There they are intensified, stretched and contorted by eddies until magnetic (Maxwell) stresses ($\propto B^2$, B = magnetic field strength) come into balance with Reynolds stresses when $B \approx 0.1 - 0.2$ T. Magnetic flux densities averaged over several granules are $\approx 0.0001 \rightarrow 0.001$ T.

(XIV) Emerging into the stratified chromosphere (Fig. 1.6), these narrow magnetic concentrations extend upwards, expanding horizontally as the magnetic

Table 1.3 Representative energy budgets for chromosphere and corona

	Energy flux density (W m^{-2})		
Region	Quiet Sun	Corona hole	Active region
Photospheric radiation	6.3 10^7		
Chromosphere	1.4 10^4	1.4 10^4	$\approx$7 10^4
Spicules (surface average)	10^3	$>10^3$	...
Corona total	8 10^2	5 10^2	$\approx 10^4$
Heat conduction %	70	8	1–100
Radiation %	30	1	50
Wind flow %	<17	90	<1

Coronal data were taken from Withbroe and Noyes (1977), Withbroe (1988), chromospheric data from Anderson and Athay (1989). The latter are larger than those of Withbroe, and more realistic, because they include radiation losses from many lines of iron. Active region (plage) chromospheric data were estimated by scaling quiet Sun chromospheric radiation losses by a factor of 5 (Withbroe and Noyes 1977; Chapman 1981).

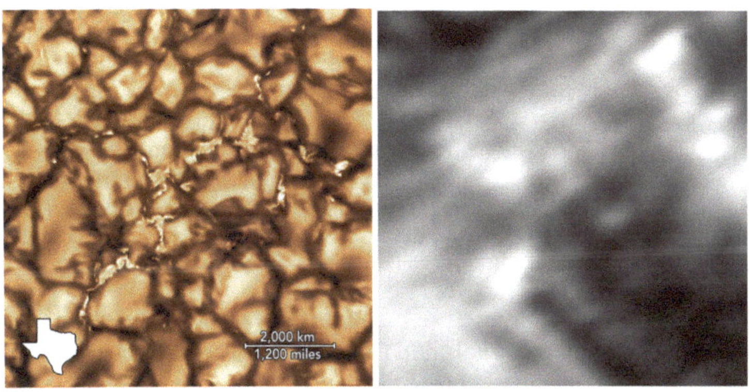

Fig. 1.5 On the left is the first image released, on January 29 2020, by the DKIST project of NSO. It shows the solar photosphere, spanning an area 8.2×8.2 Mm, $\approx 10^{-5}$ of the solar surface. Granules slightly larger than Texas (which is shown in white) have bright cores between dark lanes. Inside the lanes are small (<200 km wide) eddy-like bright features which mark areas of strong magnetic fields penetrating the photosphere from beneath. On the right is a sum of 10 images of coronal emission at 17.2 nm obtained using the Hi-C rocket payload on 11-Jul-2012, again covering 8.2 × 8.2 Mm^2, *but of a different target*, with an angular resolution of at best 0.3". These represent the highest resolution images of the solar disk photosphere and corona ever obtained. The horizontal bar shown spans 2000 km: this happens to be close to the entire vertical extent of the stratified solar atmosphere beneath the corona

stresses dominate over the rapidly falling Reynolds stresses and plasma pressure (Giovanelli and Jones 1982; Jones and Giovanelli 1983).

(XV) The brightest chromospheric and coronal structures are always associated with strong (up to 0.4 T) magnetic fields emerging from the interior, associated with sunspots. But not all strong field regions have bright coronal structures above them (see Sect. 5.6).

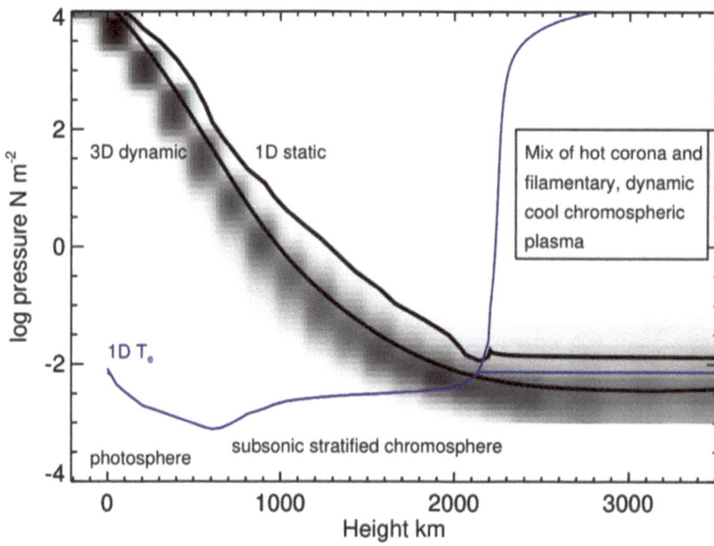

Fig. 1.6 The drop of pressure with height in a hydrostatic model is compared with a time-averaged pressure from a 3D dynamical model, shown as the probability distribution function and median value of pressures derived from the 3D Bifrost model "en024048_hion" developed for the IRIS project (http://sdc.uio.no/search/simulations, Carlsson et al. 2016). Two types of structure are present—a near-hydrostatically stratified component containing the overwhelming fraction of plasma, and dynamical extensions above heights of 2200 km, which fill just a fraction of the volume available. This figure has been adapted from an original by Judge (2015)

(XVI) Around sunspots the whole atmosphere is structured, both horizontally and vertically, by the intense magnetic fields.

(XVII) The morphology of magnetic plasma structures between photosphere and chromosphere/corona changes dramatically (Figs. 1.5 and 1.8).

1.6 First Inferences

We can already draw conclusions from these elementary inferences. Points (**III**) and (**IX**) reveal that the corona is a fully ionized plasma (with less than 1 neutral particle per million) in which free electrons are highly mobile. Thus a macroscopic electric field cannot be sustained in the plasma's frame without an imposed electric field. Any magnetic field threading the atmosphere is frozen into it on observable scales ($>$20 km), owing to the high electrical conductivity that accompanies the high electron mobility (Chap. 2). Point (**IX**) also implies that, as a first approximation, the corona is a *"simple (hydrogen-electron) plasma"*, containing less than 10% of other elements by number.

Trace elements (C,N,O,... Fe) contribute both mass and channels for loss of energy via radiation. Hydrogen and helium are poor radiators at photospheric and coronal temperatures. The corona is visible in Figs. 1.5 and 2.2 because of inelastic collisions of electrons with iron atoms with 8 or 9 electrons removed, followed by radiation emitted into space. But energy is lost mostly from the corona via downward electron heat conduction and further radiative energy losses by trace species at lower temperatures (Table 1.3).

The bulk of the chromosphere is steeply stratified by a near-balance between gravity and plasma pressure, because of (**IV**) and (**IX**). This may be surprising to some, because it is "well known" that

> the chromosphere is a highly complex, finely structured, and dynamic region ...—de Pontieu et al. (2021)

However, if we accept (**X**) and (**XI**), then it is clear that the vast bulk of the chromosphere is *not* dramatically dynamically structured (heresy for some, canon for others). The point of contention concerns not whether the atmosphere is static (it is not), but how one interprets temperature- and velocity-sensitive data of chromospheric spectral lines. We contend that the line-of-sight motions observed are almost always sub-sonic, therefore the vertical stratification cannot depart much from the hydrostatic case. There are infrequent supersonic motions seen in the strongest spectral lines formed in the highest, most tenuous atmospheric levels, which have naturally attracted much attention. In the plane-of-sky, observed motions are associated with field-aligned flows, or waves (see the morphology of the image of the chromosphere in Fig. 1.8). These structures draw a lot of attention because of the high contrasts of variations across the magnetic field (examples include Athay 1974; Damé 1985; Athay 1986; Tsiropoula et al. 1994; Bruls and Von der Lühe 2001; Judge et al. 2011; Carlsson et al. 2019; Henriques et al. 2020), which arise from the highly anisotropic energy transport in magnetized chromospheric plasma (see Sect. 2.6). They can reflect both fluid motion and apparent motion from phase coherent wave fronts. The chromosphere sets the lower boundary conditions for the corona. Therefore it is important to understand the meaning of this difference in opinion. The argument is as follows.

From the 1970s, the High Resolution Telescope and Spectrograph (HRTS) instrument measured limits on Doppler shifts and widths of optically thin emission lines in the chromosphere (Bartoe and Brueckner 1975; Athay and Dere 1991). Doppler motions were only occasionally supersonic, so that gas plasma pressures $p = nk_BT$ dominated over Reynolds stresses (components of the Cartesian tensor $\rho u_i u_j$, the product of fluid density ρ and components of the fluid velocity u_i). Figure 1.6 shows the structure of a 1D, hydrostatically- stratified atmosphere, including the sub-sonic but ad-hoc "microturbulent" pressure component, from Vernazza et al. (1981). The distributions of temperatures and pressures are also rendered, taken from a reference 3D dynamical model of Carlsson et al. (2016), built using the BiFROST computer code (Gudiksen et al. 2011). BiFROST solves time-dependent equations of motion including sub-photospheric convective motions which buffet magnetic fields inserted throughout the calculation. Yet, as is clear from the run of $p(h)$, the two calculations exhibit the same behavior. These "contradictory" perspectives are reconciled noting that the chromosphere spans many scale heights, of which only the highest—and therefore lightest plasmas—become supersonic in response to evolving hydro-magnetic forces. Of the components identified in Fig. 1.4, we noted above that the transition region contains the smallest mass. Viewed in this way, the "transition region" has so little inertia that forces from above or below, for instance those generated by magnetic reconnection, will have far larger effects than similar forces elsewhere in the atmosphere. This fact might also account for its obvious and highly cited "explosive" behavior (Mariska 1992; de Pontieu et al. 2021). It is even possible that a permanent transition region may not even exist (this is a new but mostly a heretical view, see Sect. 5.12).

The argument that the typical chromosphere has a very intermittent rise in temperature in time (Carlsson and Stein 1995) applies to weakly-magnetic regions which show large amplitude 3 minute intensity oscillations, quite unlike more magnetic regions which appear more constant and brighter (Lites et al. 1993). In some publications one can gain the impression that the entire chromosphere is highly dynamic (e.g. Carlsson et al. 2019). Yet both observations and calculations show that supersonic dynamics occur mostly above the bulk of the chromosphere, which, while not static, moves sub-sonically, and mediates the drop in pressure from photosphere to corona from 10^4 to 10^{-2} N m^{-2}. In the limited sense that the average pressure drop in the dynamic model is close to that of the 1D static atmosphere over so many scale heights, we may say that the average chromosphere is approximately given by a hydrostatic stratification. This kind of stratification is encountered in all numerical experiments which do not expressly deal with large-scale forcing or chromospheric reconnection (e.g. Arber et al. 2007).

Given (**II**), (**III**) and (**XVII**), we can infer that photospheric and lower chromospheric magnetic fields are forced by sluggish fluid motions (notice characteristic eddies in Fig. 1.5). Higher in the chromosphere, and in the corona, the fluid is forced by magnetic fields, evidenced by fibril-like morphology in the upper chromosphere (Fig. 1.8) and corona (Fig. 2.2). Not only does the ratio of fluid to magnetic stresses (plasma β) change from > 1 to < 1 within chromospheric plasma (supported by measurements of, e.g. Metcalf et al. 1995), but when $\beta \ll 1$, the coronal magnetic field has a dominant large-scale laminar component to which the plasma is frozen. The natural conclusion is that the corona must be controlled by electrodynamic coupling, not just mechanical coupling[3] through fluid dynamics, to and through the photosphere. Figure 1.7 introduces the electrodynamic coupling picture frequently referenced throughout this text.

Above we argued that the *chromosphere* must concern us greatly (see Washinoue et al. 2022, for a similar sentiment, but formulated in a different sense). We will discuss evidence that, without first understanding the chromosphere, we will forever be left asking, "what heats the corona?". The chromosphere is a plasma with a dilemma—how is it to handle the awkward transition from fluid dominance below

[3] A mechanical coupling would require $\beta > 1$ throughout (Ionson 1982).

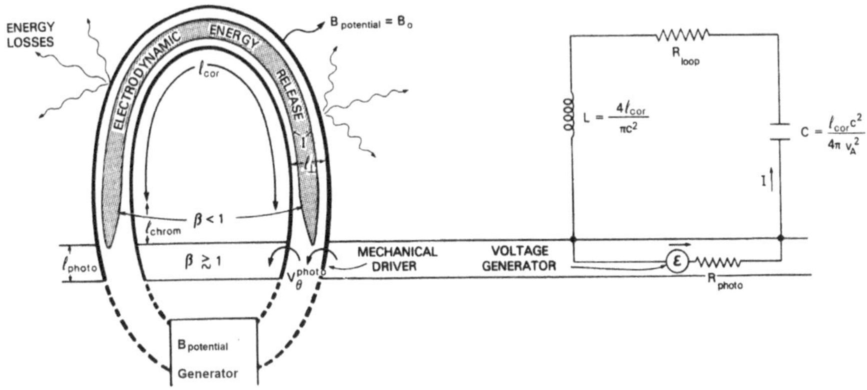

Fig. 1.7 The problem of magnetic heating of the solar corona is represented as a sketch and with the "LRC circuit" analogy (Ionson 1982). Mechanical energy under the photosphere generates a long-lived potential magnetic field, which is perturbed by photospheric (and chromospheric) motions, injecting stochastic magnetic energy via a variable Poynting flux with a component directed along the time-averaged potential field. The Poynting flux is converted to irreversible heating (represented by resistor R) and to induced kinetic energy of motions (fluid flow and electric currents), represented by inductor L and capacitor C. Ionson (1984) later extended the analogy to include slow, "direct current", and non-periodic variations far from the LRC circuit resonance frequency

(plasma $\beta > 1$), to magnetic dominance above ($\beta < 1$), while maintaining the global condition of frozen-fields-and-plasma (Chap. 2)? How does it mediate the supply of mass to the solar wind and corona, while at the same time handling non-LTE radiation transport in three dimensions, ion-neutral physics, and departures from ionization equilibrium? It is one of the most challenging parts of the solar atmosphere (Judge 2010; Carlsson et al. 2019). Perhaps the existence of the pesky chromosphere is a huge reason why we still face a "coronal heating problem". Some problems particular to the chromosphere are discussed in Sect. 3.9 (Fig. 1.8).

1.7 The Nature of the Heating Problem

We declare the heating problem for the corona to have been solved when credible mechanism(s) for the generation, transport and irreversible dissipation of energy have survived refutation against critical, information-rich experiments.

The challenge of this heating problem that we face today, is that *many* mechanisms have yet to be refuted by experiment, to the extent that at least one solar physicist has asked, impishly, "with so many mechanisms, why is the corona so

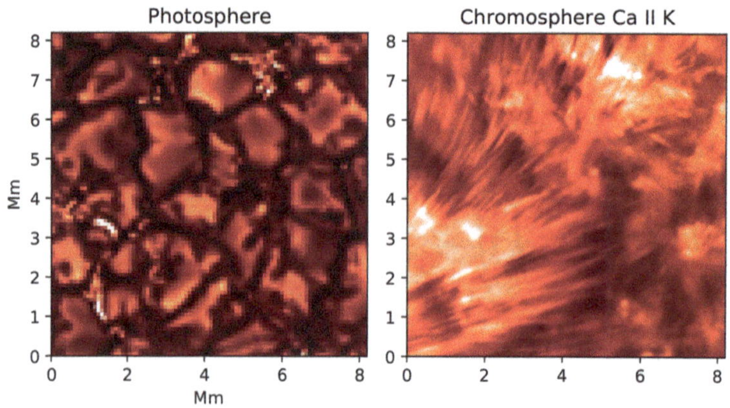

Fig. 1.8 A 8.2×8.2 Mm region of quiet Sun acquired by V. Henriques using the 1 m Swedish Solar Telescope (SST), on 25 May 2017. The dramatic change in magnetic structure is seen between the photosphere (left) and chromosphere (right). The chromospheric image was obtained through a narrow wavelength band centered at 20 km s^{-1} to the short wavelength side of the Ca II *K* line at 393.3 nm. Owing to non-LTE radiation transport, the image does not represent the brightness of plasma moving upwards at 20 km s^{-1}

cool?[4]" Naturally the Sun knows of no heating problem. To illustrate the nature of this problem, the following quotes are extracted from abstracts of refereed articles published from 2017 to 2022:

> *S* may play a role in heating the coronal plasma
> *T* are likely to be closely related to the heating of the upper solar atmosphere
> the *U* process is sufficient to account for the observed coronal heating
> The success of *V* shows the central role that this process plays for the structure, dynamics, and heating of the corona.
> Both experimental estimates and theoretical considerations still leave room for a variety of very extravagant ideas.
> ...magnetic energy release during *W* suggests that *X* can also power the quiet corona.
> *Y* may be heating the hot plasma component of these active regions
> *Z* corresponds to an energy flux larger than that needed to heat the corona. Although no direct conclusions can be drawn, it seems...

where $S \ldots Z$ represent different physical mechanisms. The final quotation reveals the honest and essential truth. Not only do multiple proposals for heating remain unfalsified, but firm conclusions are rare.

The inability to reach firm conclusions has many parallels in the history of science. In one example, to address an increase in "pseudo-science" (particularly in psychology) in the early twentieth century, Karl Popper (1902–1994) famously proposed that for a theory to be scientific, it must be *falsifiable*. While there is no general agreement on methodologies in science, including any requirement

[4] The answer chiefly lies in the dominance of electron heat conduction, which is non-linearly dependent on electron temperature in coronal plasmas, e.g. Spitzer (1956), Braginskii (1965).

for falsification, it seems that the heating problem might benefit from Popper's approach. Therefore in Chap. 4 we will review commonly applied methods from Popper's perspective. But first we ask in Chap. 3 why, after eight decades of research with huge advances in technology, we still face such problems. We will also review limitations inherent in remote sensing, extreme ranges of scales, challenges presented by non-linear and non-local processes, difficulties in dealing with the chromosphere, ill-posed problems, and stochastic behavior. In spite of these issues, the Sun remains our most accessible "laboratory" for studying enormous, magnetically-controlled hot plasmas across the universe.

However, even the scales at which ordered energy is dissipated as heat are yet to be identified, either from first principles, laboratory or numerical experiments, or observations. The answers will depend on the nature of particle distribution functions within the plasma, which we certainly do not know a priori. Mean free paths as listed in Table 1.2, derived assuming distribution functions which are small perturbations to Maxwellian distributions (Spitzer 1956; Chapman 1956; Braginskii 1965), are $\sim$200 km under coronal conditions. For a coronal scale height of length 50 Mm the Knudsen number (the ratio of particle mean free path to macroscopic length) is 0.004. But such short mean free paths apply only to particles with energy $\varepsilon \lesssim k_B T$ where T is a measure of the width of the distribution, assumed to be near-Maxwellian. These particles constitute just the low-energy "cores" of the distribution functions where most of the particles reside in velocity space. But charged particles have cross sections varying as impact energy ε^{-2}, mean free paths as ε^2. Any particle with an energy of $\gtrsim 5 k_B T$ will have a Knudsen number > 1: the particle is almost collisionless along the entire loop. Under these conditions kinetic physics would become critical, but for the particles in the core of a distribution MHD may apply. Herein lies the difficulty— it is a particularly nasty case of Joseph Heller's "Catch 22" (Heller 1961). To determine the appropriate physical regime to use in the dissipation process, we must know something of the distribution function. But the dissipation process may well determine the distribution functions. All we can say at this juncture is that, *if* there is no large population of high-energy particles in the coronal plasma, *and* compressive motions along the total magnetic field are induced, *then* dissipation scales may be as large as 100 km, MHD scales (Appendix D). If not, then it is likely that dissipation occurs on kinetic scales, on the order of 10 m (the ion inertial length, Table 1.2).

Progress is nevertheless being made in this challenging area through numerical experiments. Matthaeus et al. (2020) have reviewed how one might describe the transport of energy from MHD to kinetic scales by combining arguments from MHD turbulence with numerical experiments in the collisionless Boltzmann equation (Vlasov equation). The authors highlight terms in the Vlasov equation and identify "pathways to energy dissipation" in a hydrogen plasma. Instead of frictional heating by particle collisions converting bulk kinetic energy into heat, "collisionless heating" involves the speeding up of particles via electric fields, frequently called "wave-particle" interactions. This alternative view of the journey of energy from large to small scales will be discussed in Chap. 3. This remains an active research area (see sections 11 and 14 of Schekochihin 2022). Given this situation, we will later turn the problem around, and seek signatures of tails in distributions of particles from sensitive observations of radiation from high energy electrons (Sect. 4.5), and in the far wings of emission lines of coronal ions (Sect. 5.9).

With the advent of powerful computers, many advocate for a third pillar of science through "numerical experimentation". But like observations, these too suffer from a limited resolution and range of scales, and natural magnetized plasmas generally requiring three dimensional calculations. The ratio of largest to smallest scales currently span a factor of $\lesssim 10^3$, not solar values of 10^8, in 3D numerical MHD experiments extending from photosphere to corona (e.g. Przybylski et al. 2022, used a uniform grid of $504 \times 504 \times 840$ cells, spanning $24 \times 24 \times 16.8$ Mm). In short, observations cannot trace the flow of magnetic energy close to dissipation scales, and computers cannot, without special efforts, trace the flow over more than 3 decades in scale length. Our tools are too blunt, and will remain so before the advent of quantum computing, for direct refutation of models using experiments. Nevertheless, earlier work has yielded answers to aspects of the general heating problem (Chap. 4), encouraging further research which is hopefully better-tailored to fit the problem at hand. For example:

- Based upon observed chromospheric spectral line widths, and assuming near-hydrostatic stratification (Sect. 1.6), acoustic (sound and low-β slow mode) waves do not carry sufficient energy (Athay and White 1978). Nor can they describe observed large variations in coronal heating between strong- and weak-magnetic regions.
- Using analytical theory, Ionson (1978, 1982, 1984, 1985b,a) followed the flow of energy injected by photospheric motions into the inhomogenous corona in a unified analytical approach analogous to an LCR electrical circuit (Fig. 1.7). Ionson showed that heating rates resulting from surface waves living within internal surfaces of gradients in local Alfvén speed $c_A = B/\sqrt{\mu_0 \rho}$ (see page 48 for a definition), are essentially independent of unknown (and unobservable) dissipation rates. We will elevate this independence to a *conjecture*, in order to expand its application to all heating mechanisms on page 55. The same independence is assumed in the spontaneous formation of current sheets (Parker 1994, Sect. 4.7), and numerical work (Gudiksen and Nordlund 2005).
- At and above routinely measurable scales ($\gtrsim$0.7) Mm, coronal plasmas have been statistically related to magnetic and velocity measurements in the photosphere, in the hope of relating measurable Poynting fluxes to heating rates. For example, Fisher et al. (1998), Mandrini et al. (2000) pitted such scalings for tens of parameterized models. Such studies have revealed, however, that observable properties (i.e. large scale) tend to yield answers contingent upon assumptions, and which fail to discriminate between different scenarios. Fisher et al. (1998) found that X-ray properties all "highly correlated with all of the [macroscopic] global magnetic variables", such as coronal loop lengths, plasma densities, temperatures. Similar work compiled by Aschwanden (2001) listed 27 different qualitative scenarios, but concluded that more detailed models were required. The models were loosely defined and heavily endowed with assumptions. Also,

the Poynting fluxes were estimated using photospheric data, assuming in essence that the chromosphere always passes the same fraction of Poynting flux to the overlying corona. Nevertheless, several otherwise "reasonable" scenarios appeared to be incompatible with extant observations, so some progress has been made.

- Insight from numerical experiments has led to new and useful phenomenological descriptions of MHD turbulence (Schekochihin 2022). For example Rappazzo et al. (2008), augmented Parker's (1994) theory of coronal heating, which was based upon the general non-existence of continuous equilibrium solutions, by following the dynamical evolution towards discontinuous solutions (internal current sheets, reviewed by Pontin and Hornig 2020). Their work used the convenient framework of *reduced MHD*, ("RMHD" Kadomtsev and Pogutse 1974; Strauss 1976), an approximation which assumes dominance of a guide field $\mathbf{B}_0$, with consequent simplifying assumptions enabling the more accurate study of magneto-plasma configurations where scales parallel to $\mathbf{B}_0(\mathbf{r})$ greatly exceed those perpendicular to it. RMHD is intrisically an incompressible approximation.
- Numerical modelers have argued nevertheless that direct, full MHD numerical simulations from sub-photosphere to corona (beginning with Gudiksen and Nordlund 2005) are sufficient to identify energy transport mechanisms. In such calculations with large numerical diffusion coefficients ($\gtrsim 10^9$ m^2 s^{-1}, cf. Table 1.2), dissipation occurs rapidly, and generally it is believed that the dissipation will occur no matter the sub-grid physical processes (again, see the conjecture on page 55). In low-β plasmas with strong guide fields $\mathbf{B}_0$, anisotropic MHD turbulent velocities and magnetic fields in the plane perpendicular to $\mathbf{B}_0$ (Montgomery and Turner 1981; Goldreich and Sridhar 1995) are invoked to transport energy down through the inertial range to arrive at dissipation scales (Rappazzo et al. 2007, 2008; Einaudi et al. 2021). Einaudi et al. (2021) note that

> Since the energy input is dependent on both the external forcing and the internal dynamics, the corona is a self-regulating forced system.

The potentially important implications of this view will be explored throughout this text. The question remains, though, whether these numerical calculations, missing important sub-grid processes, survive challenges from future information-rich data, or if other sub-grid dynamical interactions must be invoked to account for the entire journey to dissipative scales. After all, this forced system illustrated graphically and schematically in Figs. 2.4 and 2.4 has a reactive as well as resistive branch. In the particular calculations of Einaudi et al. (2021), the numerical setup with strong magnetic guide fields anchored at the ends of a 3D rectangular box, with periodic boundary conditions along the other two directions, tend to suppress large-scale dynamics which might disrupt and redirect the flow of energy away from ever smaller dissipative scales.

Reconnection is one frequently-invoked dynamical process where plasma topology changes with consequences at large and small scales (Biskamp 2005; Priest and Forbes 2007; Kulsrud 2011). If it occurs below scales accessible to

MHD, it has consequences for the dissipation processes and in observations of associated plasma motions (see Sect. 5.8). Another is the recently uncovered helicity barrier, both will be discussed later. A fascinating argument for self-regularization of the corona was presented by Uzdensky (2007), based upon different speeds of reconnection in fluid and kinetic regimes.
- Techniques from artificial intelligence have yet to be widely applied to the heating problems. Among many possible applications to the coronal heating problem, they offer the possibility of coupling the dynamics across enormous ranges of scales. Significant advances have been made which explicitly coupled large and kinetic scales (e.g. Daldorff et al. 2014; Makwana et al. 2017, 2018).

These examples raise more questions than they answer, a healthy situation for fruitful future research on this old problem.

1.8 One Sun, Multiple Coronae

The reference model of Sect. 1.5 is quite inadequate in capturing the diversity of phenomena observed in the solar atmosphere. Here we briefly review the diverse nature of coronal structures in the remaining two space dimensions and time.

1.8.1 Quiet Sun and Coronal Holes

The Sun's quiet visible surface is organized mostly by the sub-surface convection, punctuated and threaded by highly intermittent magnetic fields. It exhibits rotation at the equator systematically faster than at the poles by about 40%. The overlying chromosphere is statistically similar across the entire quiet Sun. The corona, however, clearly is organized into two large but separate types of structure: *the quiet corona and coronal holes.* Modest changes in chromospheric intensities under coronal holes can be seen with some difficulty in spectral lines such as those of helium, whose formation is unusually sensitive to energies of tens of eV, more characteristic of coronal than chromospheric plasma (e.g. Kahler et al. 1983; Pietarila and Judge 2004). Magnetic fields emerging through the quiet solar photosphere plasma are believed largely to return to the solar surface forming loops of various scales. In coronal holes, a much larger fraction of photospheric magnetic fields appear to extend into space (e.g. Linker et al. 2017), areas where radial magnetic fields are, locally, predominantly of one sign. Within both these different oceans of coronal plasma lie small ($\lesssim 5''$) bright knots of emission first resolved into loops by Walker et al. (1988), they can be considered as small multi-polar active regions.

In both quiet Sun and coronal holes, differences in the driving motions appear minimal, as a working hypothesis many assume that the striking differences in the coronal plasma lie only in whether magnetic fields return locally to the Sun or extend

Table 1.4 Typical transition region and coronal pressures

Region	Plasma pressure p (N m^{-2})	Corona[a] $\log T_e$ (K)
Quiet Sun	0.01	6.1
Coronal hole	0.01	5.9
Plages/active region	0.05	6.5

[a]From Noci (2003), except that a rough value for plages is listed

into space, drawn outwards by the solar wind. The solar wind itself is a consequence of the heating whose mechanisms we seek, the interstellar medium being unable to halt the expansion of over-heated plasma along the "open" magnetic flux bundles (Parker 1958). Table 1.3 lists the power needed to sustain quiet corona and the corona and wind within a coronal hole. The two types of structures have been recognized for decades and various conclusions concerning heating mechanisms have been drawn (e.g. Axford and McKenzie 2002; Verscharen et al. 2019). We will return to these after discussing some more basic challenges in the next chapter.

In representative 1D models the increase in T_e between chromospheric and coronal plasma occurs over ~100 km, smaller than pressure scale heights. Thus a useful reference is the average pressure p of observed transition region plasmas. Pressures commensurate with typical properties in Tables 1.2 and 1.3 are listed in Table 1.4.

These pressures are important also because they also represent the pressure of the coronal plasma contributing most to its radiative losses, and the EUV and X-ray emission that we observe. These pressures can be directly used in investigations from crude scaling laws (e.g. Eq. (2.23)) to sophisticated quantitative calculations of heating processes (e.g. Ionson 1978).

1.8.2 Active Sun

A third class of corona is associated with the sunspots, bipolar and/or multipolar concentrations of magnetic flux which come and go on an 11 year cycle. In the photosphere sunspots appear dark. Outside of the dark spots is a less dark penumbra, itself surrounded by brighter "plages". Measurements of the Zeeman effect in photospheric spectra reveals their darkest regions ("umbrae") to be intense bundles (or "tubes") of magnetic flux (Hale 1908), which emerge as the unipolar legs of large (> 30 Mm) magnetic bipoles. Plages are brighter regions with lower densities of magnetic flux of mostly the same magnetic polarity of the nearest spots. The entire atmosphere of a sunspot is strongly perturbed by the magnetic forces, and even static representative models reveal major differences between active and quiet conditions (Maltby et al. 1986). The darkness of spots arises because heat-carrying convection is suppressed by the intense magnetism.

Images of coronal emission lines reveal plasma loops between umbrae, penumbrae and plages, and occasional flares. Plasmas above sunspots have temperatures between 10^6 and almost 10^7 K, expanding above and between the photospheric spots. Remarkably, the surface brightness in X-rays was found in SKYLAB data to be independent of the size of the active region. Essentially complete "loops" were observed over limb active regions, from traditional "transition region" lines ($10^{4.3}$ to $10^{5.7}$ K), clearly inconsistent with the naïve picture (Fig. 1.4) of a thin transition region (e.g. Foukal 1975, 1976, 1977; Bray et al. 1991). "Movies" of active region coronal plasma obtained since the launch of SoHO (Domingo et al. 1995) also show the evolution of the corona over periods from tens of seconds to days, in response to the emerging and evolving magnetic flux. Like quiet Sun and coronal hole plasmas, active coronal plasmas appear to glow steadily but more brightly from the excess plasma heating. However, unlike the quiet Sun, they sometimes undergo rapid dynamical changes in response to changes in topology on large scales through magnetic reconnection. The most energetic such changes lead to flares and coronal mass ejections. For reference, Figs. 1.9 (upper sketch) and 2.3 show the original model of flares, in which oppositely-directed fields of Sweet (1958) and Parker (1957) reconnect. The intense current system formed as the oppositely directed fields approach one another is dissipated as heat via the electrical resistance of plasma in the current layer. The dynamic plasma emerges at close to the Alfvén speed in the previously absent reconnected fields that are subject to the newly unbalanced magnetic tension force.

While a useful pedagogical tool, current sheets like the Sweet-Parker configuration can be unstable to the tearing mode instability in resistive MHD (Furth et al. 1963). The plasmoid instability is the special case of the tearing instability where the thickness a is set to the steady-state Sweet-Parker width $a = \delta_{SP} = LS^{-1/2}$ for arbitrary length L. If such a sheet could develop along length L (i.e. two magnetic flux systems were brought into contact along $L \approx 150$ Mm), the growth rate would be $10^{7/2}/140 \sim 20$ per second. Given the times of several minutes over which such structures are buffeted by dense plasma below, this implies that such long and narrow sheets cannot exist in the corona, they are rapidly disrupted by plasmoids. Figure 1.10 shows in panels (a)–(d) a numerical MHD calculation of this kind of structure and evolution, from Huang and Bhattacharjee (2010). The initially thick current sheet naturally thins from (a) to (b), slowly, whence the tearing instability begins in panel (c) and enters a non-linear phase in (d), with "X"-points and the beginnings of secondary current sheets. Panels (e) and (f) are examples of the "plasmoid" instability, accompanied by reconnection and the formation of plasmoids within the secondary sheets. Evolution from (d) to (e) is much faster than the primary tearing instability by a factor near 10^3 in this particular calculation made with $S = 3\ 10^6$, above the critical level for plasmoid formation of $\sim 10^4$.

Noting the peculiarly narrow geometry and sparse distribution of bright active region loops, Litwin and Rosner (1993) argued that whatever is heating the active coronal plasma has a high threshold. Otherwise active regions should be full of hot plasma loops. Whether such a threshold occurs in the driving motions at the top of the chromosphere, and/or within the corona itself is unclear. Litwin and Rosner noted that, in soft X-rays, the "switch-on times" for a loop to brighten is

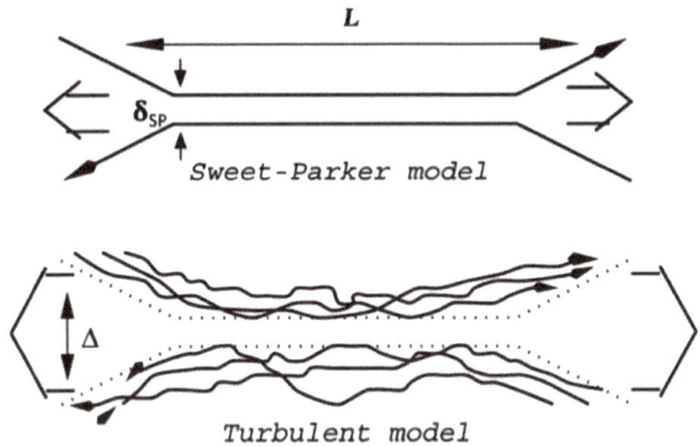

Fig. 1.9 A Sweet-Parker laminar current sheet (top) is contrasted with a current sheet consisting of wandering field lines arising naturally from frozen-in and evolving fluid turbulence. Magnetic fields frozen-in can rapidly reconnect breaking the frozen-in condition, leading to further turbulent motions, and so forth. Both the parallel and transverse fields form on small scales leading to rapid reconnection independent of resistivity. Instead of forming the thin Sweet-Parker layer through Ohmic diffusion, Richardson diffusion (Richardson 1926) dominates the cross-field dynamics and reconnection layer. From Lazarian et al. (2020)

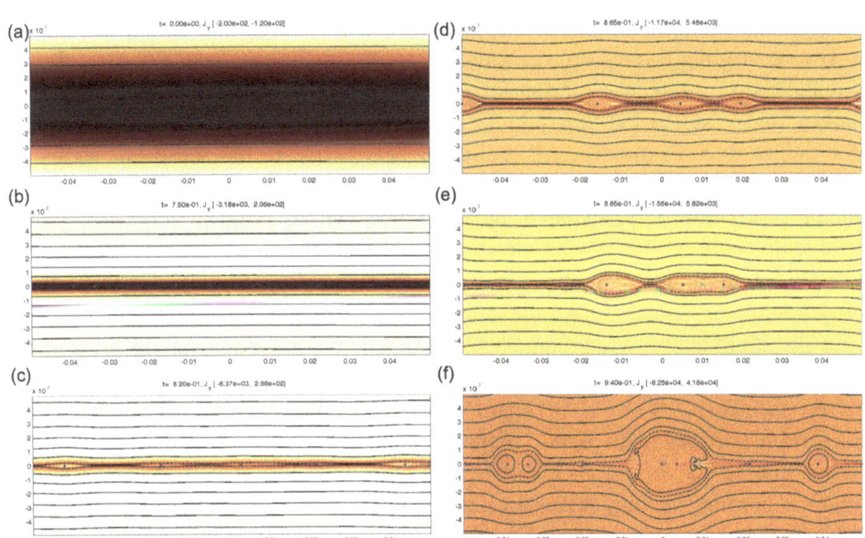

Fig. 1.10 The development of islands resulting from the tearing instability is shown in panels (a)–(d), as computed in an island-coalescing system driven by a low level of random noise is shown, from Huang and Bhattacharjee (2010). Panels (e) and (f) illustrate the rapid development of the plasmoid instability, including reconnection. The lines are magnetic lines of force and the image densities are proportional to electric current density. The vertical axis is stretched by a factor of ten, the initial sheet has an aspect ratio of about 40:1. This calculation was made with $S = 3\ 10^6$

"consistent with a δ-function input, followed by energy redistribution [...] without further energy input", in data sampled at time scales of $\approx$100 s (Golub 1996 see also Shimizu et al. 1992). In Sect. 5.6 we re-examine the question of why is so little of the magnetic field emerging into the corona is loaded with hot, bright plasma (see Fig. 2.2).

1.8.3 Further Inferences: Quiet vs. Active Corona

What might we infer about heating mechanisms from the characteristics of these different coronae? The quiet corona appears as a diffuse, statistically steady plasma. It sits above chromospheric and transition region plasma, which, unlike the corona, are both significantly brighter in a network pattern (Hale and Ellerman 1904; Gabriel 1976), associated with photospheric supergranular downflows (Leighton et al. 1962). This network harbors stronger-than-average photospheric magnetic fields. Remarkably, unlike the corona, the brightness of the chromosphere (and transition region) is almost indistinguishable between quiet Sun and coronal hole, yet the coronal hole plasma is dominated by unipolar flux. It is only in Doppler shifts of UV lines where the differences become clear (Hassler et al. 1999).

The far brighter active corona forms in response to the emergence of large fluxes of magnetism from beneath the surface (Vaiana et al. 1973; Liggett and Zirin 1985; Toriumi and Wang 2019). It exhibits a richer variety of phenomena, with apparent lifetimes of days to months. All regions appear active in the sense that loops of hot plasma come and go as emerged flux penetrates the atmosphere, and as the emerging flux adjusts to its environment. Some regions seem to achieve a quasi-steady state over days, glowing steadily like the quiet corona but brighter. Others reflect the continuing emergence of new flux by expanding, brightening with newly formed loops of plasma. Often the continually emerging magnetic energy exceeds some threshold across which large-scale magnetic reconnection can occur, leading to the dynamic evolution or coronal plasma (characterized by evolution at near-Alfvén speeds of 10^6 m s^{-1}), manifested along with non-thermal particles and radiation associated with flares, magnetic field reconfigurations and ejection of plasma. These active regions clearly entail more than just steady plasma heating, although coronal heating seems to pervade these systems despite the dynamical changes. Steady coronal heating continues, largely unaltered, right up until a flare or CME is triggered. Observational signatures of an imminent flare have been sought for decades. Significantly, the warning signs of flaring are elusive (Panos and Kleint 2020), and appear to be independent of the heating nearby. After a large flare, "post flare loop systems" (e.g. Bray et al. 1991; Qiu et al. 2012) live for many hours, far longer than dynamical and plasma cooling times (Table 1.2), requiring continuing heating.

While this diverse behavior is frequently described in terms of the "complexity" of the surface magnetic fields (Toriumi and Wang 2019), physically it results from the interaction of emerging flux, pre-loaded with varying degrees of internal

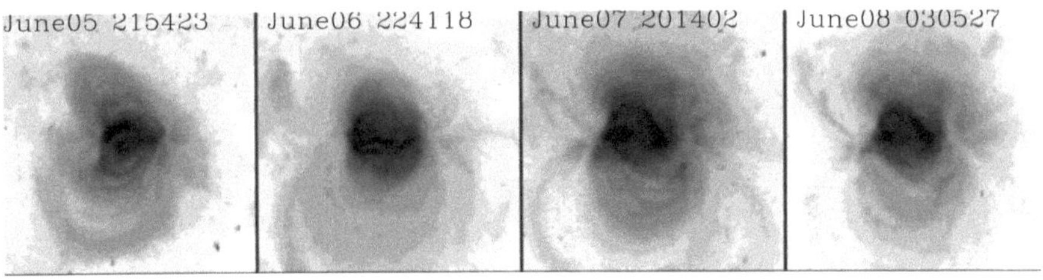

Fig. 1.11 The corona over active region NOAA 7968, which emerged during the very quiet period in June 1996 is shown evolving slowly in time over 3 days in filter images centered at 28.4 nm from the EIT instrument on the SoHO spacecraft (Domingo et al. 1995), from Deng et al. (1999). The field of view is $250'' \times 250''$

free energy, with pre-existing hydromagnetic structures. Newly emerged magnetic bundles naturally interact with both quiet and active Sun, depending on the history of interior and surface magnetic fields. Bundles can emerge with a variety of associated sizes and fluxes, from $\approx 10^{11}$ Wb ("pores") to $\approx 10^{15}$ Wb, carrying within them different embedded current systems. (An example of a system developing such current systems in-situ is shown in Fig. 1.12). Emergence of a $B \approx 0.2$ T bundle of flux into a previously quiet region ($B \approx 0.001$ T), whose coronal magnetic field initially knows nothing of the emerging flux, is likely to be an unspectacular event. One example might be the study of the life and death of NOAA 7968 during the solar minimum of 1996, shown in Fig. 1.11 (Deng et al. 1999). Deng et al. (1999) conclude that the interaction between emerged and pre-existing magnetic fields led to small ("B-class") flares, and that the emerged fields themselves were close to a potential state. Even in this relatively "simple" active region, new flux emerged untidily through the solar atmosphere over a week. Yet throughout this evolution, steady heating persisted. Perhaps then steady coronal heating has little to do with the evolution of large-scale coronal structure that accompanies flares? Instead we might conjecture that *heating mostly arises from processes merely guided and propagated passively along large-scale fields* (Sturrock 1999), which may themselves be very close to a potential state. The free energy for heating could readily be carried by magnetic and fluid velocity perturbations in the planes perpendicular to the large-scale fields (page 56), with little to do with the non-potentiality or reconnection of observable large-scale fields.

Emergence of the same flux bundle into a pre-existing 0.2 T bundle would have to be unusually well-ordered and well-aligned to avoid forming large-scale and energetic current systems. Continued emergence would then generally lead to large-scale dynamic evolution. An example of a dynamically evolving region of similar size to NOAA 7968 is shown in Fig. 1.12. NOAA 7968 exhibited changing magnetic connections especially following the launch of a CME on February 12 (McKenzie and Canfield 2008). In both of these active regions, *coronal heating persists before and after rapid, even dramatic evolution* of the large-scale structure (Fig. 1.12).

The above examples represent just two regions chosen for their relative magnetic "simplicity". Most modern research is more concerned with large-scale dynamics

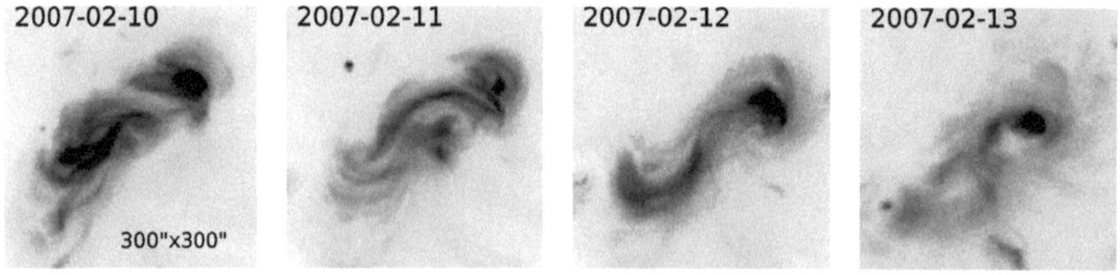

Fig. 1.12 A small active region is shown in soft X-ray data from the XRT instrument on the Hinode spacecraft, transiting the solar disk early in February 2007. This underwent a large magnetic reconnection just after the image shown for 12 February (McKenzie and Canfield 2008). The slanted "S-" or sigmoid shape seen on 12 February is associated with eruptions and coronal re-arrangements owing to an excess of large-scale magnetic free energy (Rust and Kumar 1996). The field of view is $300'' \times 300''$

than heating mechanisms, and so is also concerned with active regions far more complex than simple bipolar fields emerging through the photosphere. To put this discussion in context, examples of large magnetically complex active region include NOAA 12192, appearing in October 2014 near the peak of magnetic solar cycle 24 (Sun et al. 2015). While NOAA 12192 was prolific in flares, it produced few CMEs. In contrast, NOAA 5395 was the origin of multiple CMEs as well as flares (Feynman and Hundhausen 1994), including the CME of March of 1989 which took down the power grid in Quebec. Bobra and Couvidat (2015) have examined a list of 13 physical parameters related to production of large-scale dynamics similar in spirit to those examined by Mandrini et al. (2000), which were targeted towards to coronal heating. This pursuit of understanding of the diverse dynamic phenomena draws the attention of most solar physicists, with around 10^3 articles produced every year.

Lastly, we should draw attention to a curious, yet seemingly obscure, fact: differences between quiet, coronal hole, and non-erupting active region plasmas are greater than variations within each type of plasma. We will expand on this issue in Sect. 5.4.

Chapter Summary

For better or worse, we have boiled down the huge literature on coronal heating to the following central issues

Definition: the coronal heating problem will have been solved when *credible mechanism(s) for the generation, transport and dissipation of energy have survived refutation against critical, information-rich experiments.*

Observational differences between quiet, coronal hole, active region plasmas between eruptions are greater than variations within each plasma. Satisfactory mechanisms must account for both statistical properties.

(continued)

The finest quality polarimetric observations of photospheric magnetic fields now far outstrip the quality of observations of just the brightness of coronal plasma. Novel instrumentation should enhance coronal observations.

Coronal heating will not be fully understood without first understanding the magnetic state of the upper chromosphere, and perhaps even the capricious transition region.

We encountered two ideas which will be treated as conjectures in later chapters: once small scales develop, the free energy leads to irreversible dissipation no matter the dissipative mechanism(s); magnetic fields on observable scales play only passive roles in coronal heating.

References

Allen, C.W.: Astrophysical Quantities. Athlone Press, University of London (1973)

Anderson, L.S., Athay, R.G.: Chromospheric and coronal heating. Astrophys. J. **336**, 1089–1091 (1989)

Arber, T.D., Haynes, M., Leake, J.E.: Emergence of a flux tube through a partially ionized solar atmosphere. Astrophys. J. **666**, 541–546 (2007)

Aschwanden, M.J.: An evaluation of coronal heating models for active regions based on Yohkoh, SOHO, and TRACE observations. Astrophys. J. **560**, 1035–1044 (2001). https://doi.org/10.1086/323064

Athay, R.G.: Chromospheric Fine Structure. Reidel, Dordrecht/Boston (1974)

Athay, R.G.: Chromospheric Fine Structure. Physics of the Sun. Volume 2, vol. 2, pp. 51–69 (1986)

Athay, R., Dere, K.: Chromospheric and transition region diagnostics using emission-line intensities. Astrophys. J. **379**, 776–787 (1991)

Athay, R.G., White, O.R.: Chromospheric and coronal heating by sound waves. Astrophys. J. **226**, 1135 (1978)

Axford, W.I., McKenzie, J.F.: Coronal heating. Adv. Space Res. **30**, 505 (2002). https://doi.org/10.1016/S0273-1177(02)00328-9

Bartoe, J.D.F., Brueckner, G.E.: New stigmatic, coma-free, concave-grating spectrograph. J. Opt. Soc. Am. **65**, 13–21 (1975)

Biskamp, D.: Magnetohydrodynamic Turbulence. Cambridge University Press, Cambridge (2003)

Biskamp, D.: Magnetic Reconnection in Plasmas. Cambridge University Press, Cambridge (2005)

Blandford, R.D., Znajek, R.L.: Electromagnetic extraction of energy from Kerr Black Holes. Mon. Not. R. Astron. Soc. **179**, 433–456 (1977). https://doi.org/10.1093/mnras/179.3.433

Bobra, M.G., Couvidat, S.: Solar flare prediction using SDO/HMI vector magnetic field data with a machine-learning algorithm. Astrophys. J. **798**, 135 (2015)

Braginskii, S.I.: Transport processes in a plasma. Rev. Plasma Phys. **1**, 205–311 (1965)

Bray, R.J., Loughhead, R.E.: The Solar Chromosphere. Chapman and Hall, London (1974)

Bray, R.J., et al.: Plasma Loops in the Solar Corona. Cambridge Astrophysics Series. Cambridge University Press, Cambridge (1991)

Bruls, J.H.M.J., Von der Luhe, O.: Photospheric fine structure: an observational challenge. An analysis of radiative transfer effects on the visibility of small-scale structures. Astron. Astrophys. **366**, 281–293 (2001)

Burbidge, E.M., et al.: Synthesis of the elements in stars. Rev. Mod. Phys. **29**, 547–650 (1957)

Carlsson, M., Stein, R.F.: Does a nonmagnetic solar chromosphere exist? Astrophys. J. **440**, L29–l32 (1995)

Carlsson, M., de Pontieu, B., Hansteen, V.H.: New view of the solar chromosphere. Ann. Rev. Astr. Astrophys. **57**, 189–226 (2019)

Carlsson, M., et al.: A publically available simulation of an enhanced network region of the Sun. Astron. Astrophys. **585**, A4 (2016)

Cattaneo, F.: The solar dynamo problem. In: Pijpers, F.P., et al. (eds.) Solar Convection and Oscillations and Their Relationship, pp. 201–222. Kluwer, Dordrecht (1997)

Chapman, S.: The Electrical Conductivity of the Ionosphere: A Review. N. 4 del Suppliemento al Vol.4 Serie X, del Nuovo Cimento, pp. 1385–1412 (1956)

Chapman, G.A.: Active regions from the photosphere to the chromosphere. In: Orrall, F.Q. (ed.) Solar Active Regions: A Monograph from Skylab Solar Workshop III, pp. 43–82 (1981)

Charbonneau, P.: Dynamo models of the solar cycle. Living Rev. Sol. Phys. **17**, 4 (2020)

Cliver, E.W., et al.: Extreme solar events. Living Rev. Solar Phys. **19**, 2 (2022). https://doi.org/10.1007/s41116-022-00033-8

Cooper, J.: Plasma spectroscopy. Rep. Prog. Phys. **29**, 35–130 (1966)

Daldorff, L.K.S., et al.: Two-way coupling of a global hall magnetohydrodynamics model with a local implicit particle-in-cell model. J. Comput. Phys. **268**, 236–254 (2014)

Damé, L.: The chromospheric dynamic of fine structures. In: Schmidt, H.U. (ed.) Theoretical Problems in High Resolution Solar Physics, pp. 241–243. MPA/LPARL Workshop. Max-Planck-Institut für Physik und Astrophysik MPA, München 212 (1985)

de Pontieu, B., et al.: A new view of the solar interface region from the interface region imaging spectrograph (IRIS). Sol. Phys. **296**, 84 (2021)

Deng, Y.Y., et al.: The evolution from birth to decay of a short-lived active region. Astron. Astrophys. **349**, 927–940 (1999)

Domingo, V., Fleck, B., Poland, A.I.: Soho: the solar and heliospheric observatory. Space Sci. Rev. **72**, 81–84 (1995). https://doi.org/10.1007/BF00768758

dos Santos, L.A., et al.: The solar twin planet search. IV. The Sun as a typical rotator and evidence for a new rotational braking law for Sun-like stars. Astron. Astrophys. **592**, A156 (2016)

Eddy, J.A.: The Sun, the Earth and near-Earth space: a guide to the Sun-Earth System. Nasa (2009)

Edlén, B.: Die Deutung Der Emissionslinien Im Spektrum Der Sonnenkorona. Z. Astrophys. **22**, 30–64 (1943)

Einaudi, G., et al.: Energetics and 3D structure of elementary events in solar coronal heating. Astrophys. J. **910**, 84 (2021)

Feynman, J., Hundhausen, A.J.: Coronal mass ejections and major solar flares: the great active center of March 1989. JGR **99**, 8451–8464 (1994). https://doi.org/10.1029/94JA00202

Fisher, G.H., et al.: Coronal heating in active regions as a function of global magnetic variables. Astrophys. J. **508**, 885–898 (1998)

Foukal, P.: The temperature structure and pressure balance of magnetic loops in active regions. Sol. Phys. **43**, 327–336 (1975)

Foukal, P.: The pressure and energy balance of the cool corona over sunspots. Astrophys. J. **210**, 575 (1976)

Foukal, P.: Magnetic loops, downflows and convection in the solar corona. Astrophys. J. **223**, 1046 (1977)

Foukal, P.: Solar Astrophysics. Wiley, New York (1990)

Furth, H.P., Killeen, J., Rosenbluth, M.N.: Finite-resistivity instabilities of a sheet pinch. Phys. Fluids **6**, 459–484 (1963). https://doi.org/10.1063/1.1706761

Gabriel, A.: A magnetic model of the solar transition region. Philos. Trans. R. Soc. Lond. **281**, 339 (1976)

Giovanelli, R.G., Jones, H.P.: The three-dimensional structure of atmospheric magnetic fields in two active regions. Sol. Phys. **79**, 267–278 (1982). https://doi.org/10.1007/BF00146244

Goldreich, P., Sridhar, S.: Toward a theory of interstellar turbulence. II. strong Alfvenic turbulence. Astrophys. J. **438**, 763 (1995)

Golub, L.: The solar X-ray corona. Astrophys. Space Sci. **237**, 33–48 (1996). https://doi.org/10.1007/BF02424425

Golub, L., Pasachoff, J.M.: The Solar Corona, 2nd edn. Cambridge University Press, Cambridge (2009)

Gudiksen, B.V., Nordlund, Å: An *ab initio* approach to the solar coronal heating problem. Astrophys. J. **618**, 1020–1030 (2005)

Gudiksen, B.V., et al.: The stellar atmosphere simulation code Bifrost. Code description and validation. Astron. Astrophys. **531**, A154 (2011)

Gustafsson, B.: Is the Sun a Sun-like Star? Space Sci. Rev. **85**, 419–428 (1998)

Hale, G.E.: On the probable existence of a magnetic field in Sun-spots. Astrophys. J. **28**, 315–343 (1908)

Hale, G.E., Ellerman, F.: Calcium and hydrogen flocculi. Astrophys. J. **19**, 41–52 (1904)

Hassler, D.M., et al.: Solar wind outflow and the chromospheric magnetic network. Science **283**, 810 (1999). https://doi.org/10.1126/science.283.5403.810

Heller, J.: Catch 22. Simon & Schuster, New York (1961)

Henriques, V.M.J., et al.: Umbral chromospheric fine structure and umbral flashes modelled as one: the corrugated Umbra. Astron. Astrophys. **642**, A215 (2020). https://doi.org/10.1051/0004-6361/202038538

Huang, Y.M., Bhattacharjee, A.: Scaling laws of resistive magnetohydrodynamic reconnection in the high-Lundquist-number, plasmoid-unstable regime. Phys. Plasmas **17**, 062104–062104 (2010). https://doi.org/10.1063/1.3420208

Ionson, J.A.: Resonant absorption of Alfvénic surface waves and the heating of solar coronal loops. Astrophys. J. **226**, 650–673 (1978)

Ionson, J.A.: Resonant electrodynamic heating of stellar coronal loops - an LRC circuit analog. Astrophys. J. **254**, 318–334 (1982)

Ionson, J.A.: A unified theory of electrodynamic coupling in coronal magnetic loops - the coronal heating problem. Astrophys. J. **276**, 357–368 (1984)

Ionson, J.A.: Coronal heating by resonant (A.C.) and nonresonant (D.C.) mechanisms. Astron. Astrophys. **146**, 199–203 (1985a)

Ionson, J.A.: The heating of coronae. Solar Phys. **100**, 289 (1985b)

Ionson, J.A., Kuperus, M.: Black Hole accretion disks - coronal stabilization of the Lightman-Eardley instability. Astrophys. J. **284**, 389–393 (1984). https://doi.org/10.1086/162418

Jaynes, E.: Probability Theory in Science and Engineering. Colloquium Lectures in Pure and Applied Science. Socony Mobil Oil Co. Field Research Laboratory (1959)

Jeffreys, H.: Theory of Probability, 3rd edn. Clarendon Press, Oxford (1961)

Jones, H.P., Giovanelli, R.G.: Magnetic canopies in unipolar regions. Sol. Phys. **87**, 37–42 (1983). https://doi.org/10.1007/BF00151157

Judge, P.G.: The chromosphere: gateway to the corona? ...Or The Purgatory of solar physics? Mem. S. Astr. Italia **81**, 543 (2010)

Judge, P.G.: UV spectra, bombs, and the solar atmosphere. Astrophys. J. **808**, 116 (2015)

Judge, P.G.: Solar magnetic fields: history, tragedy or comedy? In: Almeida, J.S., González, M.J.M. (eds.) Cosmic Magnetic Fields, Canary Islands Winter School of Astrophysics, pp. 13–46. Cambridge University Press, Cambridge (2018). https://doi.org/10.1017/9781316160916

Judge, P.G.: Magnetic connections across the chromosphere-corona transition region. Astrophys. J. **914**, 70 (2021)

Judge, P.G.: The mercurial Sun at the heart of our solar system. In: Lara, L.M., Jewitt, D. (eds.) Planetary Systems Now, pp. 49–70. World Scientific, London (2023). https://doi.org/10.1142/q0388

Judge, P.G., Tritschler, A., Low, B.C.: Thermal fine structure and magnetic fields in the solar atmosphere: spicules and fibrils. Astrophys. J. Lett. **730**, L4 (2011)

Kadomtsev, B.B., Pogutse, O.P.: Nonlinear helical perturbations of a plasma in the Tokamak. Soviet J. Exp. Theor. Phys. **38**, 283–290 (1974)

Kahler, S.W., Davis, J.M., Harvey, J.W.: Comparison of coronal holes observed in soft X-Ray and He I 10 830 Å spectroheliograms. Sol. Phys. **87**, 47–56 (1983). https://doi.org/10.1007/BF00151159

Kulsrud, R.M.: Plasma Physics for Astrophysics. Princeton University Press, Princeton and Oxford (2005)

Kulsrud, R.M.: Intuitive approach to magnetic reconnection. Phys. Plasmas **18**, 111201 (2011)

Kundu, M.R.: Solar Radio Astronomy. Interscience Publication, New York (1965)

Kuperus, M., Ionson, J.A.: Black Hole accretion disks. Electrodynamic coupling of accretion-disk coronae and the partitioning of soft and hard X-ray emission. Astron. Astrophys. **148**, 309–311 (1985)

Laming, J.M.: The FIP and inverse FIP effects in solar and stellar coronae. Living Rev. Solar Phys. **12**, 2 (2015). https://doi.org/10.1007/lrsp-2015-2

Lazarian, A., et al.: 3D turbulent reconnection: theory, tests, and astrophysical implications. Phys. Plasmas **27**, 012305 (2020). https://doi.org/10.1063/1.5110603

Leighton, R.B., Noyes, R.W., Simon, G.W.: Velocity fields in the solar atmosphere I. Preliminary report. Astrophys. J. **135**, 474–499 (1962)

Liggett, M., Zirin, H.: Emerging flux in active regions. Sol. Phys. **97**, 51–58 (1985). https://doi.org/10.1007/BF00152978

Lin, H.: On the distribution of the solar magnetic fields. Astrophys. J. **446**, 421–430 (1995)

Linker, J.A., et al.: The open flux problem. Astrophys. J. **848**, 70 (2017)

Lites, B.W.: Remote sensing of solar magnetic fields. Rev. Geophys. **38**, 1–36 (2000)

Lites, B.W., Rutten, R.J., Kalkofen, W.: Dynamics of the solar chromosphere. I. Long-period network oscillations. Astrophys. J. **414**, 345–356 (1993)

Litwin, C., Rosner, R.: On the structure of solar and stellar coronae - loops and loop heat transport. Astrophys. J. **412**, 375–385 (1993)

Livingston, W., & Harvey, J.: The Kitt Peak magnetograph. IV: 40-channel probe and the detection of weak photospheric fields. In: Howard, R. (ed.) Solar Magnetic Fields, PP. 51–61. IAU Symposium 43. Reidel, Dordrecht (1971)

Makwana, K.D., Keppens, R., Lapenta, G.: Two-way coupling of magnetohydrodynamic simulations with embedded particle-in-cell simulations. Comput. Phys. Commun. **221**, 81–94 (2017)

Makwana, K.D., Keppens, R., Lapenta, G.: Study of magnetic reconnection in large-scale magnetic island coalescence via spatially coupled MHD and PIC simulations. Phys. Plasmas **25**, 082904 (2018)

Maltby, P., et al.: A new sunspot umbral model and its variation with the solar cycle. Astrophys. J. **306**, 284 (1986)

Mandrini, C.H., Démoulin, P., & Klimchuk, J.A.: Magnetic field and plasma scaling laws: their implications for coronal heating models. Astrophys. J. **530**, 999–1015 (2000)

Mariska, J.T.: The Solar Transition Region. Cambridge University Press, Cambridge (1992)

Mason, H.E.: Density measurements in the corona. Adv. Space Res. **11**, 293–301 (1991)

Matthaeus, W.H., et al.: Pathways to dissipation in weakly collisional plasmas. Astrophys. J. **891**, 101 (2020). https://doi.org/10.3847/1538-4357/ab6d6a

McCrea, W.H.: The hydrogen chromosphere. Mon. Not. R. Astron. Soc. **89**, 483 (1929a)

McCrea, W.H.: The mechanics of the chromosphere. Mon. Not. R. Astron. Soc. **89**, 718 (1929b)

McKenzie, D.E., Canfield, R.C.: Hinode XRT observations of a long-lasting coronal sigmoid. Astron. Astrophys. **481**, L65–L68 (2008). https://doi.org/10.1051/0004-6361:20079035

Menzel, D.H., Cillié, G.G.: Hydrogen emission in the chromosphere. Astrophys. J. **85**, 88 (1937)

Mercier, C., & Chambe, G.: Electron density and temperature in the solar corona from multifrequency radio imaging. Astron. Astrophys. **583**, A101 (2015)

Metcalf, T.R., et al.: Is the solar chromospheric magnetic field force-free? Astrophys. J. **439**, 474–481 (1995)

Mihalas, D.: Stellar Atmospheres, 2nd edn. W. H. Freeman and Co., San Francisco (1978)

Miyamoto, S.: Ionization theory of solar corona. Publ. Astron. Soc. Jpn. **1**, 10–13 (1949)

Montgomery, D., Turner, L.: Anisotropic magnetohydrodynamic turbulence in a strong external magnetic field. Phys. Fluids **24**, 825–831 (1981)

Niedner, M.B., Jr., Ionson, J.A., Brandt, J.C.: Interplanetary gas. XXVI - On the reconnection of magnetic fields in cometary ionospheres at interplanetary sector boundary crossings. Astrophys. J. **245**, 1159–1169 (1981). https://doi.org/10.1086/158891

Noci, G.: The temperature of the Solar corona. Memorie della Societa Astronomica Italiana **74**, 704–712 (2003)

Ossendrijver, M.: The solar dynamo. Astron. Astrophys. Rev. **11**, 287–367 (2003)

Panos, B., Kleint, L.: Real-time flare prediction based on distinctions between flaring and non-flaring active region spectra. Astrophys. J. **891**, 17 (2020)

Parker, E.N.: Hydromagnetic dynamo models. Astrophys. J. **122**, 293–314 (1955)

Parker, E.N.: Sweet's mechanism for merging magnetic fields in conducting fluids. J. Geophys. Res. **62**, 509–520 (1957)

Parker, E.N.: Dynamics of the interplanetary gas and magnetic fields. Astrophys. J. **128**, 664–676 (1958)

Parker, E.N.: Spontaneous Current Sheets in Magnetic Fields with Application to Stellar X-Rays. International Series on Astronomy and Astrophyics. Oxford University Press, Oxford (1994)

Parker, E.N.: Conversations on Electric and Magnetic Fields in The Cosmos. Princeton University Press, Princeton (2007)

Payne-Gaposchkin, C.H.: Stellar Atmospheres: A Contribution to The Observational Study of High Temperature in The Reversing Layers of Stars. Ph.D. thesis, Radcliffe College (1925)

Pietarila, A., Judge, P.G.: On the formation of the resonance lines of helium in the Sun. Astrophys. J. **606**, 1239–1257 (2004)

Pontin, D.I., Hornig, G.: The Parker problem: existence of smooth force-free fields and coronal heating. Living Rev. Solar Phys. **17**, 5 (2020). https://doi.org/10.1007/s41116-020-00026-5

Pottasch, S.R.: On the interpretation of the solar ultraviolet emission line spectrum. Space Sci. Rev. **3**, 816–855 (1964)

Priest, E., Forbes, T.: Magnetic Reconnection. Cambridge University Press, Cambridge (2007)

Przybylski, D., et al.: Chromospheric extension of the MURAM code. Astron. Astrophys. **664**, A91 (2022). https://doi.org/10.1051/0004-6361/202141230

Qiu, J., Liu, W.J., Longcope, D.W.: Heating of flare loops with observationally constrained heating functions. Astrophys. J. **752**, 124 (2012). https://doi.org/10.1088/0004-637X/752/2/124

Rappazzo, A.F., et al., Coronal heating, weak MHD turbulence, and scaling laws. Astrophys. J. Lett. **657**, L47–L51 (2007). https://doi.org/10.1086/512975

Rappazzo, A.F., et al.: Nonlinear dynamics of the Parker scenario for coronal heating. Astrophys. J. **677**, 1348–1366 (2008). https://doi.org/10.1086/528786

Richardson, L.F.: Atmospheric diffusion shown on a distance-neighbour graph. Proc. R. Soc. Lond. Ser. A **110**, 709–737 (1926). https://doi.org/10.1098/rspa.1926.0043

Rimmele, T.R., et al.: The Daniel K. Inouye Solar Telescope - observatory overview. Sol. Phys. **295**, 172 (2020). https://doi.org/10.1007/s11207-020-01736-7

Rust, D.M., Kumar, A.: Evidence for helically kinked magnetic flux ropes in solar eruptions. Astrophys. J. Lett. **464**, L199 (1996). https://doi.org/10.1086/310118

Saha, M.N.: Ionisation in the solar chromosphere. Philos. Mag. **40**, 472 (1920)

Schad, T.A., et al.: Vector magnetic field measurements along a cooled stereo-imaged coronal loop. Astrophys. J. **833**, 5 (2016)

Schekochihin, A.A., MHD turbulence: a biased review. J. Plasma Phys. **88**, 155880501 (2022). https://doi.org/10.1017/S0022377822000721

Schmitt, J.H.M.M.: 50 (38) years of stellar X-ray astronomy . Mem. Soc. Astr. Italia **84**, 532 (2013)

Schwarzschild, K.: On the equilibrium of the Sun's atmosphere. Nachrichten Königlichen Gesellschaft Wissensch. Göttingen. Math.-Phys. Klasse **195**, 41–53 (1906)

Seaton, M.J.: The spectrum of the solar corona. Planet. Space Sci. **12**, 55 (1964)

Shimizu, T., et al.: Transient brightenings in active regions observed by the soft X-ray telescope on YOHKOH. Publ. Astron. Soc. Jpn. **44**, L147–L153 (1992)

Simon, G.W., Leighton, R.B.: Velocity fields in the solar atmosphere. III. Large-scale motions, the chromospheric network, and magnetic fields. Astrophys. J. **140**, 1120 (1964)
Spitzer, L.: Physics of Fully Ionized Gases. Interscience Publishers, New York (1956)
Stencel, R.E., Ionson, J.A.: Stellar magnetic fields: the role of a magnetic field in the peculiar M Giant, HD 4174. Publ. Astron. Soc. Pac. **91**, 452–459 (1979). https://doi.org/10.1086/130519
Stix, M.: The Sun. An Introduction, 2nd edn. Springer, Berlin (2004)
Strauss, H.R.: Nonlinear, three-dimensional magnetohydrodynamics of noncircular Tokamaks. Phys. Fluids **19**, 134–140 (1976). https://doi.org/10.1063/1.861310
Sturrock, P.A.: Chromospheric magnetic reconnection and its possible relationship to coronal heating. Astrophys. J. **521**, 451–459 (1999). https://doi.org/10.1086/307544
Sun, X., et al.: Why is the great solar active region 12192 flare-rich but CME-poor? Astrophys. J. Lett. **804**, L28 (2015). https://doi.org/10.1088/2041-8205/804/2/L28
Sweet, P.A.: The neutral point theory of solar flares. In: Lehnert, B. (ed.) Electromagnetic Phenomena in Cosmical Physics, vol. 6. IAU Symposium, p. 123 (1958)
Thomas, R.N.: Some Aspects of Non-Equilibrium Thermodynamics in the Presence of a Radiation Field. University of Colorado Press, Boulder (1965)
Toriumi, S., Wang, H.: Flare-productive active regions. Living Rev. Solar Phys. **16**, 3 (2019). https://doi.org/10.1007/s41116-019-0019-7
Tousey, R.: Solar work at high altitudes from rockets. In: Kuiper, G.P. (ed.) The Sun, pp. 658–676. Chicago University Press, Chicago (1953)
Tousey, R., Watanabe, K., Purcell, J.D.: Measurements of solar extreme ultraviolet and X-rays from rockets. Phys. Rev. **83**, 792–797 (1951)
Tsiropoula, G., Schmieder, B., Alissandrakis, C.E.: Dynamical fine structures of the chromosphere. Space Sci. Rev. **70**, 65–68 (1994)
Uzdensky, D.A.: The fast collisionless reconnection condition and the self-organization of coronal heating. Astrophys. J. **671**, 2139–2153 (2007). https://dx.doi.org/10.1086/522915
Vaiana, G.S., Krieger, A.S., Timothy, A.F.: Identification and analysis of structures in the corona from X-ray photography. Sol. Phys. **32**, 81–116 (1973). https://doi.org/10.1007/BF00152731
Vernazza, J.E., Avrett, E.H., Loeser, R.: Structure of the solar chromosphere. I. Basic computations and summary of the results. Astrophys. J. **184**, 605 (1973)
Vernazza, J.E., Avrett, E.H., Loeser, R.: Structure of the chromosphere. III. Models of the EUV brightness components of the quiet Sun. Astrophys. J. Suppl. **45**, 635–725 (1981)
Verscharen, D., Klein, K.G., Maruca, B.A.: The multi-scale nature of the solar wind. Living Rev. Solar Phys. **16**, 5 (2019). https://doi.org/10.1007/s41116-019-0021-0
Walker, A.B.C., et al., Soft X-ray images of the solar corona with a normal-incidence Cassegrain multilayer telescope. Science **241**, 1781–1787 (1988)
Washinoue, H., Shoda, M., Suzuki, T.K.: The effect of the chromospheric temperature on coronal heating. arXiv e-prints. arXiv:2209.10156 (2022)
Withbroe, G.L:. The temperature structure, mass and energy flow in the corona and inner solar wind. Astrophys. J. **325**, 442–467 (1988)
Withbroe, G.L., Noyes, R.W.: Mass and energy flow in the solar chromosphere and corona. Ann. Rev. Astr. Astrophys. **15**, 363–387 (1977)
Woolley, R.D.V.R., Allen, C.W.: The coronal emission spectrum. Mon. Not. R. Astron. Soc. **108**, 292–305 (1948)
Young, C.A.: The Sun, 4th edn. Kegan Paul, Trench, Truebner & co. Ltd., London (1892)
Zeeman, P.: Over De Invloed Eener Magnetisatie Op Den Aard Van Het Door Een Stof Uitgezonden Lichtover De Invloed Eener Magnetisatie Op Den Aard Van Het Door Een Stof Uitgezonden Lichton the Influence of Magnetism on The Nature of the Light Emitted by a Substance. Verslagen en Mededeelingen der Kon. Academie van Wetenschappen, Afd. Natuurkunde **5**, 181–184 (1896)

Chapter 2
Magneto-Hydrodynamics

2.1 Underlying Assumptions, for Better and Worse

To proceed beyond the basic inferences of Chap. 1, we will not only need a representative thermodynamic model (Fig. 1.4), but also a theoretical framework that is simple, yet complete enough, with which to confront hypothetical scenarios with critical data. The initial framework we adopt is magnetohydrodynamics (MHD). As the name implies, this is a fluid approximation describing the plasma as a collisionally coupled single fluid of ions and electrons, and so we may not expect to apply it without great care on scales below the mean free paths. The fluid treatment seems rather safe in the low-temperature, high-density chromosphere where scale heights exceed mean free paths by six orders of magnitude. However, in the corona, mean free paths of 200 km are typical in structures whose scale heights are just 50 Mm. The associated Knudsen numbers of 0.004 for typical particles are uncomfortably large, they scale as particle energy ε^2. As we argued in Sect. 1.7, this energy dependence confounds attempts to justify a fluid treatment, without some care and fore-knowledge of the risks it entails.

From the outset we must recognize that MHD may be inadequate to describe the heating of coronal plasma, from the largest to smallest relevant scales. This is because the irreversible plasma heating ($\mathbf{j} \cdot \mathbf{E}$) occurs only through work done by an electric field $\mathbf{E}$ on charges, work which is dissipated through particle collisions, or by dissipation of fluid velocity gradients through viscosity, or by wave-particle interactions in collisionless conditions. Biskamp (2003) notes that

> "...in many dilute plasmas collisions are very rare, but there are other, collective, [electrodynamic] processes that play a similar dissipative role giving rise to effective (often called anomalous) transport coefficients."

For this reason the book is about *electrodynamic* not just magnetic coupling. In MHD collisions are accounted for through the plasma conductivity and viscosity. The conductivity σ is familiar through Ohm's law as $\mathbf{j} = \sigma\mathbf{E}$, which is simply

P. Judge, J. A. Ionson, *The Problem of Coronal Heating*, Astrophysics and Space Science Library 470, https://doi.org/10.1007/978-3-031-46273-3_2

derived using the equation of motion for the electrons moving among the heavy ions (Braginskii 1965). The rate of Joule (Ohmic) heating is $\mathbf{j}\cdot\mathbf{E}$. We will see that coronal plasmas are so close to ideal, i.e. having so few collisions and an enormous σ, that any electric field will be predominantly along the direction of only the induced electric field $\mathbf{E}' = -\mathbf{u} \times \mathbf{B}$ with $\mathbf{u}$ the velocity of the ion+electron fluid. Since $\mathbf{j} = \sum_i Z_i n_i e \mathbf{u_i} - n_e e \mathbf{u_e}$ and $n_e = \sum_i n_i e$, and on average $\mathbf{u}_i = \mathbf{u}_e$, then the induced field can do no work on the particles. With the addition of the small Ohmic component of $\mathbf{j}/\sigma$ to $\mathbf{E}'$, the heating rate becomes just j^2/σ in MHD, independent of the induced term. There is no reason to doubt the meaning of this term when σ is well-defined. But the point here is that in MHD transport coefficients like σ are generally derived *assuming* small departures from Maxwellian distributions, dominated by classical diffusion (e.g. Spitzer 1956; Braginskii 1965). For coronal heating to be driven by Ohmic heating, huge values of $\mathbf{j} \equiv \mathbf{curl}\,\mathbf{H}$ (see below) are required. With huge spatial derivatives in $\mathbf{H}$, long collision times and particle mean free paths, and unidentified but desired heating mechanisms, we should not expect to find fully relaxed, Maxwellian distributions a priori. Under collisionless conditions the particles are accelerated by the electric fields of waves, a complex, large area of research, currently being confronted with novel in-situ data of the solar wind from the Parker Solar Probe (e.g. Liu et al. 2023).

The transport coefficients such as σ can be significantly influenced by modest non-Maxwellian tails of electron energy distribution functions, whose development may depend on the unknown heating processes. Therefore, MHD descriptions of energy *dissipation* in coronal plasma should not be trusted without further justification, because dissipation depends upon the partial or total relaxation of macroscopic anisotropies (e.g. finite electric current densities, gradients in bulk ion velocities). The corona is simply not collisional enough blithely to use such expressions. Biskamp (2003, his p. 2) made the same point but then notes that "dissipation processes, independently of their nature, serve only as energy sinks which …do not affect the main turbulence scales". This assumption—that all small scale processes are dissipative—has proven useful and convenient in much numerical MHD work, because, in any case, the physical scales required to generate a large enough j^2/σ are unreachable, the conductivity being replaced by far larger *numerical* transport coefficients on the large numerical grids.

Nevertheless, because of the multiple "Catch 22" dilemmas of the closure problem of transport coefficients (Bogoliubov 1946) embedded in the problems outlined above, and for clarity of exposition, the approach taken in this book follows all but a few articles in coronal physics. We will assume that observed phenomena are dominated by near-Maxwellian particle distributions, but keep a watchful eye for departures. This approach is one of expediency. At the coronal base, the distributions of translational degrees of freedom of chromospheric plasma are likely to be described by thermal distributions, only the internal degrees of freedom (distribution of occupation numbers of bound energy states) and radiation being out of LTE. But while the chromosphere supplies mostly thermal particles for the corona, the very heating mechanisms we seek will (by definition!) generate hotter particles, potentially with far larger mean free paths.

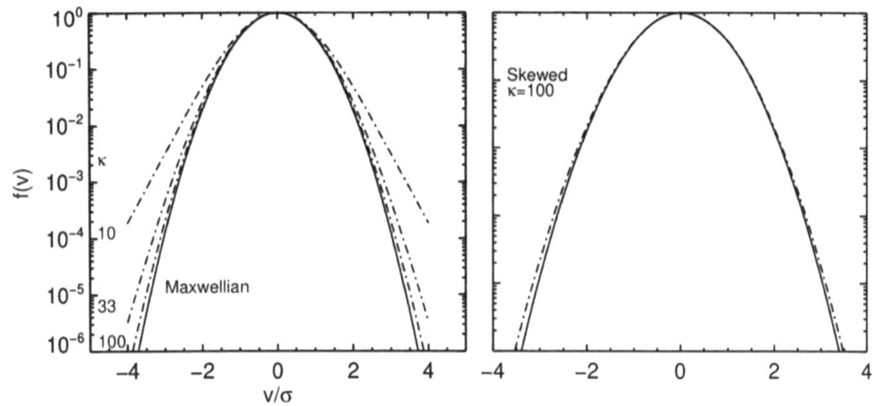

Fig. 2.1 κ distributions are shown in comparison with a Maxwellian ($\kappa = \infty$). The left panel shows standard symmetric κ distributions. The right panel shows an example of a skewed distribution—the distribution for $\kappa = 100$ multiplied by $\left(1 - 0.004\frac{v^3}{\sigma^3}\right)$—whose high energy wings ($\frac{v}{\sigma} \gtrapprox 3$) would contribute to efficient transport but whose core would not

Worse still, our ability to observe non-thermal populations of electrons, for instance, is limited, owing in part to the tendency for energized electrons to relax into the low energy "core" of the distribution functions, the large population there tending to dominate the generation of photons from the plasmas. For example, the AIA instrument (Lemen et al. 2012) on the SDO spacecraft (Pesnell et al. 2012) measures EUV coronal lines of, say Fe^{8+} and Fe^{9+}, which are generated by the inelastic collisions of the ions by electrons with energies exceeding just 2-3 times mean electron energies (see Figs. 1.3 and 5.3). Such radiative emissions are also insensitive to asymmetries in distribution functions. Yet we know that asymmetries can greatly influence transport coefficients. Thus, our observations are essentially blind to small tails in distribution functions which may dominate transport processes. A schematic example of this problem is shown in the right panel of Fig. 2.1, showing "kappa" distributions (generalized Lorentzians). These have the property of being close to Maxwellian at low energies, but with larger high energy wings, and they have been measured in low density space plasmas such as the magnetosphere (e.g. Scudder 1992).

MHD was developed by Alfvén (1950), where he expanded upon his initial ideas during the 1940s.[1] MHD treats the plasma as a single ionized and magnetized fluid. Alfvén (1942) identified a new kind of wave that now bears his name, in which transverse changes in the inertia of the frozen fluid (Sect. 2.5) were balanced by the magnetic tension force (see below). Linearization of the equations of motion around homogeneous magnetic state yields the three well-known MHD wave modes

[1] Curiously, Alfvén came to distrust MHD by the time he was awarded the Nobel Prize for this work in 1970, at least as applied to the magnetosphere.

(Chandrasekhar 1961; Ferraro and Plumpton 1966; Kulsrud 2005), and under general (inhomogeneous) conditions a rich set of instabilities.

To construct the MHD equations, Maxwell's equations are used excluding the displacement current:

$$\nabla \times \mathbf{H} = \mathbf{j} \tag{2.1}$$

$$\nabla \times \mathbf{E} = -\frac{\partial \mathbf{B}}{\partial t} \tag{2.2}$$

$$\nabla \cdot \mathbf{B} = 0 \tag{2.3}$$

$$\nabla \cdot \mathbf{E} = \rho_e \tag{2.4}$$

Neglect of the displacement current effectively sets the speed of light to infinity and disallows electromagnetic waves. In the plasma we use $\mathbf{B} = \mu_0 \mathbf{H}$ (Ginzburg 1970) and the charge density $\rho_e = 0$. These approximations are consistent with assumptions made concerning the plasma as a highly conducting, non-relativistic fluid below. Of course the plasma on small scales, below Debye lengths and other characteristic scales, consists of ions, electrons and perhaps neutral and/or dust particles. Newtonian equations of motion are included in Boltzmann's transport equation for the six-dimensional distribution function. When integrated over the three velocity components assuming small departures from Maxwellian distributions (e.g. Spitzer 1956; Chapman 1956; Braginskii 1965), the Boltzmann transport equations yield our familiar fluid equations, but which always leads to one more equation for a quantity that must be determined, to "close" the system of equations (Bogoliubov 1946). This was achieved by the above authors iteratively to yield diffusive transport coefficients from the assumption of small perturbations. These transport coefficients η and σ, for viscous and resistive dissipation on bulk velocity gradients and electric currents, determine the rates at which, in MHD, energy is dissipated as heat.

The fluid description of a plasma reduces six-dimensional distribution functions for protons and electrons (and trace elements) to fluid density, momentum density and energy density:

$$\rho(\mathbf{r}, t), \quad \rho\mathbf{u}(\mathbf{r}, t), \quad \left(\frac{p}{\gamma - 1} + \rho u^2/2 + B^2/2\mu_0\right).$$

The Lorentz force is added into the fluid momentum equation, and sources and sinks of momentum, energy and transport terms (viscosity, heat and electrical conductivity, terms Σ_i and Σ_ε below) augment the equations of motion. Another benefit of the fluid approximation is that non-linear integral collisional terms in the Boltzmann equation become just functions which are separable in density and temperature. The price for a fluid approximation is that although we can write down evolution equations for the above functions of position, new non-linearities become explicit in the equations of motion.

Experience with fluids on Earth tells us that the enormous solar plasmas must generally behave as turbulent fluids, at least in the small magnetic field (high plasma-β) regime. The associated dynamics can exhibit chaos, intermittency, attractor solutions and unpredictability. The differences with hydrodynamics arise because of free electrical charges in the plasma. Like a metal at room temperature, electrons flow relative to ions in response to an EMF induced, for instance, when the bulk plasma is moving across a pre-existing magnetic fields. These currents then enter the equation of motion through the Lorentz force, and induce further magnetic fields through Ampére's law throughout the plasma. This extra level of non-linearity is dealt with most easily using MHD.

2.2 An MHD Curiosity

The choice of rest frame and treatment of relativity underlying MHD are intriguing. Einstein (1905) famously recognized symmetries in Maxwell's formulation of electrodynamics in his development of special relativity. Michelson and Morley (1887) had established observationally that the velocity of light did not depend on the velocity of an observer. Armed with these facts, Einstein realized that time and space were linked such that Galilean transformations had to be rejected in favor of Lorentz transformations, marking the advent of special relativity. Alfvén's MHD formalism is however based on Galilean frame transformations of a fluid representation of a plasma and its electromagnetic fields correct to order u/c. Critical to Alfvén's description is the evaluation of the magnetic and induced electric fields in the frame of plasma moving with velocity $\mathbf{u}$:

$$\mathbf{B}' \approx \mathbf{B} \tag{2.5}$$

$$\mathbf{E}' \approx \mathbf{E} - \mathbf{u} \times \mathbf{B} \tag{2.6}$$

where un-primed fields are those in the frame moving with the fluid, primed fields are in our (observer's) frame. Transformations for electric current and net charge are affected by the motion only to order u^2/c^2, so that to order u/c:

$$\mathbf{j}' \approx \mathbf{j}, \tag{2.7}$$

$$\rho_e' \approx \rho_e. \tag{2.8}$$

"Ohm's law", obtained from subtracting separate equations of motions for electrons and ions, (Braginskii 1965), is

$$\mathbf{j}' - \sigma \mathbf{E}' \tag{2.9}$$

where σ is the scalar, field-aligned electrical conductivity. Eliminating $\mathbf{j}'$, $\mathbf{E}'$, $\mathbf{j}$ and $\mathbf{E}$ in favor of $\mathbf{u}$ and $\mathbf{B}$, and using vector identities yields the MHD "induction equation":

$$\frac{\partial \mathbf{B}}{\partial t} = \mathbf{curl}\,(\mathbf{u} \times \mathbf{B}) + \eta \nabla^2 \mathbf{B}, \quad \eta = 1/\mu_0 \sigma, \tag{2.10}$$

$$\equiv (\mathbf{B} \cdot \nabla)\mathbf{u} - (\nabla \cdot \mathbf{u})\mathbf{B} - (\mathbf{u} \cdot \nabla)\mathbf{B} + \eta \nabla^2 .\mathbf{B}, \tag{2.11}$$

The first three (ideal) components of Eq. (2.11) represent changes of magnetic field strength by advection, shear or stretching, and compression. The last term represents non-ideal plasmas with finite resistivities η.

In effect, Alfvén reversed Einstein's arguments to permit a more straightforward treatment of fluid dynamics using classical concepts in a Galilean frame, correct to order u/c.

2.3 Ideal MHD

The ratio of the first to second term on the rhs of the induction Eq. (2.10) is the magnetic Reynolds number:

$$\mathrm{Rm} \sim |\nabla \times (\mathbf{u} \times \mathbf{B})|/|\eta \nabla^2 \mathbf{B}| \sim u\ell/\eta \tag{2.12}$$

at length scale ℓ. Clearly, for a given (kinetic) value of η, the Reynolds number depends on characteristic speeds u and lengths ℓ. The magnetic and fluid stresses lead to characterictic speeds for a given local state, at which information tranels through the fluid. Generally speaking, then Rm is usefully regarded as a function of characteristic size ℓ within the plasma. Related is the Lunqvist number S which is simply the magnetic Reynolds number with characteristic speed $u = c_A$.

The case Rm $\gg$ 1 applies to large, highly conducting astrophysical plasmas, such as the corona, so that, in the limit Rm $\to \infty$, the second diffusive term becomes negligible. In any static plasma $\mathbf{u}$=0, fields with scale ℓ just decay resistively with a time scale $\tau \sim \ell^2/\eta$. MHD resistive time scales τ are given as a function of the size ℓ of various plasmas in Table 2.1.

The long time scales associated with the huge astrophysical plasmas suggest that, unless the plasma is riddled with far smaller structures, they will evolve essentially *as if they were ideal* ($\sigma \to \infty$, $\eta \to 0$). In the ideal limit plasma is frozen into the magnetic field ("Alfvén's theorem", Alfvén 1942). There can be no bulk motion of plasma perpendicular to magnetic lines of force, only parallel to them. On kinetic scales, the constituent ions and electrons exhibit fast gyrations around magnetic lines of force, gyro frequencies greatly exceeding collision frequencies ("the charges are magnetized", $\omega\tau \gg 1$, see Table 1.2). Averaging these motions means that the bulk plasma remains free to move only parallel to them. This "frozen-in" condition leads to a profoundly important constraint on the plasma's dynamics:

Table 2.1 Diffusion and induction in various plasmas

Object	$\log \eta$ $m^2 s^{-1}$	$\log \ell$ m	$\log u$ $m\ s^{-1}$	$\log R_m$	τ
Earth (outer core)	0.3	6	−3	3	10^4 year
Sun interior (kinetic η)	0	8	2	10	$3 \cdot 10^8$ year
Sun interior (turbulent η)[a]	10	8	2	2	3 year
Corona (ℓ = active region)	0	8	4	12	$3 \cdot 10^8$ year
Corona (ℓ =limit of observable features)	0	4	4	9	3 year
Mid-chromosphere (Spitzer η)	4	4	4	4	3 hour
Mid-chromosphere (Cowling η)	6.5	4	4	1.5	30 seconds
Laboratory liquid sodium	−1	0	1	2	10 seconds

[a] Turbulent diffusion is related to the stochastic trajectories of magnetic lines of force (Lazarian and Vishniac 1999; Lazarian et al. 2020, see Appendix C and Fig. 1.9)

> The same plasma threads the same bundle of magnetic flux, forever. Identification of plasma then identifies the magnetic bundle to which they are frozen, and vice versa. *The topology of such physical systems is thus fixed for all time.*

Figure 2.2 demonstrates one observable consequence of plasma being frozen along bundles of magnetic flux. The "loops" are extensions of photospheric flux bundles extending into the corona. The reason they are visible as complete loops of coronal plasma is that, once heated above $\approx$ 0.5 MK, electrons conduct heat very quickly along the bundles but not across them. EUV and X-ray observations trace out plasma frozen to bundles of magnetic field lines threading the corona. The nature of the thermal structure *across* the large-scale magnetic field is an important research area discussed later (Sect. 5.6).

Frozen-in plasma is an essential concept for coronal heating, because the vast bulk of the range of scales through which energy must cross occurs under almost ideal conditions. The frozen-in condition applies even below particle mean free paths, as the conductivity increases with fewer collisions and particles remain attached to their magnetic guiding centers. MHD turbulence is well described as an almost ideal cascade to kinetic scales even deep into the collisionless regime, where only then non-ideal processes occur, including turbulent electric fields and dissipation. Collisionless physics is, in the absence of strong, compressive viscous dissipation (reviewed in Appendix D), therefore critical in determining actual dissipation in the corona (Zweibel and Yamada 2009) on scales close to the ion inertial length $c/\omega_{pp} \approx 10$ m (Table 1.2). A central theme of this book, spontaneous imperfection, acknowledges the inference by Parker (1972, 1988, 1994) and others that the ideal behavior of the corona leads inevitably to the development of small scales, causing non-ideal behavior. The consequences are dramatic, they include heating by resistive dissipation through intense electric currents, viscous heating by dissipation of velocity gradients, and also magnetic reconnection. The latter efficiently converts magnetic free energy into bulk kinetic energy by permitting topological changes. In turn, these changes lead to unbalanced Lorentz forces,

Fig. 2.2 An image showing a longitudinal magnetogram (blue and orange with opposite polarities) and some associated hot plasma in the overlying corona (green). The switch from dominant "turbulent" fluid motions seen in in the photospheric magnetogram, an example of a plasma in a "high β" state, to the far more ordered coronal loop structures arising from the dominant magnetic stresses, a "low β plasma", is dramatic. But the switch occurs in the poorly understood stratified layer called the "chromosphere", not seen in this image, but lying between the visible structures. The magnetograms trace out "supergranule" cells of 20–30 Mm diameter, the entire field of view is about $100\times100\,\mathrm{Mm}^2$. Image from NASA's SDO gallery (https://sdo.gsfc.nasa.gov/gallery/main/item/18)

with dynamical time scales set by the Alfvén speed (Low 2003; Biskamp 2005; Kulsrud 2011; Pontin and Priest 2022). Reconnection naturally might explain the requirements demanded by flares (Gold and Hoyle 1960, read their words of wisdom on page 82). At least some heating must occur in response to dynamics following reconnection on observable scales (Longcope and Tarr 2012, 2015; Takasao et al. 2015), if only by adiabatic compressions and shocks. Thus we must also consider that, even if we are seeking solutions only to problems of steady heating, explicit time dependence cannot be dismissed on smaller scales a priori.

2.4 Equations of Motion

Written in conservative form to highlight the symmetry in the equations, the MHD equations of motion are

$$\frac{\partial}{\partial t}\rho + \nabla\cdot\rho\mathbf{u} = 0, \tag{2.13}$$

$$\frac{\partial}{\partial t}\rho\mathbf{u}+\nabla\cdot\boldsymbol{T}=\boldsymbol{\Sigma}, \tag{2.14}$$

$$\frac{\partial}{\partial t}\varepsilon+\nabla\cdot\mathbf{S}=\Sigma_{\varepsilon}, \tag{2.15}$$

where the plasma mass density is ρ, vector momentum density is $\rho\mathbf{u}$, and energy density is

$$\varepsilon=\frac{p}{\gamma-1}+\frac{1}{2}\rho u^2+\frac{B^2}{2\mu_0}. \tag{2.16}$$

These equations for macroscopic quantities are derived from the closure of Boltzmann equations discussed on Page 42. The system of equations can be most simply closed using a polytropic form for the equation of state with fixed γ:

$$p=\rho^{\gamma}, \tag{2.17}$$

or, more generally, through introduction of the plasma temperature T which must be solved for, along with the mass, momentum and energy densities, through

$$p=\frac{\rho kT}{\mu m_H} \tag{2.18}$$

with μ the mean molecular mass per hydrogen atom, m_H the hydrogen mass, and k Boltzmann's constant. In this case the value of $\gamma=C_p/C_v$ is either an ideal value $5/3$ or solved for from constituent equations for ionization and molecular formation. The momentum tensor is

$$\boldsymbol{T}=\left(p+\frac{B^2}{2\mu}\right)\boldsymbol{I}+\rho\mathbf{u}\mathbf{u}-\frac{\mathbf{B}\mathbf{B}}{\mu_0} \tag{2.19}$$

written in dyadic form, and the energy flux vector is

$$\mathbf{S}=\left(\frac{\gamma}{\gamma-1}p+\frac{1}{2}\rho u^2\right)\mathbf{u}+\frac{1}{\mu_0}\mathbf{E}\times\mathbf{B}, \tag{2.20}$$

where the first term is the enthalpy flux, and the last term the Poynting flux.

The terms $\boldsymbol{\Sigma}$ and Σ_{ε} represent sources and sinks of momentum and energy densities, and p is the plasma pressure. Closed forms for these terms involve transport coefficients derived assuming small departures from Maxwellian distributions (Braginskii 1965). Forces on the right hand side of the momentum equation include

$$\boldsymbol{\Sigma}=\underbrace{\mathbf{j}\times\mathbf{B}}_{\text{Lorentz force}}+\underbrace{\nabla\cdot\boldsymbol{\pi}}_{\text{viscosity}} \tag{2.21}$$

Sources in the energy equation include

$$\Sigma_\varepsilon = - \underbrace{\nabla \cdot \mathbf{q}}_{e^- \text{ conduction}} + \underbrace{\frac{j^2}{\sigma}}_{\text{Joule heating}} - \underbrace{\boldsymbol{\pi} : \boldsymbol{W}}_{\text{Viscous heating}} - \underbrace{\Phi_R}_{\text{Radiation losses}} \tag{2.22}$$

Here σ, $\mathbf{q}$ and $\boldsymbol{\pi}$ are the electrical conductivity, heat flux and viscous stress tensor. The rate-of-strain tensor $\boldsymbol{W}$, is a matrix of linear combinations of derivatives of the velocity field $\frac{\partial u_j}{\partial x_i}$ and its divergence (Eq. (D.1) of Appendix D), and $\boldsymbol{\pi}$ involves further linear combinations of $\boldsymbol{W}$. Thus viscous dissipation is a second order term in velocity gradients, dependent on the relationship between $\mathbf{u}$ and $\mathbf{B}$. For notational brevity we have ignored gravity.

All of the source/sink terms in the energy equation (except radiation losses) arise from the evaluation of work done by forces against collisions (i.e. "frictional" forces, Braginskii 1965). When, as in the corona, particles are magnetized ($\omega\tau \gg 1$), transport is suppressed in the direction perpendicular to the magnetic field. In terms of quantities in Table 1.2, we have $\mathbf{q} = -\kappa_\parallel T^{5/2} \nabla_\parallel T$ along the magnetic field, zero otherwise. Conductivities are slightly anisotropic with $\sigma_\parallel \approx 2\sigma_\perp$. Heating by viscosity is dominated by ions, and is strongly dependent on what generates velocity gradients along and across the magnetic field $\mathbf{B}$. For example in the presence of compressions along the magnetic field $\mathbf{B}$, viscosity can reach the zero-field limit described by η_0, but if these are absent, viscous heating occurs at a rate $\omega^2\tau^2$ smaller (see Appendix D). This makes the role of viscosity especially sensitive to the nature of mechanisms leading to small scales (large gradients in $\mathbf{u}$) and their relationship with $\mathbf{B}$. Radiation losses are included only in the energy equation, included as an extra sink of energy from the optically thin corona. They can be readily evaluated from the local electron density and temperature (see Fig. 3.7 in Sect. 3.8).

The purpose here is to show the origins of terms in the energy equation even if they are based on questionable premises (near-Maxwellian distribution functions). Even in the general case the heating occurs through the conversion of ordered motion into thermal motion by collisions between particles, a point made explicit throughout Braginskii's work. But the above formalism does reveal that dissipation can occur only on small physical scales, scaling with spatial derivatives of macroscopic variables to powers of 1 and 2.

Linearization of these equations around a homogeneous state gives the characteristic speed at which information travels through the MHD system, the Alfvén speed $c_A = B/\sqrt{\mu_0 \rho}$ which is of order 10^6 m s^{-1} in the quiet corona (Table 1.2). This mode is incompressible, the two other slow and fast MHD modes are compressible.

Before proceeding, we draw further attention to the properties of magnetized plasma on scales below the mean free paths. MHD transport coefficients qualitatively depend on the kinetic quantity $\omega\tau$. The coefficients are scalar quantities when $\omega\tau \ll 1$, i.e. when magnetic fields are weak and/or collisions are rapid. In the tenuous corona, collisions are infrequent so that $\omega\tau \gg 1$ for ions and electrons. $\omega\tau$

is also the ratio of collisional mean free path to gyroradius. When $\omega\tau \gg 1$ transport becomes anistropic, coefficients become tensors. Thus, perpendicular to magnetic field vectors, transport is suppressed by factors of $(\omega\tau)^n$ with $n = 1$ or 2 depending on the specific case (Braginskii 1965, p. 229. The case $n = 1$ is non-dissipative.). Such effects will be of concern later. We also note that Table 1.2 reveals an important hierarchy of properties in coronal plasma:

$$\lambda_p \gg \frac{c}{\omega_{pp}} > r_p.$$

The proton mean free path λ_p greatly exceeds both the proton inertial length (c/ω_{pp}) and gyro radius. The first inequality means that protons will evolve dynamically independently of electrons on scales of tens of meters, far below the mean free paths (which are similar for ions and electrons), i.e. in a collisionless regime, far removed from collisional MHD dynamics. Again this implies that, unless ordered energy is somehow thermalized by processes on larger scales, MHD alone, as in hydrodynamics, is insufficient to describe the dynamics and collisions responsible for irreversible heating.

Numerical physicists are obliged to use MHD algorithms applied to the Sun which cannot currently treat transport and heating explicitly using transport coefficients appropriate to solar conditions (Table 1.2). The dissipation in such calculations is numerical and may include tricks such as hyper viscosity (e.g. Biskamp 2003; Gudiksen et al. 2011; Rempel 2017; Przybylski et al. 2022, a higher order diffusive treatment of transport processes). Once again the assumption is that once the grid scales are reached, some process efficiently converts the otherwise ordered energy into heat, even if grids are far from reaching physical dissipation scales. Hyper-diffusion attempts to concentrate dissipation at the smallest possible scales, thereby "polluting" the resolved scales in the calculation as little as possible (Biskamp 2003).

2.5 Coronal Magnetism and Topology

Deep implications follow from the induction equation for highly conducting astrophysical plasmas. On macroscopic scales where MHD is valid the magnetic field evolution is dominated by the first (advection) term on the RHS of Eq. (2.10). This term is *non-linear* in variables **u** and **B** as the plasma and magnetic field evolve together, following those effects described qualitatively in Sect. 2.1. On the largest, observable scales, this term overwhelms the Ohmic diffusion term, the dynamics is essentially ideal, and the plasma and magnetic field are frozen one into the other. This condition implies that the *topology* of an ideal MHD system is fixed, there can be no leakage of plasma across any magnetic flux surfaces, such as those surrounding a tube, or bundle, of magnetic flux. The implications of a fixed topology, by definition a *non-local* constraint, on heating mechanisms are equally as important to coronal physics as are all non-linearities. The equations of motion (2.13)–(2.15) and Maxwell's equations are expressed in terms of partial differentials of *local* quantities. The ideal coronal system is therefore doubly-constrained, the local equations "know" nothing about global properties, such as the knotted or tangled inter-twining of bundles of magnetic flux along which plasma is forever frozen. It is easy to envision a state in which the equations of motion for an element of plasma demand a future trajectory which is instantaneously inaccessible because of the frozen topology. Parker (1972) proved by contradiction that, for a special but representative geometry, ideal magneto-fluid the system could not reach an equilibrium maintaining its topology, without developing tangential discontinuities. In other words, the magneto-fluid evolves dynamically towards *discontinuous* solutions, which in developing increasingly steep gradients, then become dissipative (Parker 1988, 1994). Like Shakespeare's Juliet, the very purity of ideal MHD leads to its own demise.

In Sect. 4.7 we discuss spontaneous imperfection, this idea that, independent of driving mechanisms, dissipation occurs by the inevitable development of small scales in the *almost* ideal world of solar MHD. The question of whether this interpretation constitutes a complete solution to the heating problem is addressed later.

2.6 Flux Freezing and a Useful One-Dimensional Picture

Our current understanding of the corona was framed in part using observations made during the SKYLAB mission (1973–1974). The (active) corona is built up from separate one-dimensional atmospheres of plasma frozen in to separate bundles of magnetic fields. Thus it was possible to consider transport along strong magnetic guide fields and ignore any cross-field transport. So as late as 1975, one-dimensional models of coronal heating based upon acoustic wave heating were still being constructed (McWhirter et al. 1975). Independent of the specific heating mechanism(s), Rosner et al. (1978, "RTV") re-discovered "loop scaling laws" arising from McWhirter's 1D models based only upon the assumptions that the coronal pressure is constant, and that a balance between conductive and radiative flux divergences exists. Equation 4.5 of McWhirter et al. (1975) gives the scaling:

$$T^3 \propto pL, \tag{2.23}$$

between coronal temperature T, pressure p and loop length L. This relation however became known as the RTV law. The RTV paper proposed that these plasma loops could be treated separately from one another, their large aspect ratios (Fig. 2.2) suggesting that one-dimensional approaches might be appropriate. Noting the stability of many loops, they showed that Eq. (2.23) had some empirical justification from SKYLAB observations, with L equal to the footpoint-to-footpoint lengths of the loops. By assuming that an upward energy flux $\mathcal{F}$ from the upper chromosphere is entirely deposited in the overlying corona, they also found another scaling:

$$\frac{\mathcal{F}}{L} \propto T^{7/2} L^{-2}. \tag{2.24}$$

Thus, whatever causes the heating (i.e., the derivative of $\mathcal{F}$ in the direction along the loop axis, with average value $\mathcal{F}/L$), for a given length L, $\mathcal{F}$ can have a large range of values and yet yield small ranges in the electron temperature T_e. Equation (2.24) shows that even large variations in $\mathcal{F}/L$ will lead to similar loop images at EUV and X-ray wavelengths. This property corresponds broadly to coronal observations, an example is in Fig. 1.3. We will review this picture quantitatively later (Sect. 5.4).

Later observations at EUV wavelengths revealed significant departures from Eq. (2.23). The strength of belief in the one-dimensional picture of plasma and energy transport remained unchanged, however. One option was then to replace a single atmosphere at a single temperature and pressure with an ensemble of 1D atmospheres, all unresolved but contributing to a given observed loop. By allowing for this new degree of freedom, with multiple threads each with their own history of heating and dynamic relaxation, some success was found in matching such ensembles to data (Aschwanden et al. 2000; Warren 2006; Ryutova et al. 2009; Walsh and Sarkar 2010). The basis for such proposals appears physically reasonable, but in effect it amounts to an exercise in the fitting of an empirical and multi-parameter model, based upon ad-hoc parameterizations of heating processes. Further, multi-threaded models appear to be in contradiction with results from the higher angular resolution Hi-C imaging telescope, where Williams et al. (2020), Klimchuk and DeForest (2020), and Williams et al. (2021) have argued that there is little structure below 200–350 km across the narrowest loops. Such issues will be discussed more deeply in Sect. 3.3.3. A similar (empirical) approach examined the effects of different heating rates imposed along the loop length (Priest et al. 1998, 2000). Naturally neither proposed model is unique, as will become clearer in Chap. 3.

The main conclusion here is that in the observed corona, with a strong guide field ($\beta \ll 1$) and under conditions of near-perfect plasma (frozen magnetic flux), the one-dimensional model is a useful starting point for understanding the heating and dynamics of coronal plasma.

2.7 Changing Magnetic Topology: Reconnection

Alfvén's theorem applies strictly only to ideal fluids (Sect. 2.3). On large scales it can provide a very good approximation for plasma dynamics and magnetic field evolution. However, the effects of electrical resistance depend on the scales. If small scales can be generated somewhere in a large coronal plasma threaded with magnetic fields, the fixed topology can change, and magnetic field lines will reconnect from one place to another, plasma belonging once to a flux bundle now

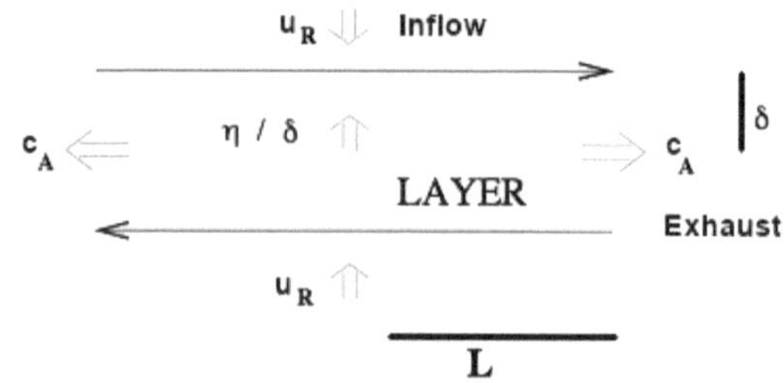

Fig. 2.3 The Sweet-Parker layer. In a steady state the magnetic diffusion speed η/δ balances the incoming reconnection speed v_R, and mass conservation requires $v_R L = v_A \delta$. The plasma outflow speed is v_A for adiabatic conditions, determined by equating the work done on the diffusion region and the kinetic energy of the outflowing plasma

inhabits another which may, for example, open into space. In order to explain the puzzling rapidity of flares, Sweet and Parker imagined bringing two flux systems together, pushing fluid aside, and bringing them into contact. Generally speaking, independent flux systems are not perfectly aligned. Then a narrow magnetic sheet, can form between the systems, which if misaligned must contain an electric current **j** through $\mathbf{j} = \mathbf{curl}\,\mathbf{B}/\mu_0$. The simplest such configuration, a "Sweet-Parker" layer with anti-parallel magnetic fields , is shown in Fig. 2.3 (Sweet 1958; Parker 1957).

The Sweet-Parker layer yields a reconnection speed of $u_R = c_A/\sqrt{S}$ with Lundqvist number $S = c_A L/\eta$. It was immediately evident that the enormous increase in speed over diffusion on scales of L at which topology could change via this mechanism was still too slow to account for time scales of minutes for flares to occur. Petschek (1964) re-examined the Sweet-Parker configuration, noting three things: diffusion across the entire length L of the diffusing layer was needed to reconnect the field lines; second, because reconnection is a topological process, a model with a shorter L can be just as effective at diffusing plasma across magnetic lines of force; thirdly, he examined the possibility that waves might be used to serve instead of diffusion as a way to move the frozen-in fields at a faster rate around the magnetically neutral layer. He thus proposed a steady-state model in which slow-mode shocks permitted plasma to flow into the exhaust region around a far shorter current sheet with length $\ll L$ of the Sweet-Parker configuration. The derived reconnection speed increased from $c_A/\sqrt{S}$ to $c_A/\ln S$, almost independent of S computed for scale L! Such a reconnection speed could indeed satisfy the requirements for the observed flare dynamics. Any mechanism which causes reconnection almost independent of S is called *fast reconnection*.

In conclusion, the theory of the 1960s showed that, in principle, one of the great puzzles of the rapid changes associated with the solar corona, could find an explanation. While not directly addressing heating per se, as we will see, this is part of a more intricate story developed during the intervening decades, with many unexpected twists and turns, and of course more questions than answers.

2.8 Electrodynamic Coupling

Figure 2.4 shows the electrodynamic coupling picture, which is adopted here to address the heating problem (Ionson 1985). Ionson introduced the framework to clarify the essentials of the heating problem. Let us first imagine that a potential magnetic field, i.e. a magnetic generated only by slowly changing sub-surface electrical currents, has emerged above the solar photosphere. Unperturbed and once it has emerged, this arrangement sends no additional free energy into the corona, there is no local energy available to induce any of the phenomena in the figure. Coronal heating and dynamics begin when we consider plasma motions **u** at and beneath the $\beta \sim 1$ surface within the chromosphere, with a component of **u** moving across the magnetic field generating an electric field, so perturbing the potential field and driving a Poynting flux upwards to generate the various manifestations of free energy shown. In the electric LRC circuit, an equivalent EMF drives a current into the circuit causing heating in the resistor, and reactive elements to amplify or diminish any AC components of the electric current (such as are generated merely by connecting the battery). In the corona, the malleable nature of the conducting plasma responds dynamically, evidently with remarkable consequences (Fig. 1.1).

The separation into "resistive" and "reactive" branches is an attempt to be guided by observations of the Sun, towards an understanding of how the corona reacts to forcing from below. The component of dynamic, small-scale electric fields in the rest frame of the plasma leads to resistive Ohmic (= Joule) dissipation via collisions between particles. The induced $-\mathbf{u} \times \mathbf{B}$ component leads to the "reactive" power highlighted in the Figure. The observation of long-lasting, steady coronal heating in

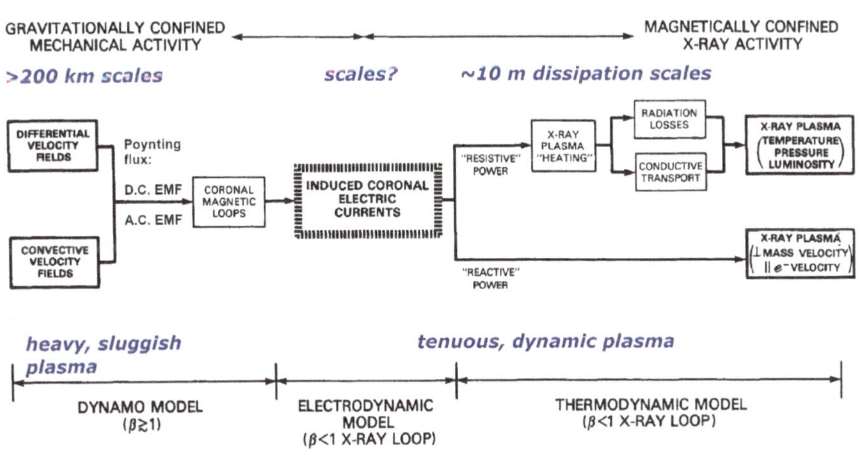

Fig. 2.4 The problem of magnetic heating of the solar corona is shown as a flow chart (modified from Ionson 1985). Energy flows from left to right, ultimately leaving the system as radiation and ejection of plasma and entrained magnetic fields from the Sun. While the figure points to coronal magnetic loops and X-rays, it applies to all electrodynamically coupled plasmas

quiet and active regions suggests that at least some of the Poynting flux is efficiently and steadily converted into heat along the resistive branch over days and weeks. The observation of flares and CMEs and other large-scale re-arrangements of active regions shows however that the Poynting flux cannot all be dissipated as heat, and that while this heating persists, free energy associated with large-scale and variable electrical currents can build up and lead to explosive behavior. Therefore while the resistive branch points to one response to a Poynting flux in which magnetic free energy is lost to heat ultimately by particle collisions, at the same time the reactive branch points to a quite different large-scale dynamic response, driven in part by dynamically fast magnetic reconnection, characterized by plasma moving with speeds a significant fraction of the Alfvén speed. Seen in this light, we must ask why the Sun is obliged to respond in these two, and perhaps only two, different ways. While this question is rarely asked, there appears to be no prior reason why such a dichotomy would have been anticipated. It would mark a significant advance if an answer were forthcoming.

The reactive branch undoubtedly is ultimately also "lossy" as the injected free energy drives coronal dynamics. The kinetic energy generated will lead ultimately to some heating as the magnetized plasmas develop small scales through, for example, shocks, or resonant surfaces (Sect. 4.6). The vast majority of the literature deals with the reactive branch, and large-scale reconnection is frequently associated with plasma heating (e.g. Shibata 1997). But can such dynamic phenomena as flares really be considered as a viable picture for steady heating (e.g. Parker 1994)?

Note that the "observables" discussed so far belong either to the extreme left (coupling of photospheric motions to photospheric magnetic fields to generate a Poynting flux), and extreme right of this figure (coronal heating and dynamics responding to the Poynting flux). On the left are observable scales, but on the resistive branch on the right we are unable to resolve directly any dynamic evolution until after dissipation has occurred. This asymmetry marks one important challenge which will be addressed in Chap. 3.

The physical region sandwiched between photosphere and corona has generally been assumed to be passive. Indeed the chromosphere receives no annotation in Fig. 2.4. No prior reason exists for presuming a passive role, but perhaps the chromosphere is generally considered too challenging, the transition region too perversely dynamic to deal with. This issue too will be identified as major weaknesses in work to date, in the next chapters.

The figure highlights the core problem of coronal heating (see also Fig. 3.1): how is kinetic energy at large scales in the sluggish (sub-) photosphere coupled through the chromosphere, to the unobservably small scales of dissipation as heat in the corona? This kind of question applies across all of astrophysics, not merely to stars (Niedner et al. 1981; Ionson and Kuperus 1984; Kuperus and Ionson 1985). On the scale of the bulk corona and scales approaching dissipation, the plasma's evolution can be understood as a perfectly conducting fluid, even below the mean free paths (Parker 2007), with electrons and ions frozen into the magnetic field down to the inertial lengths. The answers to this question in the literature include resonant processes, phase mixing, non-linear cascade of magnetic and kinetic energy downwards through an inertial range of turbulence, and the role of a fixed magnetic topology in conflict with local equations of motion. There are some conflicting ideas offering fruit for the picking. All will be discussed below.

2.9 Two Conjectures

To augment these ideas and try to pave a pathway for advancement in the future, we have found two ideas helpful. In essentially all numerical MHD work, it is assumed that the transfer to collisional kinetic scales is inevitable when Reynolds numbers are sufficiently large, $\gtrsim 10^4$ (e.g. Gudiksen and Nordlund 2005; Rempel 2017; Einaudi et al. 2021). But the reactive branch in the sequence of events epitomized by the mere existence of flares of many scales offers another dynamical route for energy transfer. To proceed, we will make this assumption explicit and call it the *"All roads lead to Rome"* conjecture ("ARLToR"):

> in the solar corona the direction of energy flow is inevitably from the large driving scales to small dissipative scales *independent of the nature of small scale processes.*

It must be classified it as a conjecture since there are situations for which it is evidently violated, but yet there may be an important class of motions in which it is essentially true, as stated in the context of turbulence by Biskamp (2003) among others. Perhaps no better counter-example is reconnection itself. While reconnection flows can ultimately be dissipated through shocks and viscosity, such dynamics is of course absent from all calculations below grid scales. Higher resolution two dimensional calculations under coronal conditions indicate that broad and shifted spectral lines would be predicted (Takasao et al. 2015) but which are not (yet) observed in steadily heated plasmas (Appendix A). Other more detailed configurations have been identified in which larger scale structures halt downward cascade in scale, such as the effect of line-tied magnetic fields. The merger of magnetic flux tubes via reconnection (Zhou et al. 2020), and the existence of a "helicity barrier" can prevent cascading to dissipation scales (Squire et al. 2022), reminiscent of the role played by conservation of cross-helicity in high-β MHD dynamos (Frisch et al. 1975; Cattaneo and Hughes 2001; Biskamp 2003).

In all numerical MHD work on the corona, energy is of course dissipated numerically at the grid scales. The calculations all are in a modest Rm regime, rarely exceeding a few times 10^4, and with Prandtl numbers around one. Rappazzo et al. (2008) studied such a forced system in detail under RMHD, showing that line-tying (i.e. the fixing of the magnetic lines of force to heavy photospheric material) inhibits any *inverse* cascade. Further, they identify the dissipative structures as narrow current sheets, the result of a non-linear cascade in a self-organizing system. The language is different from that of Parker (1994) who discusses equilibria, but the outcome is the same (Sect. 4.7). Beyond a threshold, the total dissipation becomes independent of the Reynolds numbers, in these calculations. The results clearly depend on the particular numerical problem studied, because an over-stressed loop will tend to explode in response to injection of electromagnetic energy on large scales, in particular plasma loops endowed with strong field-aligned currents during emergence (e.g. Manchester et al. 2004).

The other conjecture we introduce relates to Fig. 3.1 which gains further physical meaning upon reading the first words of an abstract of an influential paper (Sturrock 1999):

> "Coronal heating is clearly related to the coronal magnetic field. This may be due to a *passive* role of the magnetic field in modifying wave propagation and dissipation or to an *active* role resulting from the liberation of magnetic energy by reconnection or in some other way..." (our emphases).

The $B^{\flat}$ ("B-flat") conjecture[2] is that

> steady heating in the solar corona is nothing other than the journey through wavenumber space of a small magnetic field $\mathbf{b}(\mathbf{r}, t)$ on a fixed magnetic field $\mathbf{B}_0(\mathbf{r})$ which may be a potential field

The field component $\mathbf{B}_0(\mathbf{r})$ is passive. Sturrock's keen observation gains yet more clear meaning upon studying how small-scale, dissipative structures can arise, through for example resonant heating, phase mixing, formation of current sheets and MHD turbulence. In all these cases the magnetic field is envisioned as a combination

$$\mathbf{B}(\mathbf{r}, \mathbf{t}) = \mathbf{B}_0(\mathbf{r}) + \mathbf{b}(\mathbf{r}, t) \tag{2.25}$$

where $\mathbf{B}_0(\mathbf{r})$ is "the" (observable-scale) coronal magnetic field, and $\mathbf{b}(\mathbf{r}, t)$ a perturbation such that $|\mathbf{b}(\mathbf{r}, t)| \ll |\mathbf{B}_0(\mathbf{r})|$. The essence of writing (2.25) is to suggest that heating occurs without drawing energy from long-lived magnetic structures $\mathbf{B}_0(\mathbf{r})$ (see e.g., Fig. 1.11) on times exceeding dynamical time scales (Table 1.2). Clearly, if $\mathbf{B}_0(\mathbf{r})$ happens to be a potential field, this is guaranteed. The importance of the conjecture is then to try to see if it withstands scrutiny when $\mathbf{B}_0(\mathbf{r})$ is *non-potential.* In this case the free energy associated with $\mathbf{B}_0(\mathbf{r})$ may be tapped by that in $\mathbf{b}(\mathbf{r}, t)$ to trigger dynamical behavior leading to changes in $\mathbf{B}_0(\mathbf{r})$, such as by reconnection. There is of course just one magnetic field vector in space and time (away from singular points where $|\mathbf{B}_0(\mathbf{r})| = 0$). The concepts of a $\mathbf{B}_0(\mathbf{r})$ and $\mathbf{b}(\mathbf{r}, t)$ are only useful if they can guide us towards a better understanding. Thus we adopt Sturrock's idea, and test the validity and limits of its applicability with information-rich data, and try to refute it, in the hope of reaching a deeper understanding.

The formal decomposition of Eq. (2.25) only *appears* to eliminate $\mathbf{B}_0(\mathbf{r})$ from the steady heating problem, because $\mathbf{b}(\mathbf{r}, t)$ depends implicitly on $\mathbf{B}_0(\mathbf{r})$ by setting the

[2] A moniker for those interested in western music where $B^{\flat}$ denotes the note a semi-tone lower than B.

magnetic topology, the Alfvén speed and the Poynting flux traversing the corona. To lowest orders,

$$S_z = \frac{1}{\mu_0}[\mathbf{E} \times \mathbf{B}]_z \approx -\frac{1}{\mu_0}[(\mathbf{u} \times \mathbf{B_0}) \times \mathbf{B_0}]_z$$

$$= -\frac{1}{\mu_0}[\mathbf{u} \cdot \mathbf{B_0}\, B_{0z} + \mathbf{B}_0 \cdot \mathbf{B}_0\, u_z] \tag{2.26}$$

$$c_A = \frac{|\mathbf{B_0}|}{\sqrt{\mu_0 \rho}} \tag{2.27}$$

In the presence of a fixed $\mathbf{B}_0$ penetrating the solar surface, the plasma velocity $\mathbf{u}(\mathbf{r}, \mathbf{t})$ determines the Poynting flux into the corona. The conjecture represented by Eq. (2.25) focuses attention on the small-scale perturbations of $\mathbf{B}_0$ as the source of steady coronal heating, separated, for better or worse, from studies of changes in $\mathbf{B}_0$ leading to of observable (large-scale) chromospheric and coronal dynamics.

Perhaps the most important reason for proposing this conjecture is that the current literature gives the impression that magnetic reconnection on large scales (i.e. $\mathbf{B}_0(\mathbf{r})$) is the most important physical process across solar and heliospheric physics (Pontin and Priest 2022). But in terms of coronal heating, researchers should consider all scales and processes involving $\mathbf{B}(\mathbf{r}, t)$, including but not limited to reconnection, in their desire to solve the problem. It should also be noted that this picture is not new, MacKay et al. (2000) speculated that

> "...the Soft X-ray Telescope on Yohkoh showed that there are two distinct parts to the coronal heating mechanism. First there is a local time-dependent, impulsive part which occurs on short time scales and secondly a global steadier part which acts on a much larger scales (Acton 1996; Tsuneta 1996). This may indicate that different mechanisms are working together to heat different parts of the corona in different ways."

Chapter Summary

MHD is a reasonable first approximation to begin following energy from observable to smaller scales in the corona. However,

MHD cannot be justified a priori to follow the flow of energy to dissipation scales in coronal heating.

At scales below mean free paths and for supra-thermal particles, kinetic physics is important.

However, the fixed magnetic topology in ideal MHD still plays a major role in dynamic evolution towards dissipative scales.

Is coronal heating inevitably controlled by *spontaneous imperfection*, part of a self-controlling system independent of the nature of the driver?

ARLToR conjecture: The corona adjusts itself to dissipate free energy injected into it on all scales, without regard the final dissipation process.

(continued)

$B^{\flat}$ conjecture: The macroscopic, measurable magnetic field plays only a passive role in heating the steady corona (Sturrock 1999).

New approaches and methodologies will be proposed in Chap. 6, guided by these conclusions and conjectures.

References

Acton, L.W.: Comparison of YOHKOH X-ray and other solar activity parameters for November 1991 to November 1995. In: Pallavicini, R., Dupree, A.K. (eds.) ASP Conference Series 109: Cool Stars, Stellar Systems, And the Sun, vol. 9, 45 (1996)

Alfvén, H.: Existence of electromagnetic-hydrodynamic waves. Nature **150**, 405 (1942)

Alfvén, H.: Cosmical Electrodynamics. International Series of Monographs on Physics. Clarendon Press, Oxford (1950)

Aschwanden, M.J., Nightingale, R.W., Alexander, D.: Evidence for nonuniform heating of coronal loops inferred from multithread modeling of TRACE data. Astrophys. J. **541**, 1059–1077 (2000)

Biskamp, D.: Magnetohydrodynamic Turbulence. Cambridge University Press, Cambridge (2003)

Biskamp, D.: Magnetic Reconnection in Plasmas. Cambridge University Press, Cambridge (2005)

Bogoliubov, N.: Kinetic equations. J. Exp. Theor. Phys. **16**(8), 691–702 (1946)

Braginskii, S.I.: Transport processes in a plasma. Rev. Plasma Phys. **1**, 205–311 (1965)

Cattaneo, F., Hughes, D.W.: Solar dynamo theory: a new look at the origin of small-scale magnetic fields. Astron. Geophys. **42**, 3.18–3.22 (2001)

Chandrasekhar, S.: Hydrodynamic and Hydromagnetic Stability. Clarendon, Oxford (1961)

Chapman, S.: The electrical conductivity of the ionosphere: a review. N. 4 del Suppliemento al Vol.4 Serie X, del Nuovo Cimento, pp. 1385–1412 (1956)

Einstein, A.: Zur Elektrodynamik Bewegter KÖRper. Annalen der Physik, **322**, 891–921 (1905)

Einaudi, G., et al.: Energetics and 3D structure of elementary events in solar coronal heating. Astrophys. J. **910**, 84 (2021)

Ferraro, V.C.A., Plumpton, C.: An Introduction to Magneto-Fluid Mechanics, 2nd edn. Clarendon Press, Oxford (1966)

Frisch, U., et al.: Possibility of an inverse cascade of magnetic helicity in magnetohydrodynamic turbulence. J. Fluid Mech. **68**, 769–778 (1975). https://doi.org/10.1017/S002211207500122X

Ginzburg, V.L.: The Propagation of Electromagnetic Waves in Plasmas. International Series of Monographs in Electromagnetic Waves. 2nd revised, enlarged edn. Pergamon Press, Oxford (1970)

Gold, T., Hoyle, F.: On the origin of solar flares. Mon. Not. R. Astron. Soc. **120**, 89, 1960

Gudiksen, B.V., Nordlund, Å.: An *ab initio* approach to the solar coronal heating problem. Astrophys. J. **618**, 1020–1030 (2005)

Gudiksen, B.V., et al.: The stellar atmosphere simulation code BIFROST. Code description and validation. Astron. Astrophys. **531**, A154, 2011

Ionson, J.A.: Coronal heating by resonant (A.C.) and nonresonant (D.C.) mechanisms. Astron. Astrophys. **146**, 199–203 (1985)

Ionson, J.A., Kuperus, M.: Black hole accretion disks - coronal stabilization of the Lightman-Eardley instability. Astrophys. J. **284**, 389–393 (1984). https://doi.org/10.1086/162418

Klimchuk, J.A., DeForest, C.E.: Cross sections of coronal loop flux tubes. Astrophys. J. **900**, 167 (2020)

Kulsrud, R.M.: Plasma Physics for Astrophysics. Princeton University Press, Princeton/Oxford (2005)

Kulsrud, R.M.: Intuitive approach to magnetic reconnection. Phys. Plasmas **18**, 111201 (2011)

Kuperus, M., Ionson, J.A.: Black hole accretion disks. electrodynamic coupling of accretion-disk coronae and the partitioning of soft and hard X-ray emission. Astron. Astrophys. **148**, 309–311 (1985)

Lazarian, A., Vishniac, E.T.: Reconnection in a weakly stochastic field. Astrophys. J. **517**, 700–718 (1999). https://doi.org/10.1086/307233

Lazarian, A., et al.: 3D turbulent reconnection: theory, tests, and astrophysical implications. Phys. Plasmas **27**, 012305 (2020). https://doi.org/10.1063/1.5110603

Lemen, J.R., et al.: The atmospheric imaging assembly (AIA) on the solar dynamics observatory (SDO). Sol. Phys. **275**, 17–40 (2012)

Liu, W., et al.: The radial distribution of ion-scale waves in the inner heliosphere. Astrophys. J. **951**, 69 (2023). https://doi.org/10.3847/1538-4357/acd53b

Longcope, D.W., Tarr, L.: The role of fast magnetosonic waves in the release and conversion via reconnection of energy stored by a current sheet. Astrophys. J. **756**, 192 (2012). https://doi.org/10.1088/0004-637X/756/2/192

Longcope, D.W., Tarr, L.A.: Relating magnetic reconnection to coronal heating. Philos. Trans. R. Soc. Lond. Series A **373**, 20140263–20140263 (2015). https://doi.org/10.1098/rsta.2014.0263

Low, B.C.: Magnetic reconnection and the solar corona. In: Pevtsov, A.A., Uitenbroek, H. (eds.) ASP Conference Series 286: Current Theoretical Models and Future High Resolution Solar Observations: Preparing for ATST, pp. 335–346 (2003)

MacKay, D.H., et al.: How accurately can we determine the coronal heating mechanism in the large-scale corona? Sol. Phys. **193**, 93–116 (2000)

Manchester, W., I., et al.: Eruption of a buoyantly emerging magnetic flux rope. Astrophys. J. **610**, 588–596 (2004). https://doi.org/10.1086/421516

McWhirter, R.W.P., Thoneman, P.C., Wilson, R.: Heating of the solar corona II. A model based on energy balance. Astron. Astrophys. **40**, 63–73 (1975)

Michelson, A.A., Morley, E.W.: On the relative motion of the earth and of the luminiferous ether. Sidereal Messenger **6**, 306–310 (1887)

Niedner, M.B., Ionson, J.A., Brandt, J.C.: Interplanetary gas. XXVI - on the reconnection of magnetic fields in cometary ionospheres at interplanetary sector boundary crossings. Astrophys. J. **245**, 1159–1169 (1981). https://doi.org/10.1086/158891

Parker, E.N.: Sweet's mechanism for merging magnetic fields in conducting fluids. J. Geophys. Res. **62**, 509–520 (1957)

Parker, E.N.: Topological dissipation and the small-scale fields in turbulent gases. Astrophys. J. **174**, 499 (1972)

Parker, E.N.: Nanoflares and the solar X-ray corona. Astrophys. J. **330**, 474–479 (1988)

Parker, E.N.: Spontaneous Current Sheets in Magnetic Fields with Application to Stellar X-Rays. International Series on Astronomy and Astrophyics. Oxford University Press, Oxford (1994)

Parker, E.N.: Conversations on Electric and Magnetic Fields in The Cosmos. Princeton University Press, Princeton (2007)

Pesnell, W.D., Thompson, B.J., Chamberlin, P.C.: The solar dynamics observatory (SDO). Sol. Phys. **275**, 3–15 (2012)

Petschek, H.E.: Magnetic field annihilation. NASA Special Publication **50**, 425 (1964)

Pontin, D.I., Priest, E.R.: Magnetic reconnection: MHD theory and modelling. Living Rev. Sol. Phys. **19**, 1 (2022)

Priest, E.R., et al.: Nature of the heating mechanism for the diffuse solar corona. Nature **393**, 545–547 (1998). https://doi.org/10.1038/31166

Priest, E.R., et al.: A method to determine the heating mechanisms of the solar corona. Astrophys. J. **539**, 1002–1022 (2000). https://doi.org/10.1086/309238

Przybylski, D., et al.: Chromospheric extension of the MURAM code. Astron. Astrophys. **664**, A91 (2022). https://doi.org/10.1051/0004-6361/202141230

Rappazzo, A.F., et al.: Nonlinear dynamics of the Parker scenario for coronal heating. Astrophys. J. **677**, 1348–1366 (2008). https://doi.org/10.1086/528786
Rempel, M.: Extension of the MURAM radiative MHD code for coronal simulations. Astrophys. J. **834**, 10 (2017)
Rosner, R., Tucker, W.H., Vaiana, G.S.: Dynamics of the quiescent solar corona. Astrophys. J. **220**, 643 (1978)
Ryutova, M., Frank, Z., Berger, T.: Formation and dynamics of multi-thread arcades of coronal loops. In Lites, B., Cheung, M., Magara, T., Mariska, J., Reeves, K. (eds.) The Second Hinode Science Meeting: Beyond Discovery-Toward Understanding. Astronomical Society of the Pacific Conference Series, , vol. 415, p. 291 (2009)
Scudder, J.D.: On the causes of temperature change in inhogeneous low-density astrophysical plasmas. Astrophys. J. **398**, 299–318 (1992)
Shibata, K.: Rapidly time variable phemonmena: jets, explosive events, and flares. In Wilson, A. (ed.) The Fifth Soho Workshop, the Corona and Solar Wind Near Minimum Activity, pp. 103–112. ESTEC, Noordwijk, The Netherlands: ESA SP-404 (1997)
Spitzer, L.: Physics of Fully Ionized Gases. Interscience Publishers, New York (1956)
Squire, J., et al.: High-frequency heating of the solar wind triggered by low-frequency turbulence. Nat. Astron. **6**, 715–723 (2022). https://doi.org/10.1038/s41550-022-01624-z
Sturrock, P.A.: Chromospheric magnetic reconnection and its possible relationship to coronal heating. Astrophys. J. **521**, 451–459 (1999). https://doi.org/10.1086/307544
Sweet, P.A.: The neutral point theory of solar flares. In Lehnert, B. (ed.) Electromagnetic Phenomena in Cosmical Physics. IAU Symposium, , vol. 6, p. 123 (1958)
Takasao, S., et al.: Magnetohydrodynamic shocks in and above post-flare loops: two-dimensional simulation and a simplified model. Astrophys. J. **805**, 135 (2015). https://doi.org/10.1088/0004-637X/805/2/135
Tsuneta, S.: The dynamic solar corona in X-rays with Yohkoh. In: Tsinganos, K.C., Ferrari, A. (eds.) Solar and astrophysical magnetohydrodynamic flows. NATO Advanced Study Institute (ASI) Series C, vol. 481, pp. 85–108 (1996)
Walsh, R., Sarkar, A.: Observed flows in coronal loops generated by multi-thread modelling. In: 38th COSPAR Scientific Assembly, vol. 38, p. 3 (2010)
Warren, H.P.: Multithread hydrodynamic modeling of a solar flare. Astrophys. J. **637**, 522–530 (2006). https://doi.org/10.1086/497904
Williams, T., Walsh, R.W., Morgan, H.: The cross-sectional shape and height expansion of coronal loops: high-resolution coronal imager (Hi-C) Analysis of AR 12712. Astrophys. J. **919**, 47 (2021)
Williams, T., et al.: Is the high-resolution coronal imager resolving coronal strands? Results from AR 12712. Astrophys. J. **892**, 134 (2020)
Zhou, M., Loureiro, N.F., Uzdensky, D.A.: Multi-scale dynamics of magnetic flux tubes and inverse magnetic energy transfer. J. Plasma Phys. **86**, 535860401 (2020). https://doi.org/10.1017/S0022377820000641
Zweibel, E.G., Yamada, M.: Magnetic reconnection in astrophysical and laboratory plasmas. Ann. Rev Astron. Astrophys. **47**, 291–332 (2009)

Chapter 3
Challenges

The hurdles preventing the clear identification of specific heating mechanisms are essentially unchanged from the 1960s. Some challenges such as limited instrumental and numerical resolutions, radically improved since then, or the intrinsic non-linear nature of the governing equations of plasma motion, are easily appreciated yet still present the same qualitative problems. Others are more nuanced. The information content of remotely sensed radiation is naturally limited. Inverse problems are ill-posed, requiring additional information in order to find solutions. Such information is often subjective, ad-hoc, or philosophically more worrisome, added from numerical calculations (see Sect. 4.10, but also refer to de Pontieu et al. 2014; Carlsson et al. 2019, for an entirely different viewpoint). Other issues include both radiative transfer and heat conduction which redistribute radiation and heat in space. EUV and X-ray radiation from coronal plasma is optically thin, leading to line of sight confusion. Last but not least, we have no measurements of the evolving coronal magnetic field, the agent through which coronal heating occurs. The purpose of this chapter is to pinpoint important limitations both in models and observations to avoid potential pitfalls and identify possible routes towards progress. We begin with a brief overview of how magnetic energy is thought to be transferred from macroscopic to dissipative scales.

3.1 Disparate Scales, Diverse Regimes

The solar plasma undergoes critical regime changes between the driving and dissipation scales (Fig. 3.1). A hydrodynamically controlled, convecting high-β, partially ionized collisional plasma drives the low-β, weakly collisional magnetized plasma. Dynamical changes in the photosphere and corona propagate at wave speeds of several km s^{-1} and at Mm s^{-1} respectively. Consequently there is a "impedance mis-match" between photospheric motions and the coronal response (Fig. 1.1). The

P. Judge, J. A. Ionson, *The Problem of Coronal Heating*, Astrophysics and Space Science Library 470, https://doi.org/10.1007/978-3-031-46273-3_3

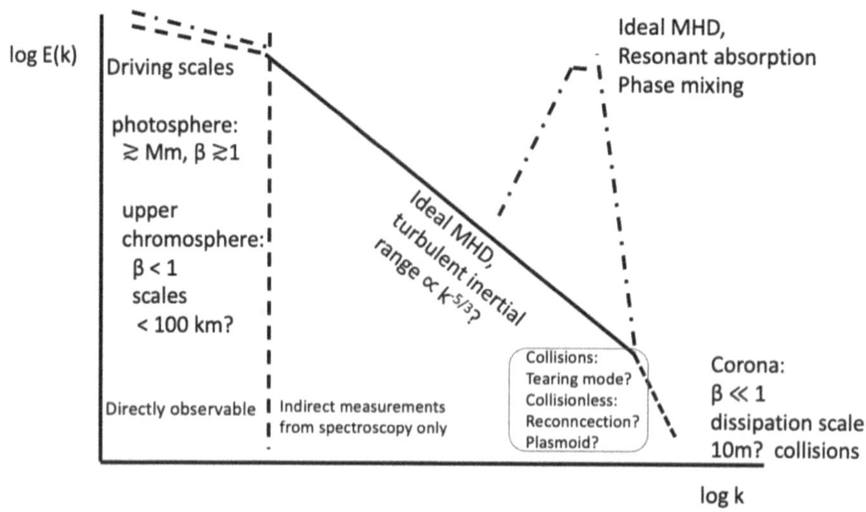

Fig. 3.1 Schematic energy spectra $E(k)$ over the huge range of scales (5 decades) required to follow energy from large to small scales are shown. A typical turbulent picture leads to $E(k) \sim E^{-\alpha}$, a direct transfer from large to small scales (e.g., via phase mixing) exhibits a peak (dot-dashed line). Highlighted in the box is a scale of particularly active research, the transition from ideal to non-ideal MHD conditions (Schekochihin 2022)

electrodynamic coupling is consequently weaker because the EMF generated by photospheric motions occurs mostly at frequencies lower than the natural oscillation frequency of the coronal LRC circuit (Ionson 1984, 1985a,b). Parker (1991a) used the same analogy in a study of heating efficiency of coronal hole plasmas by Alfvén waves. Figure 3.1 shows the enormous range of scales between the outer, driving scales of $\gtrsim$40 km under and in the photosphere (Fig. 1.5), to dissipation scales in the corona, which may approach meters. At least four orders of magnitude separate these scales. Given the short dynamical times and long observed lifetimes of heated coronal plasma outside of flares, it is reasonable to assume a steady state. With the impedance mis-match, perhaps the system undergoes a quasi-static evolution with rapid but modest adjustments to an average MHD state.

A statistically steady state is a common assumption in MHD turbulence theory. There, the energy injected at small k is steadily dissipated as heat at large k, and subsequently lost to heat conduction, radiation and plasma motions, bulk and differential (fluid compression, shear, and electric currents). Owing to the presence of the strong guide field $\mathbf{B}_0$, we will benefit from separating dynamical variables parallel and perpendicular to $\mathbf{B}_0$, with annotations $u_\parallel$ and $u_\perp$, and $b_\parallel$ and $b_\perp$ respectively.

The current phenomenological understanding of low β MHD turbulence theory (based largely on numerical work and solar wind observations) assumes that *MHD turbulence requires the non-linear interaction between two oppositely directed Alfvén waves* (Iroshnikov 1964; Kraichnan 1965; Goldreich and Sridhar 1995). This is not to say that compressible modes are unimportant, indeed they can be generated by the non-linear terms intrinsic to the MHD turbulence picture. Brandenburg and Lazarian (2014) reviewed compressible MHD turbulence, noting that the interacting Alfvén wave picture is applicable to compressible fluids. Compressibility can reduce the imbalance between Alfvén waves. Observationally, under conditions of the undisturbed quiet and active corona, subsonic line widths suggest that compressible effects are modest in the heating problem. Numerical work by Cho et al. (2002), Cho and Lazarian (2003), even under supersonic conditions, suggests that coupling between the slow, fast and Alfvén modes does not greatly affect scaling laws for the turbulent cascades of each mode. The Alfvén mode can cascade to the slow mode, but the the fast mode scalings behave as isotropic fluid turbulence. Computationally, the convenient RMHD model excludes compressive modes. Nevertheless compressive effects in heating cannot entirely be discounted, because strong MHD turbulence will generate compressions with a component along the total magnetic field vector **B**. Even if subsonic, these motions can be strongly damped by viscosity (see Sect. 5.2, Appendix D). We will follow the development of MHD turbulence below based on the incompressible case, but will discuss the role of compressive effects in dissipation as needed.

In the turbulence picture, the non-linear inertial terms dominate dynamics across the bulk of the wavenumber range. Under the influence of a driven, strong guide field on macroscopic scales, they produce the cascade of magnetic and kinetic energy from large to small scales, mostly through the interaction of motions perpendicular to $\mathbf{B}_0$ (Appendix C). We must remember that this is just one particular model of energy transfer to small scales. Even within this picture, we have seen that inverse cascades occur in natural MHD systems, for example when reconnection enables the formation of larger structures. An example is the turbulent dynamo (Cattaneo and Hughes 2001), relating to the conservation of scalar quantities "cross helicity" as well as energy (e.g. Biskamp 2003; Schekochihin 2022). But, as we shall see, energy is not obliged to flow always in the direction to smaller scales even with strong guide fields. As is well-known, steady state MHD turbulence yields power-law shapes for the distribution of magnetic and kinetic energy with scale length, corresponding to the solid straight line in Fig. 3.1 with negative exponents of wavenumber k (see Appendix C).

Also shown in the figure is a schematic power spectrum for mechanisms such as resonant absorption or phase-mixing, where large amplitude velocity fluctuations are induced directly in mildly inhomogeneous layers even from small amplitude, long-period waves associated with large-scale driving motions. Along each narrow bundle of magnetic flux, Alfvén waves propagate without knowledge of their neighbors along the real characteristics of the wave equation. This linear picture breaks down in the presence of a resonance of the wave with the local Alfvén speed (resonant absorption), or until gradients across the bundles become so steep that non-linear effects couple otherwise independent Alfvén waves (phase mixing). This is richer behavior than can be exhibited by the single LRC analog, owing to the ability of a fluid to develop its own internal dynamics, in this case internal, localized waves owing their existence to internal gradients in Alfvén speed (Ionson 1978). The corona might be considered as a continuum of LRC circuits each with their own values of L, R and C. The appeal of this picture is that gradients are guaranteed to exist in these coupled plasmas, and therefore local resonances, and enhancements of dissipation through R, must occur *somewhere* in the plasma. Provided these occur frequently enough compared to a cooling time (tens of minutes, Table 1.2), the heating problem is qualitatively "solved"! While this does not meet our criterion for a physical solution (Sect. 1.7), nevertheless, and with some justification, some consider this sufficient to declare that there is no longer a *fundamental* problem to be solved. We respectfully disagree, offering a different perspective throughout this text.

The dot-dashed lines in Fig. 3.1 represent schematically this more direct form of transport of energy to dissipative scales within the corona. Again, the velocity and magnetic field fluctuations are perpendicular to the unperturbed magnetic field, owing to the ideal condition of frozen field and plasma undergoing Alfvénic oscillations. The figure indicates that if power spectra $E(k_\perp)$ can be measured, including perhaps measurements of plasma velocities differentially as a function of the angle between line of sight vector and $\mathbf{B}_0$, then they might discriminate between the cascading MHD picture and other pictures where large-scale motions are converted more directly to small scales for dissipation.

The source of the corona's energy on large scales is a Poynting flux arising from magneto-convection and magnetic flux emergence (Fig. 2.4), modified by the dense and partially ionized chromosphere before reaching the corona. The way the chromosphere changes the nature or amount of Poynting flux available to heat the corona is almost entirely unknown. In presenting average conditions it would seem that, somehow only a relatively constant and small fraction of the energy entering the chromosphere is sent onwards to heat the corona (Table 1.3, see the discussion in Sects. 5.4 and 6.2). What might this observation imply? Already it is clear from images such as Fig. 2.2 that only some bundles of emerging magnetic fields host hot plasma, even when the underlying magnetic fields show nothing special at their footpoints. This constitutes a problem to be discussed in Sect. 5.6. Only further studies of the chromosphere and its coupling to the corona will provide answers.

While the driving scales can, to some degree, be directly observed in the dense photosphere, dissipation scales cannot. In principle, we ought to be able to determine driving scales and other conditions closer to the coronal base, at the upper levels of the chromosphere. However this has not been achieved, despite the likelihood that the chromosphere is not a passive, linear filter of upward directed energy and magnetic field (Sturrock 1999). How do we know what scales dissipation occurs on? This question is being intensely scrutinized following recent advances in understanding of the transition from MHD scales to dissipation scales (see later Sect. 4.3 and the boxed text in Fig. 3.1). In this picture, the heating problem only can be deemed as solved when we arrive at a full understanding of the journey of the energy flux across wavenumber space (Fig. 3.1) as well as position and time (Fig. 2.4).[1]

[1] This does not require that every possible process on all scales needs to be understood if it can be convincingly proven that, beneath some scale yet to be identified, dissipation is inevitable.

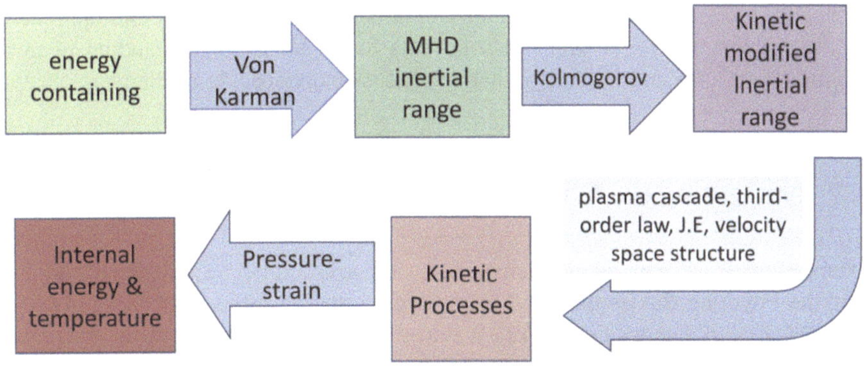

Fig. 3.2 A recent proposal for energy transport from large, MHD-scales ("Von Karman" and "Kolmogorov" arrows) to kinetic scales is shown, from Matthaeus et al. (2020). Below MHD scales, energy is transferred between various terms in the Vlasov equation ("plasma cascade" and "pressure-strain" arrows) to thermal energy increase. "Pressure-strain" terms in the Vlasov equation are similar to "pdV-work" in the more familiar fluid equations. This remains an active research area

Towards the high-k end of the inertial range of MHD turbulence, the transition from ideal to non-ideal conditions remains unclear (see the box in Fig. 3.1, and detailed discussions by Matthaeus et al. 2020; Liu et al. 2022). In the turbulence literature, questions arise as to the MHD plasmoid instability, the role of separate ion and electron dynamics, the collisionless dynamics of processes such as reconnection, and other plasma processes in accelerating electrons before they have time to relax through impacts with surrounding plasma (section 14 of Schekochihin 2022). The large mean free paths of ions and electrons ($\approx 10^5$ m in $T = 10^6$ K coronal plasma, scaling as T^2) suggest that collisionless processes are likely to occur before the classical processes of viscous dissipation, Joule and ion-neutral heating are able to convert bulk velocity gradients and currents into heat. But, and this is an important *caveat*, if compressive motions parallel to $\mathbf{B}_0$ are significant, this may dissipate energy well before kinetic scales are reached (see Sect. 5.2 and Appendix D). The novel conceptual picture introduced by Matthaeus et al. (2020) focuses on the tricky MHD-to-kinetic region (previously highlighted by the review by Tsiklauri 2009). In their heuristic model of energy transport on kinetic scales, primary "pathways", i.e. terms in the Vlasov equation, are identified, through which energy is dissipated (converted to internal energy, see Fig. 3.2). Heuristically, this amounts to a complex generalization of the Kolmogorov cascade (Kolmogorov 1941). Of the processes shown in Fig. 3.2, the pressure-strain interaction (epitomized by fluid "pdV work" but applied not to a fluid but electron and proton distribution functions) is found to balance the dissipation rate at kinetic scales, regardless of the scale separation (Yang et al. 2022).

These are just some of the general complications intrinsic to the huge range of scales involved with the coronal heating problem. The transport of energy at MHD to kinetic scales and dissipation remains an intense area of research. Our approach to this problem is to seek signs in solar data to aid in our understanding of this problem (Sects. 4.5 and 5.11). But first we discuss more specific challenges.

3.2 Navigating the Chromosphere

Even before these general problems arise, embarrassingly, we still do not know how the Poynting flux is modified as it traverses the chromosphere. Roughly 5% and 10% of this energy flux makes it into the corona, according to the averages given in Table 1.3. The range of possibilities entailed in a first-principles MHD approach to this problem is daunting, but they are being systematically explored numerically, from explorations of elementary processes by Arber et al. (2007), to attempts to construct a physically complete model, e.g. Nóbrega-Siverio et al. (2020), Przybylski et al. (2022). It has been shown convincingly that only a small fraction of the magnetic field (note, this is not the same as the Poynting flux) associated with granular dynamics reaches above the chromosphere to emerge into the corona (Rempel 2014; Lites et al. 2017), the bulk of the granular magnetic fields returning to the surface. Studies addressing chromospheric processes that change the Poynting flux available for coronal heating and dynamics are needed, both observational and theoretical.

The chromosphere is far from being a tame, benign region, a mere inconvenience for numerical work. For example, compressive slow modes in the chromosphere will tend to produce shocks and accompanying high frequency perturbations as waves propagate into lower density plasma. Reconnection is not unlikely across the whole chromosphere owing to its conductivities which are smaller than the corona (Sturrock 1999). Fast modes will generate Alfvén waves close to the $\beta = 1$ surface (Cally 2011), with β evaluated using the sum of ionized and neutral partial pressures, because of rapid ion-neutral collisional coupling.[2] In the literature, the treatment of the chromosphere varies, from little modification (implicit in, for example Cranmer and van Ballegooijen 2005), to the generation and mode conversion of Alfvén waves near the $\beta = 1$ surface, and the generation of high frequency waves (>100 Hz) resulting from fast reconnection in the upper chromosphere (e.g. Axford and McKenzie 1992; Sturrock 1999). Observations of reconnection including rapid "explosions" in the chromosphere support this picture (Dere 1994; Peter et al. 2014; Judge 2015), along with numerical calculations (Heggland et al. 2009; Wargnier et al. 2023). With new, more capable facilities such

[2] Collisions between charges and neutral atoms were needed and compiled by aeronomers (Banks and Kockarts 1973). The hydrogen-proton collision cross section for momentum-changing is unusually large, $\approx 200\pi a_0^2$, involving a resonant charge-transfer interaction at energies in the solar chromosphere.

as DKIST capable of detecting small magnetic fields and their changes, this is an area ripe for novel observational research.

Importantly, observations of plasma properties alone are not sufficient to identify the nature of magnetic fields and upward Poynting flux emerging from the chromosphere. Measurement of the evolving magnetic and velocity fields *at the base of the corona* is a multi-faceted problem, considerably more challenging than measurements in photospheric plasma (Metcalf et al. 1995). However, some fresh approaches have been proposed which might help in this particular problem (Judge et al. 2021, 2022), to augment the paucity of information available in radiation that reaches the ground. In Sect. 3.9 we will address the peculiar nature of the interface between chromosphere and corona, spicules, jets, fibrils, corrugations and all.

3.3 Remote Sensing

The richness of possible solutions to almost-ideal MHD equations that are possible under the scenario shown in Fig. 2.4 is formidable. First principles computer modeling cannot, without leaps of faith, convincingly capture the entire class of solutions to the coupled governing partial differential MHD equations, constrained to maintain the large-scale magnetic field topology.

In the words of theorist B.C. Low, "solar physics is an observationally driven science" (private communication, ca. 2010). As well as possessing native uncertainties, observations do not contain sufficient information desired to follow the sequence of events represented in Fig. 2.4, and much information from multiple spectral lines is redundant. We have no measurements of magnetic fields evolving in the corona, even though plasma heating just using classical transport coefficients requires us to estimate the rate of change of current density vector $\mathbf{j} = \mathbf{curl}\,\mathbf{B}/\mu_0$ or differentials of the fluid velocity vector, both of which require measurements in all 3 dimensions. The solar literature is replete with ingenious ways around these difficulties, usually observations are combined/compared with models to be able to assess the efficacy of a given hypothesis (see e.g. the many examples discussed by Aschwanden 2005; de Pontieu et al. 2021). At best, this methodology, driven largely by necessity, lacks a clean confrontation between modeled and observed behavior. In advocating for refutation as a pillar of scientific method, Karl Popper would probably not approve. But naturally this mix of theory and observation is reflected in almost all astronomical research. The problems facing researchers seeking answers to coronal heating are far removed from many of the solved problems involving elementary physics and astronomy, such as the successes of the theory of stellar evolution which spans so many disciplines (Iben 2012). Instead we must acknowledge that the problem is rooted in strongly-driven complex systems capable of intricate dynamical evolution (Holland 2014).

An obvious, but perhaps important limitation is that we cannot observe *all* the plasma in any region of the solar corona. The frozen-flux condition implies that we cannot assume that the corona is well-mixed across the dominant magnetic field, we must expect significant cross-field inhomogeneities. Since the brightness of optically thin coronal plasma varies as n_e^2, we must recognize that we may not see vast volumes filled with magnetic field but almost devoid of plasma. The data that we receive represent at best a biased view of what is most evidently an inhomogeneous plasma. There are entire regions devoid of sufficiently bright plasma beneath our ability to measure whose untraceable magnetic fields can support very fast MHD waves along and across magnetic lines of force. Like dark energy and matter, we must be open to the fact that we may see only part of the story when we observe the corona. We must also be aware that sometimes the *absence* of a signal is as important as those (biased) signals that we are able to see.

The deepest exposures of chromospheric and coronal plasmas sometimes reveal tracers of supersonic motions as infrequent and broad line emission. Most easily observed in spectra obtained at high angular resolution, they also appear even in stellar spectra as weak emission in the wings of "transition region" lines. These explosive events seen in the Sun and stars are believed to be related to magnetic reconnection, possible agents in coronal heating in general (e.g. Dere et al. 1989, 1991; Simnett 1994; Shibata 1997; Innes et al. 1997; Wood et al. 1997; Peter 2006; Innes et al. 2015, ... and the list goes on. The large number of references reflects in part the availability of data for easily observable Li- and Na-like ions at UV wavelengths, see page 96). It may be that under certain scenarios (e.g., nanoflares) deep exposures might set limits on the viability of certain models. Such observations illustrate the biases inherent in remotely sensed data, in that important but weak radiative signatures can be missed. Weak or even absent spectral features may well represent robust signatures from specific mechanisms, or hypotheses. As we will emphasize later (Sect. 6.1), zeros can be as important as ones, in information-rich datasets.

3.3.1 2D Observations Versus 3D Structure

The inference that magnetic fields and electrical currents are implicated in heating demands that *observations reveal information in all three spatial dimensions*. The electrodynamics demands that to describe heating mechanisms three dimensions, or at least two dimensions and a third ignorable coordinate, are needed to handle the pseudo-vectors **B** and **j**, and Poynting flux **S** (Eq. (2.26)). There is no prior reason to expect that coronal heating is amenable to treatments with ignorable coordinates, indeed current understanding implies that inhomogeneities be present to enable efficient heating. Consider the anisotropic nature of dissipative structures in MHD turbulence (Appendix C). Observations also support this position. Imagine, for example, extrapolating magnetic fields from the contorted patterns shown in Fig. 1.5 and/or the cross-field coronal plasma structures at the edges of loops (Fig. 2.2). The frozen-field condition alone demands that neighboring bundles of magnetic field are loaded with their own plasma, each with their own histories and thermodynamic states. We should not expect to address this problem successfully in fewer than three spatial dimensions.

Photons emerging from solar plasmas are measured with instruments mostly situated along the Earth-Sun line. All measurements from Earth involve integrations of a radiative transfer equation along this one line of sight (henceforth "LOS"). Thus, the kind of 3D information we usually acquire from Earth's viewpoint is asymmetric. Just a few spacecraft, in particular the STEREO spacecraft (Kaiser et al. 2008), Solar Orbiter (Marsch et al. 2005) and Parker Solar probe (Kinnison et al. 2013; Fox et al. 2016), have been able to acquire solar observations from well away from the Earth-Sun line. Therefore, to recover 3D information it is usually necessary to find ways to determine light originating from different places along each LOS. In an entertaining article entitled "Revealing the True Solar Corona" Woo (2010) discusses potential pitfalls including optical illusions such as Moiré patterns, or murmurations of starlings, in which our eye and brain is fooled into seeing patterns when there is no such underlying physical structure.

The intensities of spectral lines and continua of different ions do emerge from different locations along the LOS (e.g. see the figures in articles by Vernazza et al. 1973, 1976, 1981). This is most readily appreciated when the plasma is optically thick to a variety of spectral features with different optical depths. The emergent intensity is then given roughly by the "Eddington-Barbier relation" (refer to Rutten 2003; Mihalas 1978, for definitions of source function S_λ, τ_λ, also Sect. 1.5):

$$I_\lambda \approx S_\lambda(\tau_\lambda = 1), \tag{3.1}$$

and since

$$\tau_\lambda = \int \alpha_\lambda \, ds \tag{3.2}$$

$$= \int \kappa_\lambda \, dm \tag{3.3}$$

with s distance along the LOS and α_λ the opacity in m^{-1}, different for each spectral feature, then the source function can be retrieved as a function of s. The second definition of τ_λ uses the opacity $\kappa_\lambda = \alpha_\lambda/\rho$ in $\mathrm{m}^2\ \mathrm{kg}^{-1}$, to be multiplied by the column mass increment $dm = \rho ds$, with ρ the mass density. Since $\alpha_\lambda \propto \rho$, κ_λ becomes sensitive only to effects of elemental abundance, ionization and removes the main dependency of stratification across the atmosphere. If the spectral features (which may include radio-wavelengths) are sensitive to magnetic fields, then in principle some components of $\mathbf{B}(s)$ can be recovered (as reviewed by Casini et al. 2017). But when the plasma is not opaque (i.e. optically thin, such as in most X-ray, EUV coronal lines) then the intensity is just the sum of the emission along each line of sight, plus a contribution from background sources of radiation, this background given perhaps by Eq. (3.1) when observing the corona against solar disk. Even when optically thick, photons emerge not from a single point along the LOS but from a region typically spanning one photon mean free path Λ.[3] For stratified atmospheres $\Lambda \approx$ a pressure scale height, $\approx$100 km for the Sun's photosphere and chromosphere (Judge et al. 2015). Thus any positions along the LOS are determined only to within $\pm\Lambda/2$ reflecting the intrinsic "fogginess" of the Sun's atmosphere. The same argument applied to the two other dimensions implies that the emergent intensity arises from a volume

$$\mathcal{V}_\Lambda \approx \Lambda^3 \approx 10^6 \ \text{km}^3 \tag{3.4}$$

(see also Bruls and Von der Lühe 2001). Quantitative analyses of the LOS structure require solutions to "integral inversion" problems, which are notoriously ill-posed, i.e. multiple solutions are compatible with observational data (Craig and Brown 1976, 1986). At least one ambitious study has nevertheless applied such inversions, replete with ad-hoc procedures to add information to data (a mathematical procedure called "regularization") to study the relation between the temperatures of plasma and magnetic fields in the chromosphere of a sunspot (Socas-Navarro 2005b,a). Even if such procedures can satisfy the desire for reproducibility, they are entirely subjective but necessary to yield a mathematically unique solution, but with no justification in physics.

Plasmas which are optically thin ($\tau_\lambda \ll 1$), such as the corona observed at EUV and visible/infrared wavelengths contain no information along the LOS. In these cases, two or more different LOS are required to probe 3D structure. The STEREO mission delivered three dimensional views of coronal features (e.g. Schad et al. 2016). When new data become available from optically thin plasmas, there is no intrinsic limit to the smallest observable volumes of plasma. Tomographic work can take advantage of solar rotation to probe 3D magnetic structure, assuming that structures on the Sun do not change during rotation (Davila 1994).

A recent study has reminded us of complications in the interpretation of radiation from the optically thin corona. Malanushenko et al. (2022) used a 3D MHD numerical calculation from photosphere to corona to explore the effects of integrations along just a single LOS through optically thin coronal plasmas. Figure 3.3 illustrates that what we see as "loops" of magnetic field in images, arise from rather amorphous distributions of plasma embedded in a low-β plasma which has a simple bipolar structure. More importantly, one can see that the 2D image showing plasma "loop" structures consists of a large number of smaller plasma structures all frozen in to their own unique topology with their own plasma history of heating, cooling and dynamics. In particular, these smaller structures have their own different Alfvén speeds, with significant consequences for the evolution of magnetic and kinetic free energy perpendicular to the lines of force (i.e. in the

[3] In strong lines with scattering dominating the source functions, the appropriate value of Λ is the "thermalization length" which can be larger in physical space, see Mihalas (1978).

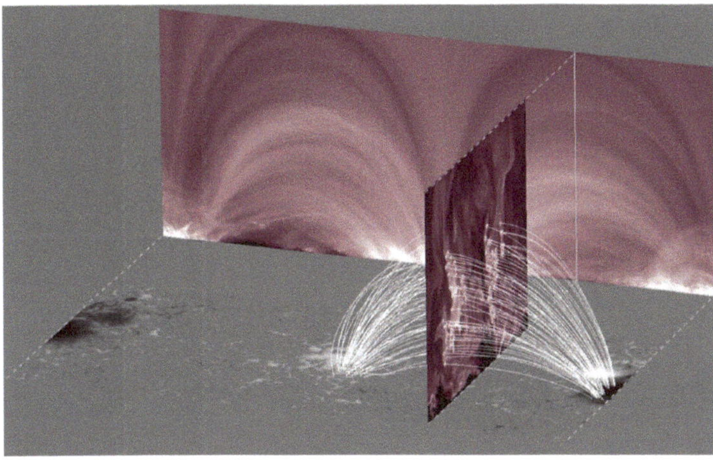

Fig. 3.3 A rendering of a snapshot of the photosphere and corona from a dynamical calculation is shown as projected onto the plane-of-sky (back-most images showing coronal loops). The apparent plasma loops are LOS sums of thermodynamic structure from plasma embedded in a bipolar field (image slicing across the magnetic lines of force), because the plasma is optically thin. The magnetic and thermal structure is from a 3D MHD calculation. The appearance of the plasma loops in the image results from multiple plasma structures all frozen along their own magnetic lines of force, all of which have their own dynamical histories and characteristic Alfvén speeds. Figure from Malanushenko et al. (2022)

darker inclined plane), and thus heating. These consequences will become clearer in Sect. 3.5. Complementing this rendering, Fig. 3.4 shows images of various spectral lines of iron ions shown in order of increasing plasma electron temperature, again from Malanushenko et al. (2022).

If this calculation reasonably represents line-of-sight effects through active region coronal structure then we must conclude that the state of the plasma within the corona is *far more highly structured than it may appear in 2D images.* Or re-phrased, the corona appears less structured, smoother and more ordered, by virtue of the line-of-sight integration of structures which are part of the same, low-β, hydromagnetic structure. It would be difficult to imagine that a calculation that assumes a hoop-like symmetry (a torus, like a pumped-up bicycle inner tube) might be applied successfully to such structures, even though our two-dimensional view from Earth suggests these might be geometrically simple structures. This calculation reveals another potentially important feature. The chromosphere does not appear to have simplified the hydromagnetic state of the coronal plasma- the cross field structure appears genuinely turbulent.

The reason that the corona appears so ordered and simple in two dimensions is that plasma is frozen into loop-like structures on large scales. Large structures are very frequently well approximated by potential extrapolations, which being governed by an elliptic partial differential equation, smooths out the magnetic field with height. So long as heating and conduction fill each bundle with low-β plasma,

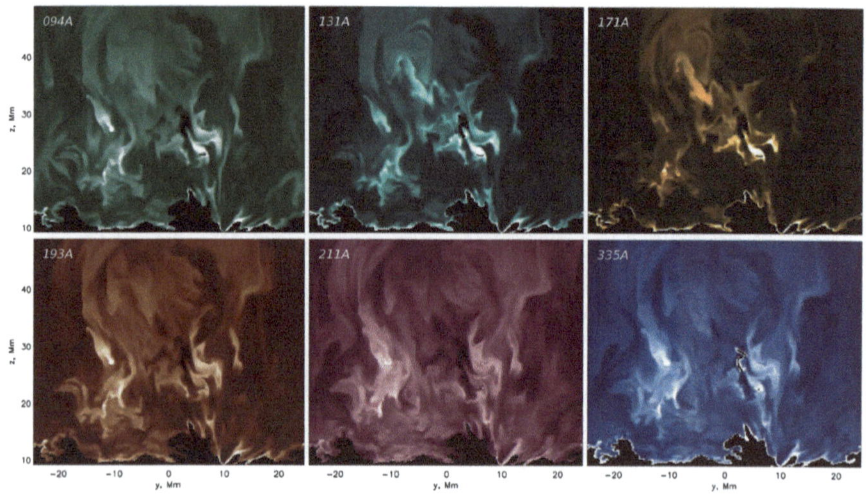

Fig. 3.4 Images of the thermal structure which might be observed at various coronal temperatures, obtained by Malanushenko et al. (2022) by integrating along the lines of sight parallel to the line joining magnetic footpoints shown in Fig. 3.3. Instead of revealing the strong "loop" fields $\mathbf{B}_0$, the images reveal the structure associated with $\mathbf{b}$, i.e. the turbulence and/or wave motions in this plane. (This point will be made clear in Sect. 2.8 and later in the book.) Typical turbulent structures are striking, and are not reported in the literature as much as loops. Our eyes are *not* drawn to see such structures within the corona, when, as usually happens, the alignment between line of sight and magnetic loops is not as good as the calculation shown. Even a modest tilt will reveal to us structures defined by $\mathbf{B}_0$, which may be close to potential, and loop-like in projection. There is *no difference in the "fuzziness" of the images as a function of temperature of line formation*, unlike in observations (cf. Sect. 5.3)

we will always be led to believe that the hydromagnetic state of an active region's corona is well-ordered, even approaching a potential state (e.g. Schrijver et al. 2005). This observation has interesting consequences to be discussed in Sect. 5.3.

In spite of such messy cross-field structure, coherent transverse wave motions have been observed above the solar limb through infrared Doppler line shifts using instruments integrating over large volumes of plasma (Tomczyk et al. 2007; Morton et al. 2015). It is perhaps in contradiction to the above picture that such signals can survive averaging, offering potential for new understanding through resolving the disagreement. However, intriguing MHD calculations are already revealing how coherent wave structures can emerge from initially complex conditions (e.g. Pascoe et al. 2011). The observations cited show the usual characteristic of smaller Doppler shifts than Doppler widths noted on page 14, by an order of magnitude. They must be at least partly coherent in time and space, the Doppler shifts otherwise cancel one another, and the phase speeds must be very similar. Does this coherence imply coherent motions of a subset of magnetized bundles or plasma which survive averaging? The coherence seems to imply conditions which seem to be rather special. By themselves the measured velocities are not inconsistent with MHD turbulence, but they raise questions as to what can survive all the averaging that might otherwise be expected. The amplitudes of the resolved motions (Doppler shifts) appear to be small, in a linear regime, and the authors identify them as transverse (Alfvén) waves. For now we speculate that either there are enough structures along the LOS that at least one might statistically exhibit such behavior even in a linear case, but it would have to dominate all emission along the lines of sight. Perhaps more likely is the idea that such waves exist on gradients in Alfvén speeds as surface modes akin to those proposed by Ionson (1978), Parker (1991b), or that they are linked in phase through second- or higher order terms in dispersion relations.

Another class of coronal wave, standing waves seen across active region loops which have received an impulse from, for example, flares (Nakariakov and Verwichte 2005) have been observed. In both cases wave speeds have been associated with transverse modes whose phase speeds vary as $|\mathbf{B}|/\sqrt{\mu_0\rho}$. With spectroscopic or other estimates of ρ, many authors have found reasonable values for magnetic field strength. Again, because of LOS issues and other special conditions which may be needed for such oscillations to be visible, the meaning of such values remains unclear. So, we shall return to coronal oscillations in Chap. 5.

3.3.2 Finite Angular Resolution

The highest resolution observations of photospheric and chromospheric plasma emerge from volumes of order

$$\mathcal{V} \approx \varpi^2 \times \Delta\ell \gtrsim 20^2 \times 100\ \mathrm{km}^3, \tag{3.5}$$

where ϖ is an (optimistically small) pixel size. $\varpi = 20$ km corresponds to the resolution limit of DKIST close to 400 nm wavelengths (most observations even with DKIST's 4m aperture will not achieve this), and $\Delta\ell$ is the path length over which the spectral feature forms. Again for the stratified atmosphere beneath the corona $\Delta\ell \approx \Lambda \gtrsim 100$ km. For optically thin lines, ℓ has little meaning except that it spans the line-of-sight scale of the entire volume from which photons emerge. In the corona ℓ can vary from perhaps a narrow width of tens of meters and up (set by the ion inertial length), and loop segments of 10 Mm, perhaps up to a solar radius (700 Mm). No matter what, the scales of irreversible dissipation are orders of magnitude smaller than ϖ and ℓ currently accessible to us. Clearly these estimates of $\mathcal{V}$ depend on the Sun's atmospheric properties ℓ as well as instrumental properties. We cannot assume then that we can reach arbitrarily small values of $\mathcal{V}$ simply by reducing ϖ alone, for when $\varpi < \Delta\ell$ then $\mathcal{V} = \varpi^2\Delta\ell > \varpi^3$.

Figure 1.8 highlights a technological mis-match: current observations of coronal plasma are an order or magnitude lower resolution, sampling volumes *at least* three orders of magnitude larger than observations of the photosphere. Of course, Murphy's Law dictates that this situation is *precisely the opposite* of what is required to study directly the development of coronal structure approaching dissipation scales. But the Hi-C instrument used (see e.g. Fig. 1.5) sampled pixels in the focal plane which are seven times the diffraction limit (Kobayashi et al. 2014). In principle much smaller pixels will yield important results if structure exists below the scales of Hi-C pixels (see Sect. 6.3.2).

The gain of angular resolution achieved by DKIST for the photosphere is a critically important advance ($\varpi \rightarrow 20$ km) in measuring the upward injection of electromagnetic energy. However, measuring the transport of this energy through the chromosphere (Fig. 1.8), into the corona where it can be stored and dissipated (Fig. 1.5) represents a severe observational challenge in many other ways (see particularly Sect. 3.3.4). It is unfortunate that, given the nature of the heating problem, instruments capable of such high resolution have not been selected for space missions. In the 1990s, the SIMURIS mission was studied but not selected for launch (Damé et al. 1991; Damé et al. 1994). Can a high resolution EUV instrument be flown? Higher resolution coronal instruments have been rejected in part because of the untested notion that existing instruments have essentially resolved elementary coronal structures (L. Golub, private communication ca. 2000).

As already noted on page 2, instrumentation was selected primarily to study only phenomena such as coronal dynamics on large scales. Perhaps instrumentalists might attempt to acquire EUV coronal data of the solar disk with an order of magnitude decrease in pixel size ϖ (interested skeptics might skip to the discussion in Sect. 6.3.1, in particular Fig. 6.1). The diffraction limit of a 20 cm telescope observing coronal plasma near 10 nm is 0.013 arcseconds, equivalent to 9 km on the Sun's surface. Many bright solar EUV lines can be found which are optically thin in coronal emitting structures. Those small-scales inherent within the magnetic structure shown in Fig. 3.3 might become visible as intense, far narrower structures which become physically separated in the plane-of-sky images. Either result, a detection or non-detection of finer structures would be of immediate interest for the heating problem (Sect. 6.3.1)! A high intensity requires a large emission measure $n_e^2 s$ (defined on page 93), and if s is of order 10 km or 20 times smaller than the current limit of 200 km, the electron density n_e would need to be 5 times higher, less so if multiple 10 km structures were present in the field of view. Given innate difficulties in assessing the relationship between average and peak electron densities using spectroscopy (Judge 2000), this picture cannot be immediately dismissed. Alternatively, one might adopt the unjustified idea that we have already resolved elementary structures in the corona at a pixel scale of 200 km. Only higher resolution data can decide this issue, which we return to in Sect. 6.3.1.

Figure 3.1 identifies regions that are observable at small values of wavenumber k. The values of k accessible to current observation vary from scales of $1 R_\odot$ to about 20 km (for DKIST), so that, with $k < \frac{1}{2}$ (pixel size)$^{-1}$,

$$10^{-9} \leq k \leq 3 \cdot 10^{-5}$$

spanning over 4 orders of magnitude. However, the scales for dissipation are closer to $0.1\ \mathrm{m}^{-1}$, yet the upper limit for k at EUV wavelengths is currently 200 km,

$$k \leq \frac{1}{2}\,\frac{1}{2\,10^5}\cdot \leq 3\cdot 10^{-6}\mathrm{m}^{-1},\ \text{at EUV wavelengths (Fig. 1.5)}.$$

3.3.3 Challenges for Spectroscopy

Independent of angular resolution, there are well-known limits to how we can dissect the photon spectrum to diagnose properties of the emitting plasma encoded in the spectra. The goal of plasma spectroscopy (Griem 1964; Cooper 1966; Fujimoto 2004) is to infer plasma properties by the influence of the plasma and its associated magnetic and electric fields on the radiation emitted by otherwise isolated atoms and atomic ions. Within the corona, researchers routinely derive volume-averaged electron temperatures, moments of fluid velocity distribution functions, distribution of electrons with temperature (emission measures, see page 93), electron densities from appropriate spectra. Such quantities, weighted averages of properties within volumes $\mathcal{V}$, naturally form the observational basis for essentially all analyses of the chromosphere and corona (e.g. Pottasch 1964; Dupree and Reeves 1971; Foukal 1975; Jordan 1992; Fisher et al. 1998; Mariska 1992; Mandrini et al. 2000; Aschwanden 2005, and many more). Both electric and magnetic fields make their presence known though the Stark, Zeeman and Paschen-Back effects, although spectro-*polarimetry* is a more powerful technique for measuring the relatively weak fields of the chromosphere and corona.

Precise impacts of the intrinsic volume-averaging associated with observations on solving the heating problem have yet to be properly understood. But some hints are found in the literature. We do know that any inhomogeneities within $\mathcal{V}$ skew derived quantities (such as plasma temperature, density) away from volume averages (Jefferies et al. 1972a,b; Almleaky et al. 1989; Brown et al. 1991; Judge 2000, 2020) that have been used to investigate heated coronal structures. Given that in MHD, intense heating is intrinsically linked to the development of small-scales within the plasma (Fig. 3.1) and that electron heat conduction is highly anisotropic (when $\omega\tau \gg 1$ Table 1.2), there is again, no prior reason to expect that homogeneity can be assumed within observable volumes $\mathcal{V}$. Currently observable plasma volumes all exceed expected dissipation scales by orders of magnitude, with the one possible exception involving viscous dissipation of particular compressive motions assessed later (Sect. 5.2, Appendix D).

Spectroscopy has some potential to address important questions in energy dissipation. While a major thrust of spectroscopy has been to derive bulk thermal properties of coronal plasma, such as through emission measure and line ratio analyses, such methods have failed to deliver definitive results concerning heating mechanisms. Augmented with studies of coronal emission line profiles, this kind of work forms the *starting point* for studies of processes behind coronal heating (e.g.

Jordan 1992; Mariska 1992). Unfortunately, it is evident that such *thermodynamic* quantities are in fact the *end result* of coronal heating mechanisms. In starting from such derivative quantities (based upon a host of assumptions), seeking coronal heating mechanisms is almost akin to inferring the mechanisms which evolved the human eye from the finished organ. This provocative analogy was chosen partly in jest because the eye seemed to deny a possible explanation by mutation and natural selection, even according to Darwin himself. But high spectral resolution spectroscopy has been used to reveal properties of ion distribution functions through measurements of line profiles. Resolved line profiles can in principle limit the number of compatible mechanisms, whether in the fluid or kinetic regime, without the need to use elaborate inversions schemes to solve for thermal properties. Some ingenious published examples are discussed in Sect. 3.3. Questions which might be addressed might be, do the assumptions behind single-fluid MHD break down before reaching dissipation scales? Are collisionless processes (wave-particle interactions, plasma turbulence, fast reconnection) needed to determine the mode by which energy is dissipated? Or does it even matter (Sect. 3.1, the ARLToR conjecture)? We will be obliged again to try to identify such processes through properties which survive averaging in space (and time needed to acquire data), to seek unique signals of their presence in the processes of coronal heating. These must hinge on spectroscopy sampling a number N_λ of spectral wavelengths, or spectropolarimetry (see Sect. 3.3.4), sampling at least $N_\lambda \times 4$ wavelength and polarization states (Landi degl'Innocenti and Landolfi 2004), given that images contain just one number per spatial pixel.

Returning to existing observations, by 1973, SKYLAB's instruments had barely sufficient angular resolution to establish the essential inhomogenous (or "fine structure") nature of plasma *across* magnetic field vectors within the corona (Rosner et al. 1978). This observation has been amply verified by later spacecraft instruments. Frozen-in fields and field-aligned heat conduction are so dominant that the Sun's corona can be usefully thought of as an ensemble of independent one-dimensional flux bundles (the RTV picture of Sect. 2.6). But further comments on the true scales of any cross-field inhomogeneities are warranted because of their implications on energy transport and dissipation. As noted earlier, some believe that data from the two successful Hi-C flights have resolved some cross-field structures already at 200 km (see a subset of these data in Fig. 1.5, Williams et al. 2020; Klimchuk and DeForest 2020; Williams et al. 2021). However, others have argued in favor of thermal sub-structure using spectral signatures which depend on various powers of electron density (e.g. Dere et al. 1987), but again see the findings by Judge (2000), or the empirical multi-thread picture (Aschwanden et al. 2000; Warren 2006; Ryutova et al. 2009; Walsh and Sarkar 2010). The methods by which sub-resolution image structure has been inferred face difficulties of non-unique interpretations, sometimes amounting to rank extrapolation, with no information provided by the data analyzed. In the highest quality datasets, such as reported by Rachmeler et al. (2019), the authors note that "Hi-C 2.1 resolution is between 0.3 [220 km] and 0.47 arcsec [340 km] in images that are not affected by motion blur." The narrowest coronal structures are certainly optically thin, so that superposition along the LOS is a likely source of image confusion and degradation (Malanushenko et al. 2022). These problems might be ameliorated using high spectral resolution observations, in separating emission from plasma that is moving at different speeds along the lines of sight. Judge (1994) applied such ideas to density-sensitive line pairs in the $\mathcal{R} \approx 70{,}000$ UV spectra of red giant stars, spatially unresolved, from the Goddard High Resolution Spectrograph (Heap et al. 1995) on the Hubble Space Telescope. However solar spectra have neither the photometric quality nor high spectral resolution to enable equivalent studies, except perhaps for a paper by Judge et al. (1998, see Table 5.1).

We conclude that the evidence that the corona has elementary structure only above 200 km is not strong, particularly given LOS problems highlighted in Fig. 3.3. However, more recent work from the HRI instrument on the Solar Orbiter (Halain et al. 2017) has observed scales similar to those of Rachmeler and associates (Chitta et al. 2022), with an instrument with a similar spatial resolution, at a far closer distance from the Sun. Again, the smallest structures observed are close to the instrumental angular resolution.

In relation to such scales, a theoretical study of the resonant absorption of Alfvén waves (Davila 1987) suggested that the ion viscosity in magnetized plasmas (Braginskii 1965) can in principle lead to resonance layers within the corona of 250 km width, a result contingent on the presence of compressive modes aligned with $\mathbf{B} = \mathbf{B}_0 + \mathbf{b}$. 250 km is also comparable to the mean free path of protons and electrons (Table 1.2) as well as current observational limits. To examine this possibility generally requires detailed dynamical calculations. However, the phenomenology of MHD turbulent motions allows us to estimate these components, it will be studied in Sect. 5.2 (see also Appendix D) to predict dissipation rates specific to MHD turbulence. We will examine processes below mean free paths of order 250 km, and their influence on dissipation in Sect. 5.11. In contrast, Davila noted that shear motions ($\frac{\partial u_i}{\partial x_j}$, $i \neq j$) are far less dissipative, owing to their scaling with $(\omega\tau)^{-2}$, refer to Sect. 2.4. Shear motions would require far narrower layers to supply the necessary rate of dissipation.

The temperatures T_e and T_i of electrons and ions are well defined when each particle species reaches its own equilibrium (time scales τ_{ee} and τ_{ii}), and are equal when the electron-proton collisional equilibration time τ_{ei}^{ε} ($\approx$80 seconds in the quiet Sun) is shorter than typical dynamical times in the plasma. Viscosity heats primarily the ions, current dissipation heats the electrons (Braginskii 1965). One would expect then, that determining the ion and electron temperatures would be a prime target for those interested in coronal heating. More attention has been paid to electron temperatures than ion temperatures because the electrons directly cause the emission of radiation, and they control the states of ionization, excitation and radiative emission of ions (e.g. Cooper 1966). Ion temperatures can only be probed remotely using widths of emission lines and other special techniques such as Doppler dimming, where the brightness of light from coronal ions, Doppler-shifted in the accelerated wind plasma, is reduced because bright incident disk light is red-shifted out of resonance (Hyder and Lites 1970; Kohl et al. 1998). But such methods are inherently non-unique (see, for example Raouafi and Solanki 2004).

Measurements of T_e are frequently reported in the literature, but how useful are they in the heating problem? This may seem an odd question, but one must remember that the plasma temperature is the end result of a dynamical adjustment of the plasma to electrodynamical forcing. Such measurements are meaningful when the distribution functions are Maxwellian, but they are also useful in other cases where one needs information on the typical width of the distribution function. Yet we have already observed that, as far as energy transport and dissipation is concerned, particles in the high energy wings can be significantly more important. In a low-β plasma, the plasma pressure nk_BT plays little role in the equation of motion, instead the electrons and ions adjust to the changing sources and sinks mostly in the energy equation along with charge conservation, an equation of state and, under some circumstances, Ohm's law. There are many methods by which T_e can be estimated from observations, for example by using brightness ratios of coronal emission lines measured spectroscopically. Two more points can be made regarding the relation of measurements of T_e to coronal heating:

- The accuracy of T_e from emission line intensities is fundamentally limited because emission lines are excited efficiently only over typically a factor of 2 in T_e (Pottasch 1964; Cooper 1966; Judge 2010). Figure 3.5 shows one of the main factors determining the emission rate of various coronal emission lines as a function of electron temperature. *Changes* in T_e, related to the precision not accuracy of the methods used, are better detected. For example, the ordering of the curves as a function of temperature in Fig. 3.5 is always correct, even if the shapes of them differ between different calculations. Priest et al. (1998, 2000) used differential estimates of temperature at various distances s along coronal loops to constrain heating rates as a function of s. The results, based upon broad-band X-ray measurements were, as we might have anticipated, quite ambiguous (MacKay et al. 2000).
- Recognizing the dominance of heat redistribution by conduction in the energy equation for the plasma (see also Sect. 3.8), we can use the well-known approximate solution of the steady-state energy equation for a loop of length L from RTV to find a relation between T_e and the total heating flux $\mathcal{F}$:

$$\mathcal{F} \propto T^{7/2} L^{-1} \quad \text{or} \tag{3.6}$$

$$d \log_{10} T = \frac{2}{7} d \log_{10} \mathcal{F} - d \log_{10} L, \tag{3.7}$$

For a fixed L, a change in $\log_{10}\mathcal{F}$ of ± 0.5 leads to a change in $\log_{10} T_e$ of ± 0.14, which is of order the typical half-width at half-maximum of each line's dependence of emission on T_e (Fig. 3.5). Judge (2010) asked, how broad does a distribution of plasma temperatures need to be before one can tell that it is not isothermal (i.e. a δ-function in T_e)? The answer is that any plasma with a Gaussian distribution of T_e of full width at half maximum below 0.3 in $\log_{10}T_e$ cannot be distinguished from isothermal. This result can be contrasted with that of Landi and Feldman

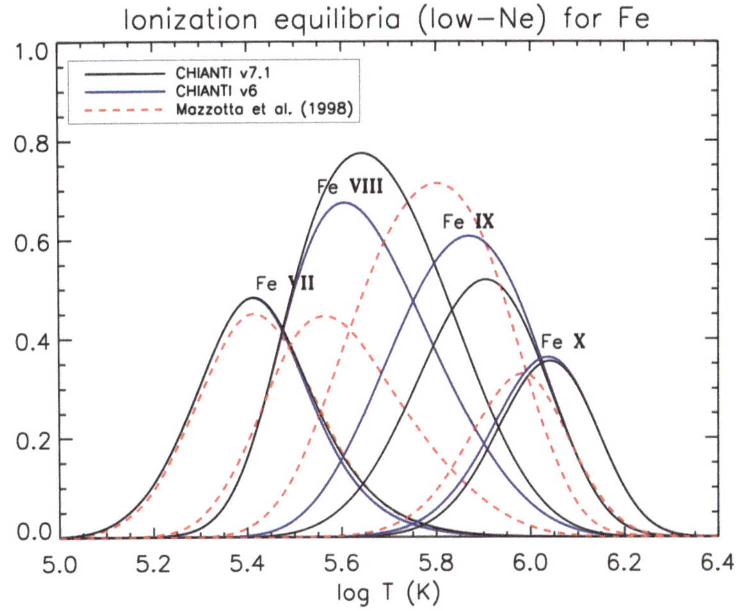

Fig. 3.5 Ion populations of some important coronal ions of iron in equilibrium, from Del Zanna and Mason (2018). The Roman numerals identify the spectra- Fe I is the first (neutral) spectrum, Fe X the tenth (nine-times ionized Fe^{9+}) spectrum. For EUV lines these curves dominate the variation of line emission rates with T_e. Note that the full-width at half maximum of emission over is typically 0.3 dex, spanning a factor of two in electron temperature. The peak of each curves is at an ionization equilibrium temperature T_{ie}

2008, table 3), who quote measured log T_e with three significant figures. Notice also that the three different calculations shown in Fig. 3.5, if used as diagnostics of electron temperature using line ratios, would yield very different answers (see the three different ionization fractions computed for Fe^{7+}, labeled "Fe VIII"). We must conclude that *electron temperatures derived from coronal emission lines are ill-suited to estimating* T_e, $\mathcal{F}$ and hence heating rate $\mathcal{F}/L$, with any degree of precision or accuracy.

Lastly, spectroscopy is sufficient to determine magnetic fields through the Zeeman effect only when magnetic fields are sufficiently strong to split a spectral line into Zeeman patterns. This can only be done within the intense umbrae of sunspots, at long wavelengths in narrow lines formed in the dense, cool photosphere (Penn and Livingston 2006). The broad lines in the corona, coupled with far weaker magnetic fields, mean that the Zeeman effect can only be applied in weak field limits using spectropolarimetry.

Such are some difficulties associated with the interpretation of radiation from optically thin coronal plasma. Radiation emerging from optically thick chromospheric plasma is influenced by non-local scattering of photons across the entire chromosphere, not just the regions from which photons finally escape into space.

Observations there are poorly equipped to reveal small dissipative structures, such as current sheets (Rappazzo et al. 2008).

A more elementary problem concerns the use of spectroscopy, and/or spectroscopic images, of the corona. By observing just the plasma properties, and not the properties of coronal magnetic fields, the current set of instruments on board solar observatories in space observe just the *aftermath* of the heating processes. The ingenious use of oscillations, fortunate accidents in atomic physics, and witnessing the changing magnetic morphologies offset this to some degree, but these data cannot yet be regarded as "information rich" in their application to the problem of coronal heating as defined on page 21.

3.3.4 Challenges for Spectropolarimetry

Like spectroscopy, plasma *spectropolarimetry* seeks to understand properties of the plasma, *and* electromagnetic fields permeating the plasma (Stenflo 1994; Landi degl'Innocenti and Landolfi 2004; Casini and Landi Degl'Innocenti 2008). The light emitted by most natural sources, including the Sun, is anisotropic, which on interacting with atomic ions or electrons generates polarized light (e.g. van de Hulst 1957). But the magnetic fields in magnetized plasmas such as coronae (again, where $\omega\tau \gg 1$), are oriented by processes quite different from those generating polarized light. The magnetic fields perturb the atomic ions through the magnetic moments, to cause space quantization, a preferred cylindrical symmetry along $\mathbf{B}$ in which the magnetic substates are most simply represented. The splitting of the substates leads to the Zeeman effect, and the interaction of anisotropic light with these states leads to the Hanle effect (Hanle 1924). The Hanle effect is sensitive to magnetic properties when the atomic system relaxes on time τ, for example by spontaneous emission of radiation, happens to be $\approx$ the inverse gyro-frequency of the ion, ω^{-1}. At optical wavelengths, spontaneous emission through permitted transitions occurs on time scales $\tau \approx 10^{-8}$ seconds. Then, magnetic field strengths of about 10^{-2} T can be revealed through this effect. These and other effects (Landi degl'Innocenti and Landolfi 2004) are routinely used in solar physics to measure components of $\mathbf{B}_0$ within volumes $\mathcal{V}$ in the photosphere and low chromosphere (Lites 2000). At radio wavelengths, Bremmsstrahlung emission is modified to produce weak polarization in the free-free emission in the presence of magnetic fields (Gelfreikh 1994) most readily understood in terms of magneto-ionic theory (Gelfreikh and Lubyshev 1979). Gyro-resonant emission from electrons also is valuable when $|\mathbf{B}_0| > 0.02$ T (Casini et al. 2017), but it is not easy to determine where, geometrically within a hydromagnetic structure the radiating, resonant surfaces are. While large-scale electric fields cannot be sustained in plasmas, the thermal motions of electrons and protons generate variable electric fields with a mean of zero, over scales of Debye-lengths. In turn, these cause the Stark effect (line shifts and broadening), and other, secondary signatures in the emitted spectra (such as the reduction of the ionization potential, Cooper 1966) which can have observational consequences.

Even in the photosphere where the largest observable magnetic fields exist, the energies induced by electric and magnetic fields in atomic ions are a small fraction of the energies in atomic Hamiltonians. Polarization signatures of the magnetic fields, as revealed by polarized spectral lines formed across the solar atmosphere, are frequently small, particularly in lines formed above the photosphere. In the photosphere they can be significant fractions of the intensity. The spectral lines must be "chopped up" not only in wavelength but also into at least four "polarization states" to recover components of **B** (see chapter 1 of Landi degl'Innocenti and Landolfi 2004). Consequently the sensitivities needed to achieve accurate measurements of the electromagnetic fields are difficult to achieve, the Sun is simply not bright enough (Landi Degl'Innocenti 2013)! Landi Degl'Innocenti showed that, because signal-to-noise ratios for intensity using pixels of size ϖ matched to the diffraction limit, even the bright photosphere generally requires exposure limes long compared with the evolution of features ($\approx\varpi/c_S$, $c_S =$ sound speed) within the field-of view. This constraint comes only from constants of nature for a star of solar age and mass. Practical work-arounds are to observe with pixels of size $N > 1$ times the critical sampling size of the diffraction limit. N is the desired increase in signal-to-noise ratio, accepting the lower angular resolution. The original resolution may be restored by moving a given instrument N times closer to the Sun, a very expensive option recently realized by the SO/PHI instrument on the Solar Orbiter mission, where $N \approx 2.5$ (Gandorfer et al. 2011). The problem is compounded for *coronal* spectropolarimetry because, at visible wavelengths the global corona is a million times dimmer than the solar surface, requiring the solar disk to be successfully occulted by the moon or an occulting disk. Also, magnetic fields are correspondingly weaker, emission lines are broader, and Zeeman signals are weaker (Casini et al. 2017).

Practical applications of solar spectropolarimetry tend to adopt the *inverse* methodology in seeking physical solar parameters from measured polarized light. In part this is owing to the fact that the atmosphere below the corona is highly stratified (Fig. 1.6), so that different spectral lines form at various heights or mass columns (Eqs. (3.2) and (3.3)). In turn this requires radiative transfer calculations ("forward" calculations of spectra from magnetic atmospheric models) to be "inverted" to extract magnetic information. Large efforts to develop computer codes for such inversions have been undertaken under various physical approximations (reviewed in detail by del Toro Iniesta and Ruiz Cobo 2016). Such work occupies a major part of research in solar physics. But it must be remembered that, like the challenges in spectroscopy, inversions are not a panacea. All are dependent on the assumptions behind, and quality of, the models upon which they are based. Also, not all problems can be cast in inverse form. This is in part why so much effort has been placed in comparing dynamical models directly with solar data in recent years (e.g. Carlsson et al. 2016; Rempel 2017; Carlsson et al. 2019; de Pontieu et al. 2021; Przybylski et al. 2022) where inversion schemes are not easily applied.

New data from DKIST offer the opportunity for us to study the injection of electromagnetic energy to a higher precision owing to its advanced imaging capabilities, large aperture, both of which enable us to evaluate the upward Poynting flux in the photosphere and (potentially) chromosphere. The latter will be crucial for our understanding the change in amplitude and nature of the Poynting flux into the corona after crossing the chromosphere. The first results support the $B^{\flat}$ conjecture. No correlations were found between any measurable magnetic or plasma parameters on scales between several Mm down to the angular resolution of 0.15 Mm (Judge et al. 2024), in stark contrast to analyses of observations on larger scales (Fisher et al. 1998). Therefore the possibility of following the magnetic energy from chromosphere to dissipation in the corona by direct measurements of $\mathbf{B}_0$ remains an important target for future research.

Facing such extreme challenges, it is easy to understand why following the energy from ordered to disordered forms has rarely been directly pursued (notable exceptions being studies by Socas-Navarro 2005b,a; Vissers et al. 2022). Instead, researchers have tried to extract conclusions from a blunt toolset which includes analysis of real and numerical experiments, in the hope of identifying, through statistical and other methods, causes of the observed effects. Arguably, as well as suffering from methodological problems (highlighted at the start of this chapter, also see Sect. 3.7), this approach has proven too fertile for us to make real progress. Hundreds of articles have been written in comparing observations and modeled data in the hope of identifying cause and effect, only to end with conclusions which are contingent upon a number of conditions which may, or may not, ultimately be refutable. A remedy might be to strive to refute certain pictures, which will require the community accepting an abundance of research articles with negative results. A strategy to address this problem is summarized in Sect. 6.1.

3.4 Stochastic Behavior

A central tenet of physics is that we can relate effects to underlying cause(s). However, non-linear dynamical systems (such as MHD) can exhibit sensitivity to initial conditions, stochastic behavior (unpredictability), chaos and intermittency, with the system evolving between "attractors" in the space of solutions (readable short introductions are given by Smith 2007; Holland 2014). Stochastic behavior was anticipated already by Gold and Hoyle (1960), in the context of flares:

> The requirements of the theory can therefore be stated quite definitely. Magnetic field configurations must be found that are capable of storing energy densities hundreds of times greater than occur in any other form, and that are stable most of the time. A situation that occurs only a small fraction of the time must be able to lead to instability in which this energy can rapidly be dissipated into heat and mass motion......

Litwin and Rosner (1993) arrived at essentially the same argument to explain why only certain bundles of flux are loaded with coronal plasma over an active region undergoing steady heating (Fig. 2.2). Flares are of concern to the coronal heating problem not only by sharing a common energy source, but also, and *precisely because* they require some part of the magnetic free energy within the corona to be stored, not efficiently dissipated. We have already argued that, right up until the sudden release of a flare, coronal heating in the associated structures continues with little indication that a flare is imminent. In the language of our B$^\flat$ conjecture, it must be stored in a system where $\mathbf{B}_0$ in the corona contains free magnetic energy, i.e. supporting large and long-lived electric currents $\mathbf{j}_0 = \mathbf{curl}\, \mathbf{B}_0/\mu_0$. Whatever heats the corona steadily cannot sap energy completely from $\mathbf{j}$. In other words, heating must be sufficiently effective between flares to heat the plasma yet at the same time be sufficiently *ineffective* to permit free energy to build and be released as flares. We have not encountered this perspective in the literature, most likely we have simply missed it, but a genuine absence might be a result of *confirmation bias* (see Chap. 4, and refer to Barker Bausell 2021), the lack of an effect being seen as a negative, and perhaps unpublishable, result.

Magnetic reconnection on macroscopic scales lies at the heart of the "standard model" of solar flares and CMEs. A gradual buildup of free energy within coronal plasmas is released unpredictably with the onset of a subsequent change in large-scale topology (Carmichael 1964; Sturrock 1966; Hirayama 1974; Kopp and Pneuman 1976). The model predicts plasma moving with bulk velocities approaching the Alfvén speed of the reconnected field, and, at least some of the time, the consequent development of shocks, accelerated particles and hard X-ray radiation. This stochastic behavior appears to be independent of the steady heating. It can be analyzed in terms of the B$^\flat$ conjecture: flares involve free energy associated with $\mathbf{B}_0$, steading heating involves just $\mathbf{b}$. The conjecture then finds itself in disagreement with at least two pictures of steady heating. Firstly, MHD turbulence couples $\mathbf{B}_0$ and $\mathbf{b}$ down in scale through a cascade that is local in wavenumber space, leading to a power law for magnetic and velocity fluctuations as a function of wavenumber. Secondly, a major candidate for "steady" coronal heating is the "nanoflare" proposition (Parker 1988), asserting that all heating is related to small flares generated as a result of stochastic relaxation of magnetic fields stressed as a result of the incompatibility between local forces and global topology (Parker 1994, 2007; Bogachev et al. 2020). If a nanoflare is merely a scaled-down flare, then any separation into a $\mathbf{B}_0$ and a $\mathbf{b}$ has no real meaning meaning, there being a continuum of measured flare sizes.

As already noted, disagreements generally provide avenues for scientific progress. These two challenges to the B$^\flat$ conjecture will be pursued in Sects. 3.5 and 4.8. What evidence is there for MHD turbulence in steadily heated plasmas? Can an avalanche from the smallest scales responsible for apparently steady plasma heating in the corona lead to the enormous inverse cascade to release 10^{25} J in a few minutes?

Stochasticity arises in most, if not all possible models of steady heating. Resonant absorption is naturally stochastic in that as one region in resonance is heated, the heat deposited will move the heated plasma away from resonance. The plasma evolves as resonant regions come and go as Alfvén wave resonant conditions continually change (Ionson 1978, see also Sect. 3.1).

In analogy with weather prediction on Earth and other non-linear systems such as traffic, while causes of individual effects may not be predictable, statistics of many realizations of observational and numerical experimental studies may of course be predictable and useful in the application to coronal heating and dynamics, as well as in large flares (e.g. Campi et al. 2019; Leka et al. 2019a,b). Stochasticity must therefore be embraced as part of the attempts to solve the heating problem as formulated. Refutation of cause and effect is clearly more difficult in stochastic than deterministic systems.

3.5 Turbulence

As illustrated in Fig. 3.1, a solution of the problem at hand requires us to identify the nature of the energy flow from large scales to far smaller, unobservably small dissipation scales within the corona. One solution is potentially provided by *MHD turbulence*. While the term "turbulence" is frequently used broadly in solar and astrophysics, this kind of turbulence describes a specific kind of dynamic evolution of magnetic field and conducting fluid, described below.

Fluid turbulence—disordered motions encompassing a large range of length and/or time scales—arises when the non-linear advection term dominates all others in the fluid's momentum equation (Eq. (2.14)–(2.19) with $\mathbf{B} = \mathbf{j} = 0$). The "disorder" is stochastic by its very nature, until the kinetic energy of turbulent motions ultimately reaches scales of viscous dissipation, when it is completely disordered (thermalized). MHD turbulence is not yet as well understood as hydrodynamic turbulence. Excellent introductions to MHD turbulence are Biskamp's book (Biskamp 2003), and chapter 11 of Kulsrud (2005). Steady theoretical developments since then now enable researchers to study specific observable predictions in the context of the heating problem (Schekochihin 2022), forming a basis for later sections of the present book (see Chaps. 4 and 5). As in hydrodynamic turbulence, energy injected at large scales is, through dynamical non-linear interactions, transferred to smaller scales where there is a sink of energy through a cascade that is characterized by interactions between "eddies", but only between neighboring scales. There are unique characteristics of MHD turbulence in the presence of a strong guide field $\mathbf{B}_0$, which introduces anisotropic dynamical effects into turbulent motions (as already indicated in the $\|$ and $\perp$ symbols in the "reactive" branch in Fig. 2.4). In the 1960s, it was realized that incompressible MHD turbulence should be understood as the non-linear interaction of field-aligned, but oppositely directed, Alfvén waves (Iroshnikov 1964; Kraichnan 1965). Discussion of compressive motions is postponed until Sect. 5.2. So, when "MHD turbulence" is discussed in the context of the solar atmosphere, it has a specific meaning, characterized qualitatively by the above properties, including frozen-in plasmas. As in hydrodynamic turbulence, MHD turbulence also is characterized by specific (power-law) spectra of turbulent power as a function of scale. The subject is briefly reviewed in Appendix C.

Is MHD turbulence inevitable in the corona? Is there observational evidence for MHD turbulence in heated coronal plasmas? It would seem likely to occur in the presence of oppositely-directed Alfvén waves, where second and higher order terms in perturbations are retained in the equations of motion. Small amplitude Alfvén waves have been observed (with amplitudes in Doppler shift $\sim 10^{-3} c_A$, Tomczyk et al. 2007; Tomczyk and McIntosh 2009; Morton et al. 2015) propagating along coronal loops, in both directions along loops (recall that such "loops" may be artifacts of our view from a single viewpoint, see Figs. 3.3 and 3.4). The linewidths reported are always in excess of Doppler shifts. But whereas widths always increase with variations of plasma velocity along each line-of-sight, the latter can decrease. These observed conditions are in principle compatible with the generation of MHD turbulence as envisaged. Thus the answer to the first question is a "definite maybe".

The answer to the second question posed above is less clear, in part because observable predictions have changed with steady advances in understanding. There is at least one example of what appears to be freely decaying MHD turbulence within the solar atmosphere (as opposed to wind) of which we are aware. Lacatus et al. (2017), after a process of elimination, showed that peculiar post-flare line profiles of the resonance lines of Mg^+ behaved in space and time as decaying MHD turbulence might predict, the kinetic energy of post-flare motions decaying along with observed and inferred radiation losses from the Mg^+ lines themselves. But this system differs significantly from steadily heated coronal plasma. So we are left with, as usual, an answer contingent upon better understanding of theory and/or the availability of information-rich data.

It should be emphasized that if the answer to the second question turns out to be "no", there is no special problem, it merely means that MHD turbulence is not dominating the dynamics within the heated plasmas. Examples of non-turbulent modes of energy transfer from large to dissipative scales include both resonant absorption and phase mixing. In these cases there is no role for energy carried at intermediate scales, and so no "power law" dependence of power on wavenumber (Fig. 3.1). In one case reported on page 88, the power spectrum is far from a power law. Alas again, this case has its own special interpretive challenges.

It is well known that power laws are encountered in the variation of energy densities with scale size in unmagnetized fluids. Kolmogorov (1941) argued that a (statistically) steady energy flow can be identified across a range of scales from large "driving" scales at small $k = k_0$ to those far larger values of k where energy losses due to viscosity (losses $\propto k^2$) dissipate the small scale kinetic energy. This "cascade" from small ($\approx k_0 = 1/L$, L the outer scale) to large k occurs through non-linear interactions, the strongest of which occur between perturbations (or "eddies") of similar scales, ignoring interactions between motions on very different scales. The cascade spans an "inertial range", covering orders of magnitude in scale from Rm $\gg 1$ (k_0) to Rm ≈ 1 ($k = 1/$dissipation scale). Kolmogorov (1941), assuming that the motions are *isotropic*, found that the kinetic energy spectrum followed his eponymous "$-5/3$ law":

$$E_k \propto \varepsilon^{2/3} k^{-5/3}, \tag{3.8}$$

which is shown schematically as the spectrum of the inertial range of Fig. 3.1. ε is the [outer scale] energy injection rate.

But what are the power law distributions of MHD turbulence? Numerical and analytical work, entertainingly reviewed by Schekochihin (2022), has led to predictions concerning the anisotropic nature of MHD "eddies", described by $k_\perp$ and $k_\parallel$ which denote wavenumbers of coherent fluid scales perpendicular and parallel to $\mathbf{B}_0$ (see Appendix C, which draws heavily from Schekochihin's review). Then for a given outer scale L, at each $k^2 = k_\parallel^2 + 2k_\perp^2$ the turbulent motions are predicted to be *highly anisotropic* in regions where MHD is valid (Goldreich and Sridhar 1995):

$$k_\parallel \propto k_\perp^{2/3} L^{-1/3}. \tag{3.9}$$

Kinetic energy spectra perpendicular and parallel to $\mathbf{B}$ are

$$E_{k_\perp} \propto \varepsilon^{2/3} k_\perp^{-5/3}, \tag{3.10}$$

$$E_{k_\parallel} \propto \frac{\varepsilon}{c_A} k_\parallel^{-2}, \tag{3.11}$$

which is an anisotropic form of the Kolmogorov spectrum (Eq. (3.8)). The picture of Goldreich and Sridhar (1995) underwent an important modification in the subsequent decades in the realization of the importance of the geometric alignment between turbulent velocity and magnetic vector components in the plane perpendicular to $\mathbf{B}_0$ (Boldyrev 2005, 2006): leading to

$$E_{k_\perp} \propto \varepsilon^{2/3} k_\perp^{-3/2}, \tag{3.12}$$

which is supported by numerical experiments (Schekochihin 2022). The phenomenon is consistent with the dynamics seeking to reduce non-linear interactions at these scales. However, Schekochihin warns that roles of anisotropy, intermittency, stability of current sheets and other issues are still actively debated, and the theory itself is still guided by diverse solar wind observations. For example, Squire et al. (2022) point to observations from high in the corona and in the solar wind, which demand a kinetic treatment

> ...the difficulty is that most theories predict that, in the strongly magnetized limit relevant to the solar corona (the low$-\beta$ limit), ...such low-frequency, high-$k_\perp$ structures dissipate to heat predominantly electrons. Other low-frequency plasma motions, such as compressive waves, generally cause parallel heating of ions. Both possibilities are inconsistent with observations [of the solar wind, e.g. Marsch 1990]...

The basic challenge presented by lack of a complete theory even of MHD turbulence is offset, to some degree, by some emerging robust conclusions (Schekochihin 2022). *If dominant in driving large to small scales needed for coronal heating, MHD turbulence should be readily testable through observations of power spectra, anisotropies of spectral line broadening, and perhaps by finding the smallest observed scales across magnetic fields,* in spite of the incompleteness of the theory.

As applied specifically to the coronal heating problem, multiple numerical calculations have explored specific aspects of MHD turbulence, some in the reduced MHD approach ("RMHD", Rappazzo et al. 2007, 2008; Perez and Chandran 2013; van Ballegooijen et al. 2017; Einaudi et al. 2021), others using 3D MHD calculations, (Gudiksen and Nordlund 2005; Dahlburg et al. 2016; Rempel 2017; van Ballegooijen et al. 2017; Howson et al. 2020; Pontin et al. 2020). Applications of RMHD to the corona have relaxed assumptions in some earlier numerical experiments (e.g. van Ballegooijen et al. 2011; Perez and Chandran 2013; van Ballegooijen and Asgari-Targhi 2017). For example, in permitting gradients of density perpendicular to $\mathbf{B}_0$, important effects such as phase mixing and resonant absorption have been included (van Ballegooijen et al. 2017). These calculations all assume a dominant $\mathbf{B}_0$ which is perturbed by photospheric motions, by moving conducting fluid under the solar corona perpendicular to $\mathbf{B}_0$. If $\mathbf{B}_0$ penetrates the photosphere at both ends of a loop-like structure, the system is called "line-tied". If penetrating only once, the field is considered (locally) open and turbulence will require existence of downward propagating Alfvén waves (in homogeneous, incompressible background conditions), or, more generally, "uni-turbulence" (Magyar et al. 2019, see page 141).

The initial motions are classified as either wave-like or quasi-static depending on whether the product of driving frequency and wave crossing time is >1 or <1. Accordingly, after initial linear phases, the non-linear dynamics that develops is characterized with names such as phase mixing, resonant absorption, topological dissipation, field-line braiding, tearing mode, plasmoid formation, magnetic reconnection, and turbulence. In a general sense these descriptive words are just projections of the non-linear dynamics onto some familiar terms under different, nearly-ideal MHD or plasma dynamic responses to different drivers. These terms form a convenient basis for mutual understanding and communication between researchers based upon more idealized pedagogical examples deliberately simplified to highlight physical mechanisms (van Ballegooijen et al. 2017). Calculations for quasi-static evolution have been reviewed by Wilmot-Smith (2015). In most of these cases an over-riding theme seems to emerge: the non-linearities themselves determine to a significant degree the state of the coupled plasma and magnetic field. Heating mechanisms do not simply depend on the underlying driving motions. Like the decay of whirling eddies in rivers whose kinetic energy originates from changes in an external gravitational potential, internally-generated dynamics also controls the evolution of coronal plasma (Uzdensky 2007; Einaudi et al. 2021). This important point has received recent observational support (Judge et al. 2024).

$\mathbf{B}_0$ influences the turbulence only by introducing anisotropy into the fluctuations (Eqs. (3.9)) to (3.12), and providing tension supporting the propagation of Alfvén waves across the system, whose phase speed varies linearly with $|\mathbf{B}_0|$. This injected energy by construction forces the direction of a cascade of energy towards small scales where dissipative processes await to rid the system of energy, yielding the above energy spectra upon reaching a steady state. An example of typical structures which develop within such computations is shown in Fig. 3.6 (note the compressed scale in the z-direction along $\mathbf{B}_0$). The derived phenomenological understanding of MHD turbulence can be tested, to some degree, observationally. Some spectra

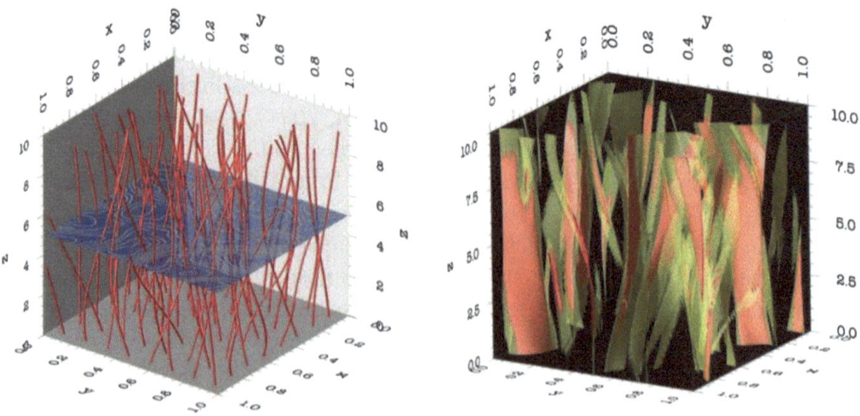

Fig. 3.6 Magnetic lines of force (left) and current densities (right) are shown from a 3D calculation of MHD turbulence driven, following Parker, by a slow re-shuffling of footpoints. The state of the calculation is shown after 18 Alfvén crossing times. Magnetic field lines are weakly entangled on large scales, because an efficient energy cascade has already occurred. The rendering has been compressed tenfold along the z-direction, so that all current sheets have aspect ratios >50:1. From Rappazzo et al. (2008)

E_k have been already been measured (e.g. Dere 1989), albeit for lines of C^{3+} formed in transition region plasma, which are slippery data to interpret (Sect. 1.8.2). Nevertheless, ostensibly the results, would seem to contradict the MHD turbulence prediction (Dere 1989):

> Essentially all of the power resides at the higher wave numbers, and the spectrum is flatter than predicted for the inertial range of a turbulent fluid.

The observations show this directly: on resolved scales ($\gtrsim$700 km), the rms Doppler shifts of lines are about 6 km s^{-1}, whereas the Doppler widths are between 10 and 35 km s^{-1}. These measurements of Doppler widths potentially include a modest opacity broadening (Pietarila and Judge 2004). However, it is not clear if an inertial range was reached in Dere's observations on resolvable scales. Further, the transition region plasma has so little inertia that it is readily powered by processes both from beneath and above (heat conduction and enthalpy), and so the meaning of such results for coronal heating is debatable. Nevertheless this would seem to be an area worth exploring further. We re-visit such a discussion in Sect. 5.1.

A second test of MHD turbulence might involve the anisotropy of fluctuations in **u** and **b**. Fluctuations in **b** will be unobservably small. The numerical work of, e.g. Rappazzo et al. (2008) shows that, when aspect ratios (loop lengths/correlation lengths of photospheric motions) are ~50, fluctuations are predominantly in **b** and not **u**, owing to the tying of magnetic fields to the photospheric boundaries. However, larger aspect ratios increase the kinetic energy over magnetic energy of the turbulence. If the system can evolve to such slender structures (reminiscent perhaps of the empirically proposed unresolved strands in the corona, Aschwanden et al.

2000?) then the effects of line-tying are reduced and any anisotropic turbulence may become evident in anisotropic widths of emission lines. In an enlightening paper, Pontin et al. (2020) found broadened, asymmetric profiles of coronal lines computed from a line-tied MHD calculation. However, their results are dependent on magnetic Reynolds numbers, which are far smaller than in the solar corona. Center-to-limb variations in the widths of coronal emission lines have not previously been sought. So we do not know if they reveal anisotropies. We do know that lines are optically thin and subject to much statistical averaging. Measurements of line widths during transits of coronal loops across the solar disk might set limits on any anisotropy. Alternatively, if observations of active region loops near disk center which are demonstrably oriented with parallel and perpendicular components systematically changing along their apparent lengths, then the anisotropies expected can be probed.

In terms of the basic challenges presented by systems undergoing steady MHD turbulence, the news is not all bad. The good news is that the probability of observations which might refute MHD turbulence may be small, but it is not zero.

3.6 Theoretical and Numerical Challenges

The main challenges facing **analytical** models involve the frequent need to adopt unphysically high degrees of symmetry and/or approximation, such as linearization of the equations of motion, or some part of them. This is not a problem for the resonant absorption picture, in which linear waves of frequency ω and wavenumber k propagate freely, unless the local condition $\omega/k = c_A$ occurs whence perturbations grow and convert linear wave motion into heat (Ionson 1978). An early problem concerning the use of ideal MHD equations to infer eventual dissipation (Lee 1980) was resolved confirming that dissipation is indeed expected (e.g. Davila 1987), satisfying with the ARLToR conjecture (page 55). But in another more cautionary example, Parker (1991b) not unexpectedly examined the physics behind phase mixing as developed analytically by Heyvaerts and Priest (1983). Here, a strong guide field $\mathbf{B}_0 = (0, 0, B_0)$ threads a density profile which varies only in the Cartesian x-direction $\rho(x)$. Alfvén waves are propagated along the z axis, and the y coordinate is ignored. The waves propagate along their own wave characteristics centered at x and $x + dx$ at different Alfvén speeds $c_A(x) = B_0/\sqrt{\mu_0 \rho(x)}$. Neighboring wavefronts become increasingly out of phase, quickly setting up steep, rapidly varying gradients in $\mathbf{u}$ and $\mathbf{b}$, enhancing dissipation. But Parker pointed out that inhomogeneities along the 3rd dimension (y) can qualitatively alter the outcome of the wave dynamics. Instead of phase-mixing, the waves can excite a locally coordinated wave mode, even in the linear wave equation, which has a fixed phase speed and therefore does not develop steep gradients. The truth perhaps lies between these two cases, recent assessments of the efficacy of phase mixing suggest that a variety of outcomes are possible depending on inhomogeneity scales within the corona (e.g. Howson et al. 2020; Pagano et al. 2020). But the message from Parker is clear: imposing prior symmetry can be risky.

This concern can be leveled also at the significant amount of applied mathematics devoted to understand waves along highly symmetric *tubes*, cylinders of magnetic flux (reviewed by Nakariakov and Verwichte 2005). Few of the results of such studies appear to be relevant to the study of waves in less homogeneous media, studied numerically for example by Howson et al. (2020), Pagano et al. (2020). These works found, not expectedly, that the heterogeneous densities and Alfvén speeds greatly enhance dissipation of wave energy in otherwise homogeneous models.

Analytical models present formidable challenges when, as well as local non-linearities in the equations of motion, they are subject to global topological constraints. Ingenious non-algebraic thought experiments have been developed to handle such challenges. Parker's 1994 monograph discusses the formation of current sheets which provide a natural reason why the corona exists, inevitably from the conflict between non-linear evolution subject to a fixed magnetic topology. This perspective has recently been reviewed with further insight by Low (2023), it will be discussed in Sect. 4.7.

Numerical experiments naturally offer a powerful, and increasingly popular, alternative to analytical approaches. A colleague expressed many years ago that "a good numerical physicist should be able to model anything" (P. Charbonneau, ca. 1995). Almost 20 years have passed since Gudiksen and Nordlund (2005) performed 3D numerical experiments on coronal heating with Lundqvist numbers $S_L = c_A L/\eta_N \approx 10^3$, where η_N is the numerically-determined coefficient of magnetic diffusion. Today, computations can model directly the slow build-up of magnetic energy for many hours to the point where dynamic evolution such as flares occur (Chen et al. 2022). In simplified large-scale geometries, one can examine in detail small-scale dynamics with magnetic Reynolds numbers above 10^4 (e.g. Xia and Keppens 2016; Fan 2020; Einaudi et al. 2021). In Sects. 4.3 and 4.10 we will discuss philosophical differences between numerical experiments and numerical simulations.

Ultimately the heating problem is essentially that of understanding coupling between the largest and smallest scales in the magnetic structures within the solar atmosphere. In a perceptive and interesting article entitled "Missing bits of the solar jigsaw puzzle: small-scale, kinetic effects in coronal studies", Tsiklauri (2009) reminds us:

> ...It is important to realize that reproducing or mimicking the reality in (mostly) numerical models does not necessarily mean that we have an understanding of physical mechanisms in action.

This lends support to our assertion that *refutation* should now become a dominant methodology in the particular problem of interest.

3.7 Ill-Posedness

The problems of chromospheric and coronal heating are both *ill-posed*, not just in the mathematical sense (Sect. 3.3.1). Both are components of the atmosphere which require minute fractions of the outward convective energy flux just beneath the photosphere. Just 2 parts in 10,000 are needed to sustain the chromosphere, and typically just 8 parts in 100 of that is needed in heating the corona. These small fractions imply that small, unobservable changes in local energy fluxes can have order zero effects on the overlying atmosphere. This extreme sensitivity is reflected in publications concerning chromospheric and coronal heating. Many observational studies report "effects", such as spicules and supersonic "explosions" with no ability to access underlying causes. When the causes are sought, frequently one reads (these words are paraphrased, see Sect. 1.7 for quotations):

> we conclude that there is sufficient energy in [spicules, waves, turbulence, reconnection, jets,...] to heat the corona.

Energy budgets which are orders of magnitude smaller than the energy available unfortunately provide a *weak constraint, a necessary but not sufficient condition* to accept a specific proposal. These energy budgets have changed over time, as listed in Table B.1 of Appendix B, which should be considered in assessing different claims about the success of heating mechanisms published over the decades. The fact is that unobservably small changes in the upward Poynting flux associated with photospheric observations lead to overwhelming changes in the corona. This situation must partly be responsible for the sheer number of claims and models which *can* satisfy energy requirements that are considered as viable solutions (e.g. Mandrini et al. 2000; Aschwanden 2005). The familiar logical ideas of necessity and sufficiency provide more justification for adopting refutation as a working methodology. If various pictures can be refuted, however incrementally, eventually an increasingly smaller set of them can be considered, until just those that appear to be necessary remain. Of those, given the diversity of observed coronal phenomena, it may eventually become clear that each phenomenon (e.g. steadily heating active region loops) can reasonably be identified with one physical cause. This might be sufficient to claim that a heating mechanism is known, the other physical pictures having been rejected, for each phenomenon. This claim is then open to further investigation.

In an attempt to address this type of ill-posedness, a relatively new methodology has been applied. Observations of effects (using, e.g., data from the IRIS instrument (Lemen et al. 2011), SDO (Pesnell et al. 2012), SoHO (Domingo et al. 1995), and many others) are combined with computer "simulations" in order to try to make progress, replacing unobserved (perhaps even unobservable) signatures of causes, such as changes in Poynting fluxes near the photosphere with those computed, assuming that the numerical algorithms and physics of the driving motions and magnetic fields is well understood and modeled from first principles. Specific heating processes might then be mapped to specific observed effects (de

Pontieu et al. 2021). However, there is a serious concern which must eventually be addressed. The methods intrinsically mix models with observations, refutation becomes always contingent on the physical credibility of the numerical models, the initial and boundary conditions, and the unknown and untested relevance of a particular calculation to a specific set of observations.

To "fill in" missing data, the method of "data assimilation" has gained considerable popularity, addressing to some degree the latter concern. The technique was developed as a need to improve forecasting (Asch et al. 2016), with roots in early work by L. F. Richardson. The idea is to incorporate sparse observations directly into boundary conditions for the governing equations of motion of complex systems. The equations require stipulation of variables everywhere, not merely at locations where and when observations happen to available. The art of its application lies in beginning dynamical calculations with initial and boundary conditions, which optimize dynamical variables for calculations to advance in time in a weighted and controlled fashion. In this way, over-sensitivity to initial conditions, a clear form of ill-posedness, is kept at bay. A forecast is built upon a series of measurement-time integration cycles, with a confidence estimate of the forecast state. Its application to solar physics appears mostly in the context of forecasting phenomena (e.g. Kitiashvili and Kosovichev 2008), and it may be tempting to apply this to the heating problem. However, it is a statistical method to look for differences in the state of a non-linear system and use those to update a set of new initial conditions. The differences, resulting from classical over-sensitivity (the "butterfly effect") are not generally used to test scientific ideas, merely to improve deviations from predicted physical behavior. As such, assimilation is poorly suited to refutation in natural science, it should not be applied without great care. One of the cornerstones of scientific methodologies is violated—*hypothesis testing becomes highly contingent on the credibility of the methods used*, and to some extent on the eye of the beholder. We regard this as swimming in dangerous waters, urged on by sirens offering the comfort of confirmation bias.

The other, far more familiar and ill-posed problems include "inversions" of sets of integral equations in which source terms (computed from a given atmospheric state) are derived using observed spectra (Sect. 3.3.4, also see Craig and Brown 1986 for a readable overview). These might include the determination of thermal structure, the nature of non-thermal distributions of particles, each as a function of depth into the atmosphere and/or time. One glaring problem is that to invert an integral, the quantity modeled must be the solution to a problem cast in a form that can be inverted. It is not obvious that this applies to a dynamically evolving atmosphere.

While well-recognized as basic difficulties in remote sensing, it is worth reminding readers of commonly encountered examples related directly to the heating problem. "Inversions" of radiation transfer (integral) equations have become a standard way to derive thermal and magnetic properties of a stratified atmosphere as well as the entrained vector magnetic fields (see del Toro Iniesta and Ruiz Cobo 2016, for a review). In essence, one seeks solutions to integral equations relating measurements of all components of polarized light to a stratified thermal model, along with models for signals of magnetic fields such as those created by the Zeeman effect and/or atomic polarization (Hanle and other effects). The ill-posedness is clearly seen simply by comparing the typical number n_{ind} of independent measurements with n_{mod} model parameters. In practice,

$$n_{mod} \gg n_{ind}$$

so that at most n_{ind} model parameters can be derived. This translates into a set of inferences for the magnetic field vector at just 2–3 depths (nodes) in the atmosphere, along with thermal parameters at these nodes. The role of the assumptions is then clear- to reduce the number of possible mathematical solutions by imposing draconian restrictions on acceptable solutions. Enforcing maximum entropy on the atmospheric structure will, for example, yield the least structured solution that is compatible with data, an appealing approach based on the wisdom of William of Occam in medieval times. But basic questions remain. Non-linear and non-local dynamical systems can explore a solution space with multiple, quasi-stable solutions (attractors), in which case one can ask, what does the "simplest answer compatible with the data" really mean? Should such answers be considered just one of many answers? Would multiple realizations of an ensemble of such solutions found by the solar plasmas (i.e. observations) or numerical experiments be sufficient to reject certain mechanisms?

Another example further illuminates ill-posedness, but it much more broadly encountered in coronal literature, especially in articles concerned with coronal heating (e.g. Gabriel 1976; Jordan 1980; Cally 1990; Jordan 1992; Mariska 1992; Golub and Pasachoff 2009). The widely used "differential" emission measure $\xi(T_e)$ is a solution to a re-casting of integral of the equation of radiative transfer for optically thin emission lines, into a function of T_e. With some simple factorization (Lang et al. 1990), the radiative transfer equation (or simply the conservation of radiative energy under optically thin conditions) gives emergent intensities, sitting on top of a continuum spectrum formed much deeper in the atmosphere, from a coronal line labeled "i":

$$I_i = \int_s n_e^2(s) G_i(T_e(s))\, ds \equiv \int_{T_e} \xi(T_e) G_i(T_e)\, dT_e \qquad \mathrm{W\,m^{-2}\,sr^{-1}}. \tag{3.13}$$

Each spectral line i radiates $G_i(T_e)$ W m^{-3} sr^{-1} (electron density)$^{-2}$, a function of T_e only under the assumptions that two body collisions dominate (the "coronal approximation"), that ion populations are controlled by electrons thermalized at temperature T_e, and that ions have time to reach ionization equilibrium (Seaton 1964; Cooper 1966; Lang et al. 1990). These assumptions cannot be justified *a priori* (Chap. 3), or a posteriori (Sect. 5.12 reviews such challenges). Nevertheless, with these common assumptions the leading dependence of $G_i(T_e)$ on T_e follows the curves of Fig. 3.5: each $G_i(T_e)$ has a width comparable to its peak temperature, i.e. far from a Dirac δ-function. $\xi(T_e)$ is a solution to a system of integral equations, so that solving for $\xi(T_e)$ is the solution to an inverse problem (a, given uncertainties

δI_i in measurements I_i, account of this particular problem is given by Craig and Brown 1986). Ill-posedness arises because $\xi(T_e)$ *lies under the integral over* T_e. Any function $\psi(T_e)$ can be added to $\xi(T_e)$ without changing the agreement with data, if

$$\left| \int_z \psi(T_e) G_i(T_e) dz \right| < \delta I_i \quad \forall i. \tag{3.14}$$

$\xi(T_e)$ can only be determined uniquely if there are no functions $\psi(T_e)$ satisfying this, yet it is readily seen that rapid oscillatory functions in T_e form a infinite set for which Eq. (3.14) is true. Typically, positivity and/or smoothness of $\xi(T_e)$ is imposed to "regularize" the procedure effectively so that the number of independent measures of $\xi(T_e)$ does not exceed the number of lines n (Craig and Brown 1986). Positivity of the solutions is required, because physically, $\xi(T_e)$ is to be understood as a positive-definite distribution function, fixed by the plasma thermal structure alone such that plasma between T_e and $T_e + dT_e$ which has density $n_e(T_e)$ and a thickness $dz(T_e)$ is summed along each line of sight s:

$$\xi(T_e) dT_e = n_e^2(T_e) ds \tag{3.15}$$

(compare Eq. (3.13)). $\xi(T_e)$, encumbered with critical assumptions and ill-posedness, has nevertheless served as a foundational rock against which theories of coronal heating have crashed (Appendix B), and largely survived. It has also prompted alternative ideas for energy transport (e.g. Rabin and Moore 1984; Antiochos and Noci 1986; Cally 1990; Roumeliotis 1991; Ashbourn and Woods 2001). C. Jordan has turned the problem around, and presented interesting work on the inference of properties of heating mechanisms, beginning from the form of $\xi(T_e)$ (Jordan 1976, 1992). For some reason, this work was regarded by some as controversial (Mariska 1992).

In summary, we find that again, in regards to coronal heating, results are conditional upon unknown assumptions and conditions. There does exist reason to suspect that assumptions are not always reliable, since the solutions derived for $\xi(T_e)$ do not account consistently for all spectral lines (Jordan 1975; Judge et al. 1995; Landi et al. 2002; Pietarila and Judge 2004) even using modern, accurate atomic data. Feldman (1993) provides an entertaining account of atomic properties, assumptions and the solar atmosphere.

3.8 Changing the Scene of the Crime

Direct signatures of coronal heating, such as increased coronal radiative emission or increased plasma ionization, are intrinsically sanitized or even hidden from our view by the plasma itself. It is as if each potential culprit has their work mopped up by a partner-in-crime: the plasma itself acts like a felon, altering the crime scene.

We have already seen that electrical currents dissipate their energy mostly into the electrons (Braginskii 1965). The collisions between them are so rapid (see τ_{ee} in Table 1.2) that heat is rapidly transported away from the deposition site by field-aligned heat conduction which rapidly removes spatial variations in electron temperatures. Inelastic collisions between these electrons and ions is the primary mechanism by which coronal emission line radiation is excited. The time required for a bright EUV line (e.g. of Fe^{11+} at 19.5 nm) to respond to an increase in temperature is roughly the inverse of the inelastic collisional transition probability per unit time, given by

$$n_e \langle v\sigma(v) \rangle \approx n_e \sqrt{\frac{kT_e}{m_e}} \frac{\pi a_0^2}{Z^2} \approx 1/10 \text{ per second} \tag{3.16}$$

Where $\langle\rangle$ represents averaging of the velocity components of the particle distribution function. Here $Z = 10$ for 9 times ionized ions, $Z = 1$ is a neutral atom, a_0 is the Bohr radius, and $n_e \approx 3\ 10^{14}$ m^{-3}. Over 300 elastic collisions occur between each such inelastic collision (Table 1.2), during which electrons travel a distance $\sqrt{300}\lambda_e \approx 2$ Mm before losing excess energy to radiative decay of the excited ion. The ratio of macroscopic radiative energy loss time to that for conduction in the plasma is about 40 (Table 1.2), consistent with these microscopic arguments.

In 10^6 K coronal plasma, hydrogen and helium are fully ionized, and so can contribute little to plasma radiation losses via inelastic collisions with electrons (Fig. 3.7). Among the next most abundant elements (C, N, O, Ne), H- or He-like

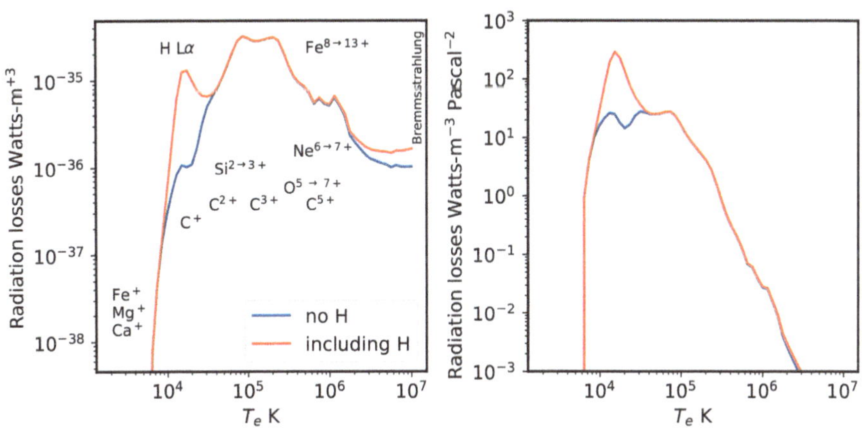

Fig. 3.7 Radiative losses from an optically thin plasma with solar elemental abundances are shown. The function in the left panel is to be multiplied by n_e^2 to obtain losses in W m^{-3}, that in the right panel by electron pressure^2. Under equilibrium conditions the Sun's atmosphere is close to a constant pressure above the chromosphere ($T_e < 10^4$ K, Fig. 1.8), so the losses follow the curves on the right. Curves with and without transitions of hydrogen represent rough limits of the functions which are affected by radiative transfer. Ions emitting strong lines are indicated at their formation temperatures across the chromosphere-corona

ions are most numerous when $T_e \approx 10^6$ K. However, they too are modest radiators because each inelastic collision requires a change $\Delta n \geq 1$ in principle quantum number, for example from $1s$ or $1s^2$ electronic shells to a $n = 2, 3 \ldots$ shell. The excitation energies required to excite the $n = 2$ levels of hydrogen-like and helium-like ions are:

$$k_B T_{ex} = E_{2p} - E_{1s} \approx \frac{3}{4} R_\infty Z^2 \equiv 1.1\ 10^5 Z^2 \text{ K}, \tag{3.17}$$

whereas the temperatures kT_{ie} at which each ion is abundant are typically smaller. Balancing two-body radiative recombination with electron impact ionization (Seaton 1964) we obtain the crude estimate $T_{ie} \sim 1\ 10^4 Z^2$. This approximation is not appropriate for ions with complex sub-shells (e.g. typical coronal ions of iron, silicon, etc.), for which the process of dielectronic recombination is important, where a rough guide is $\log_{10} T_{ie} \sim 3.8 + 0.24Z$ (below $Z = 12$ or so only, from inspection of Fig. 3.5). In all cases, $k_B T_{ex} \gg k_B T_{ie}$, the radiation losses vary as $\exp(-T_{ex}/T_{ie}) \ll 1$.

In contrast, Li-like and Na-like transitions of the type $ns - np$ with no change in principal quantum number n ($\Delta n = 0$) have smaller excitation energies varying as Z not Z^2 (Dufton and Kingston 1981). They are therefore prominent in the solar spectrum between 60 and 160 nm. These include coronal lines of Ne^{7+} and Mg^{9+}. Similarly, $\Delta n = 0$ transitions of elements of the iron group contribute most to radiation losses at coronal temperatures, examples are EUV lines of Fe^{8+} to Fe^{13+}. In short, the corona is an inefficient radiator.

The rightmost plot of Fig. 3.7 shows that for a given pressure, the plasma near 10^5 K radiates a thousand times more efficiently than at 10^6 K or higher. This implies that a pulse of heating in coronal plasma to above 10^6 K is most readily observed as a pulse of radiation from conductively heated plasma closer to 10^5 K (see, for example, the numerical results of Hansteen 1993). As a consequence, observed fluctuations in brightness of 10^5 K "transition region" plasma cannot without care be interpreted in terms of local, non-conductive energetic processes. The transition region contains less than 10% of the mass, on average, than of the overlying corona (page 13), so that this plasma is influenced by both the chromosphere and corona. This is true even if some of the transition region emission emerges from separate, cool loops (see, for example the conclusions of Feldman 1983, 1987, 1993; Hansteen et al. 2014 in direct contrast with those of Judge 2021). Its low inertia and consequent dynamics explains perhaps the enormous (over?) attention given to "transition region" spectra over many decades (Mariska 1992, we will return to this point in Sect. 5.12).

The chromosphere too suffers from well-known *non-local radiative* transport, because scattering terms dominate source functions (S_λ in Eq. (3.1)) of the strongest lines in the optically thick, low density chromosphere (low compared to the photosphere). The source function is the amount of radiative emission integrated along a ray of length one mean free path. To illustrate this, consider the source function (Eq. (3.1)) for a two level atom (Avrett and Hummer 1965):

$$S = (1 - \varepsilon)\overline{J} + \epsilon B, \tag{3.18}$$

where $\varepsilon \ll 1$ is the fraction of photons emitted at the expense of local thermal plasma energy from an isothermal source with Planck function B, i.e. resulting from local upward electron collisions relative to the total. Then $(1-\epsilon)\overline{J}$ is the contribution to S from local and non-local radiative scattering. This form for the source function applies assuming for simplicity complete redistribution, wavelength dependence enters through the variation in optical depth τ_λ (Mihalas 1978). $\overline{J}$ is the wavelength and angle- averaged mean intensity which results from contributions from all over the atmosphere. The emergent radiation given by the local Eddington-Barbier relation (Eq. (3.1)) depends on local values of S which is however determined *non-locally* by its dominant dependence on $\overline{J}$. The famous "square root of epsilon law" ($S(\tau = 0) = \sqrt{\varepsilon}B$, Avrett and Hummer 1965), yielding the emerging line center intensity, is a direct consequence of non-local scattering, as explained by Hubeny (1987).

Thus small scales are rapidly smoothed geometrically. The partially ionized chromospheric plasma has an adiabatic index γ approaching 1 from above, versus 5/3 for a perfect gas, because of extra degrees of freedom associated with the internal energy of hydrogen, helium and other atoms (Mihalas and Mihalas 1984). Instead of increasing plasma temperatures, much of the dissipated energy is stored as latent heat of ionization. Spectral changes in response to chromospheric heating are accordingly smaller. The chromosphere is also subject to a thermodynamic hysteresis as hydrogen recombination ($e^- + p^+ \rightarrow$ H) for each proton is slow ($\approx$ 10–100 seconds for an electron density of 10^{17} m^{-3} Allen 1973[4]). This timescale is barely sufficient to produce local ionization equilibrium in the stratified region which tends to oscillate with a 3 minute period (Gibson 1973; Hansteen 1997), with slow MHD shocks producing higher frequencies.

In another example, studies of UV spectra have for decades revealed "explosive events", most readily observed in emission lines formed in plasma from $10^{4.3}$ to $10^{5.7}$ K. This appears entirely to be expected in terms of its relatively low inertia, as well as being sandwiched between two larger reservoirs of energy. Yet some analyses investigate these as dynamic signatures occurring *within* a pre-existing transition region (Dere et al. 1991; Innes et al. 1997; Dere 1994, title "explosive events in the transition zone"). Others have recognized that such explosions are statistically unlikely to occur within such little pre-existing transition region plasma (Sturrock 1999; Sturrock et al. 1999; Innes 2004). Calculations of reconnection within a stratified chromosphere by Heggland et al. (2009) revealed signatures characteristic of some explosive events *within the pre-existing chromosphere*, with nothing to do

[4] While resonant charge transfer is vastly more rapid (see the opening of the present chapter), it merely swaps an electron during a collision between a H atom and a proton, resulting in a proton and a H atom, a momentum-changing collision.

with a pre-existing and higher transition region. Galsgaard and Moreno-Insertis (2014) reported on more general calculations which focused on the interaction of locally open and closed fields with similar outcomes. Peter et al. (2014) found direct evidence that explosive events can form deep with the chromosphere, but well above the photosphere (Judge 2015). Such events are important in two related ways. First they illustrate how the chromosphere can change magnetic connections before they reach the corona, and second how it might create a (unobservable?) high frequency spectrum in the Poynting vector emerging into the corona to address difficulties for example, in powering the solar wind (Axford and McKenzie 1992; Axford and McKenzie 2002). More recent work has included novel use of optics to acquire slit-less EUV spectra of explosions including spatial information (Fox et al. 2010; Rust and Kankelborg 2019). These authors used the MOSES (Multi-Order Solar EUV Spectrograph) sounding rocket instrument with inverse methods to extract spectra with 2D spatial information, finding a diverse set of line profiles with important implications for the coupling between magnetic field rearrangements and plasma heating. Huang et al. (2017) studied explosive events at higher resolution using the IRIS instrument and supporting data from the Swedish Solar Telescope (SST, Scharmer et al. 1999). Recent studies including more complete thermal physics are reported by Wargnier et al. (2023). Analyses of IRIS and SDO data continue to be explored with interesting results compatible with chromospheric origins for explosive events at and above the solar limb (Tavabi et al. 2022; Alissandrakis and Vial 2023). But in the absence of data for chromospheric magnetic fields the conclusions remain speculative in all observational studies. All indicate that what appears to be a "transition region" phenomenon may not actually be so.

In short, *while detectives prefer an untarnished crime scene, both corona and chromosphere have ways to obfuscate the pictures appearing in our telescopes, particularly of the intervening transition region plasma.* But like a good crime novel, part of the fun is to deal with such challenges.

3.9 The Poorly Understood Chromosphere-Corona Interface

The role of the chromosphere outlined so far is by no means the only aspect of coronal heating affected by it. In the absence of measurements of evolving chromospheric magnetic fields, and some apparent success of extrapolations based upon photospheric measurements (Altschuler and G. Newkirk 1969; Démoulin et al. 1997; Semel 1998; Neukirch 2005), at least on large scales, it is tempting to conclude that perhaps the chromosphere does not influence the large scale corona. The magnetic fields emerging from the chromosphere have been probed using fibrils which are assumed to trace magnetic lines of force (e.g. Wiegelmann et al. 2005, 2008), and more recently using full Stokes polarimetry of the Zeeman effect (reviewed by Vissers et al. 2022). So far, such analyses point to the need of higher quality, dedicated observations. In short this is another justification for improving measurements of physical conditions close to the coronal base. But chromospheric observations can be complicated, and challenging to interpret. Here we attempt to condense and infer their implications for the conditions at the base of the corona.

Interest in the chromosphere grew the 1950s and 1960s during the development of the theory of non-LTE radiative transport (Thomas and Athay 1961). Later it was recognized that the chromosphere produced structures (Fig. 1.8) quite different from those upon which extrapolations of magnetic lines of force from the photosphere had been made (Wiegelmann et al. 2005, 2008). Figures 1.5 and 1.8 show that the morphology of the chromosphere is very different from the photosphere. Assuming that the chromospheric fine structure results from freezing of plasma along thermally isolated magnetic flux bundles, and that the emerging radiation traces such plasmas, we conclude that the magnetic morphology changes dramatically across the chromosphere. This conclusion is fully consistent with the plasma β being >1 low in the chromosphere, and <1 in the upper chromosphere.

In terms of the coronal heating problem, it is unsafe to consider that the chromosphere is a passive stratified layer, merely allowing pressures to drop and magnetic field to expand to fill the available volume. Its particular role(s) in setting the state of the large-scale corona remains unclear, it is a prime target for future research. Above the steeply stratified chromosphere, structures observed in "chromospheric" lines have been seen extending far higher into space in unexpected structures called "spicules" a name used by Roberts (1945) because of their narrow, straw-like appearance. Observations from the stable, seeing-free Hinode spacecraft in the light of chromospheric Ca^{+} lines revealed structures at the solar limb which, because of detrimental effects of atmospheric seeing, otherwise had been difficult to identify from the ground. It is worth dwelling on the potential importance of these findings for the coronal heating problem, because such structures obviously extend into coronal plasma.

Athay and Holzer (1982) drew attention to the potential role of traditional spicules (Beckers 1968) in supplying the corona with energy and mass. Many later articles have considered the implications of spicules on the corona (de Pontieu et al. 2007; Martínez-Sykora et al. 2011; Pereira et al. 2014; Martínez-Sykora et al. 2017, 2018, to name but a few). Unfortunately the literature is confusing, replete with speculations, and some (implicit) contradictions. de Pontieu et al. (2007) classified two kinds of chromospheric structures (type I and II) extending well above the stratified chromosphere shown in Fig. 1.4. Of the two types, type II spicules appear to eject material into the corona, type I return to the photosphere in accordance with hydrodynamic shock theory within a magnetic flux bundle (Hansteen et al. 2006).

Spicules of type II have not yet clearly been identified with spicules discussed in historical ground-based observations by Beckers (1968), the literature is confusing[5]. They are almost universally suspected as being important for the solar corona.

[5] de Pontieu et al. (2007) did not relate their spicules to traditional spicules, but explained type I spicules as generated by magneto-acoustic shocks along flux tubes perturbed by photospheric motions, but mostly those extending outwards only to 2–3 Mm lengths. Yet in Martínez-Sykora et al. (2011) we read that type I spicules are the historical spicules discussed by Beckers (1968), reaching heights of 2–9 Mm!

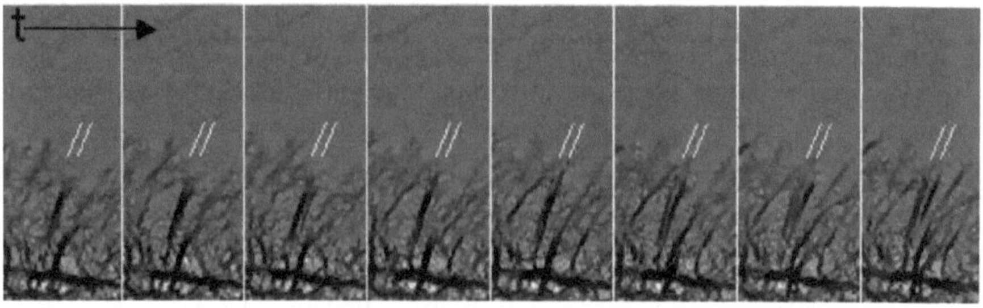

Fig. 3.8 Time series of sharpened images of type II spicules observed over the solar limb, in the chromospheric H line of Ca^+, from Suematsu et al. (2008). The cadence is 5 seconds, and the field of view roughly 7 ×18 Mm. The authors mark with two white lines the appearance of spicules rotating around one another, but this interpretation is of course not unique. The narrow black lines are spicules of type II, their typical widths of 0.35 Mm are barely resolved by the Hinode instrument. Note that such observations pick up many spicules along each line of sight because spicules of length $\ell \approx 5$ Mm contribute to the images along a path length $\Delta \approx 2\sqrt{2R_\odot \ell}$, $\approx$160 Mm extending across the solar limb

However, they have yet to be given a satisfactory physical identity, in spite of many numerical experiments (Martínez-Sykora et al. 2011; Martínez-Sykora et al. 2017). These authors claim to find features representing type II spicules and their counterparts, but the aspect ratios of these numerical structures are far too small (nearer 2:1 than over 10:1 as observed, e.g. Suematsu et al. 2008; Lipartito et al. 2014, see Figs. 3.8 and 3.9).

Of more concern is the remarkable dynamical behavior observed in type II spicules which appears inconsistent with a magnetic tube with a straw-like morphology (Judge et al. 2011, 2012; Lipartito et al. 2014). Given the propositions for coronal heating that include the interactions between apparently separate tubes of photospheric flux (Priest et al. 2002; van Ballegooijen et al. 2017), such current sheets may well have scales as large as these tubes (~200 km) down to dissipation scales. There is no physical reason to dismiss sheets, yet the novel refutation work has been curiously omitted from prominent and relevant reviews (Carlsson et al. 2019; de Pontieu et al. 2021).

Judge et al. (2011), Judge et al. (2012); Lipartito et al. (2014) suggested that line-of-sight issues can be at play, on the basis that super-Alfvénic motions were needed to explain the observed phenomena in terms of tubes. Instead they suggested that some of these features might correspond to current sheets seen edge-on, their rapid dynamics being akin to the coming and going of striations in lace curtains blowing in the wind (the later coronal work by Malanushenko et al. 2022, is of course concerned with such projections). Narrow sheets of plasma lie beyond the scope of existing 3D numerical models based upon advanced solvers of partial differential equations (Martínez-Sykora et al. 2011; Martínez-Sykora et al. 2017; Rempel 2017; Malanushenko et al. 2022) which suppress the formation of large gradients.

Observational evidence for plasma sheets is not insignificant. Lipartito et al. (2014) measured phase speeds exceeding any physical speed (except light!) in

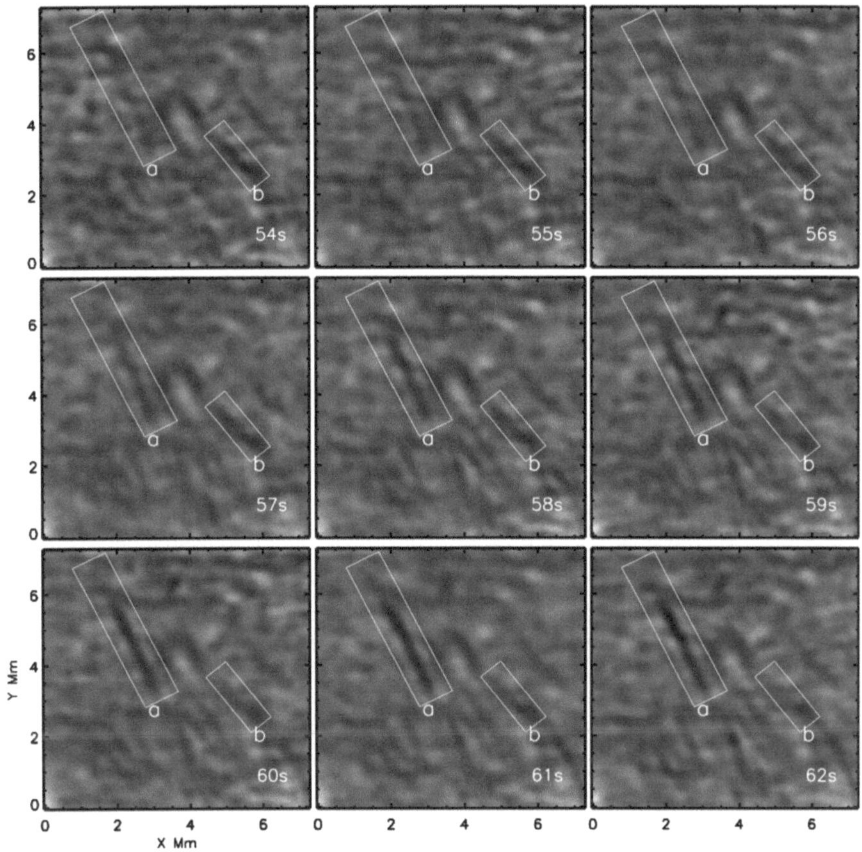

Fig. 3.9 Examples of super-Alfvénic variations from Judge et al. (2012). A 9s sequence of 1-second-cadence images in the red wing of Hα reveals the rapid appearance/ disappearance (region a/b) of dark, spicule-shaped features with phase speeds $\gtrapprox$ 1 Mm per second. Estimated Alfvén speeds of 0.05 Mm per second are expected using a field strength of 0.001 T and a particle density of 10^{17} m^{-3} (Beckers 1968)

the Hα line of disk features clearly similar to off-limb type II spicules. The data were inconsistent with fluid motions, but consistent with the line of sight effect of "sheets blowing in the wind" (Fig. 3.9). On the theoretical side, long ago Levine (1974) argued for the presence of such current sheets on the scale of and above super-granulation boundaries, precisely where these narrow, long spicules and fibrils originate. As well as interacting bundles of magnetic flux, at least two simpler physical mechanisms naturally create sheets: spontaneous formation of current sheets in nearly ideal MHD, and MHD turbulence (see the thorough discussions by Parker 1994; Schekochihin 2022). On the other hand, such structures are potentially unstable to resistive tearing mode instabilities, especially in thinner current sheets such as the Sweet-Parker configuration of Sect. 2.7 (Loureiro et al. 2007). By examining the evolution towards a Sweet-Parker layer, Pucci and Velli (2010) showed further that even before the non-ideal effects of tearing can occur, ideal effects caused by the plasmoid instability disrupt Sweet-Parker layers well before resistive effects occur. In this regard, Uzdensky and Loureiro (2016) note that:

> the recent realization that Sweet-Parker current sheets are violently unstable to the secondary tearing (plasmoid) instability implies that such current sheets cannot occur in real systems.

The statement applies to the Sweet-Parker idealization of resistive magnetic reconnection (Sect. 2.7) in which the sheet thickness has reached the tiny thickness of $S^{-1/2}$ of the length. Thicker current sheets and those with Prandtl numbers > 1 are less unstable. The intricate subsequent dynamics involves the formation of plasmoids separated by current sheets, under several regimes of Prandtl number and the wavenumber of the fastest growing modes (Schekochihin 2022, devotes considerable time to reviewing the advances). Evidence for large (>1 Mm in size) plasmoids in the solar atmosphere has been found in many observational studies (e.g. Innes et al. 2015; Mulay et al. 2023, and references therein). Evidence independent of physical size has come from polarization measurements (French et al. 2019) of the large-scale equatorial current sheet. This structure arises as trans-equatorial coronal loops are drawn out to interplanetary space by the solar wind (Smith 2001). The striking absence of linearly polarized light scattered from Fe^{12+} ions within the current sheet, compared with outside, puts demands on the nature of the emitting plasmas and the embedded magnetic fields. Given the low-β plasma and frozen-field condition on macroscopic scales, plasmoid formation offers the only natural explanation for the anomalously low polarization signatures. This particular observation requires unobserved magnetic structures on scales below the 4 Mm pixels used. Perhaps the stage is set for significant advancement concerning unresolved plasma dynamics, applied to other polarization observations, obtained for example with the DKIST.

The recent explosion in publications on the solar chromosphere (reviewed by Carlsson et al. 2019; de Pontieu et al. 2021) has resulted from a combination of new measurements from the ground and from space, with the maturation of radiation-MHD models. These first-principles models begin with an upward flux of convective energy beneath the photosphere and extend into the corona, embedded with magnetic fields. Equations of motion are evolved in time using numerical integrations of partial differential equations (see, e.g. Tóth 1996). The algorithms seek smooth solutions by the necessity to suppress well-known numerical instabilities. Yet there exist entire classes of discontinuous weak solutions to the MHD equations (Grad 1970), including well-known shocks and Parker's "tangential discontinuities" ("TDs" Parker 1994, 2007). While shocks have been captured with adaptive meshes, TDs which form throughout the MHD fluid are inaccessible to such calculations, without special techniques designed to find such solutions (e.g. Bhattacharyya et al. 2010). Reduced MHD calculations of turbulence reveal the "fettucine-like" ribbons on multiple scales (e.g. Einaudi et al. 2021; Schekochihin 2022). Such structures are absent from diffusive yet "realistic" published numerical solutions. The most recent work claiming to find type II spicules in traditional numerical MHD calculations (Martínez-Sykora et al. 2017, 2018) will therefore be unable to find an entire class of solutions which have been predicted, modeled, and perhaps already observed as spicules of type II (Lipartito et al. 2014)!

Ironically, the conclusions of Malanushenko et al. (2022) regarding line of sight interpretations were based upon the same kind of numerical calculations (Rempel 2017) which did not produce such narrow sheets (although some structures in Fig. 3.4 may be approaching 2D-like configurations). Again, the take-away point is that comparisons of computed and observed features are neither unique nor free of multiple implicit assumptions. There is a risk of finding what you are looking for (confirmation bias), and perhaps missing structures important in forming the building blocks of the radiating plasmas. There remains one nagging problem concerning the chromosphere-corona interface: is the emission from plasma we observe around 10^5 K dominated by a structure thermally and magnetically isolated from the corona (Feldman 1983)? Distributions of plasma with temperature above 10^5 K seem compatible with downward heat conduction (Jordan 1980), but a debate still rages about the 10^4–10^5 K region, discussed most recently by Judge (2021).

At stake is our zeroth-order picture of magnetic chromosphere-corona interface. For each cool spicule extending into the hot corona, a sheet presents the largest contact area, a circular tube the smallest area. The resolution of the question will prove to be critical for our understanding of the transport of mass, momentum and energy from chromosphere into the corona, and vice-versa.

Chapter Summary

All challenges in this chapter are compounded by one fact: the dominant agent responsible for coronal heating, the changing coronal magnetic field, is not directly measurable.

Scales of dissipation at kinetic scales will never be measurable using known methods. Instead we must accept unresolved 2D images mostly from integrals along the Earth-Sun line-of-sight.

Coronal magnetic field measurements have been replaced with qualitative "proxies" such as observed plasma loop morphology and kinematics, using computer models which cannot span the huge range of scales involved.

With current remote sensors, we observe just the smoking guns, i.e. the *aftermath* of the processes we wish to understand. This "crime scene" is further confused by non-local radiative transfer and heat transport by electrons.

The majority of coronal observations suffer from confusion owing to contributions from all emitters along each line-of-sight, highlighted by Malanushenko et al. (2022).

(continued)

Without measurements of evolving magnetic fields *emerging from the chromosphere*, the amplitude and frequency spectrum of the Poynting flux entering the corona will remain in doubt.

Our task for the remains of this book is to find ways around these difficulties. Next, we will assess methods of attack, to try to identify the most promising approaches for cracking the problem at hand.

References

Alissandrakis, C.E., Vial, J.C.: Explosive events in the quiet Sun near and beyond the solar limb observed with the interface region imaging spectrograph (IRIS). Sol. Phys. **298**, 18 (2023). https://doi.org/10.1007/s11207-023-02111-y

Allen, C.W.: Astrophysical Quantities. Athlone Press, University of London (1973)

Almleaky, Y.M., Brown, J.C., Sweet, P.A.: Density diagnostics and inhomgeneous plasmas I. Isothermal plasmas. Astron. Astrophys. **224**, 328–337 (1989)

Altschuler, M.D., Newkirk, Jr., G.: Magnetic fields and the structure of the solar corona I: methods of calculating coronal fields. Sol. Phys. **9**, 131–149 (1969)

Antiochos, S.K., Noci, G.: The structure of the static corona and transition region. Astrophys. J. **301**, 440 (1986)

Arber, T.D., Haynes, M., Leake, J.E.: Emergence of a flux tube through a partially ionized solar atmosphere. Astrophys. J. **666**, 541–546 (2007)

Asch, M., Bocquet, M., Nodet, M.: Data Assimilation. Society for Industrial and Applied Mathematics, Philadelphia (2016). https://doi.org/10.1137/1.9781611974546

Aschwanden, M.J.: Physics of the Solar Corona. An Introduction with Problems and Solutions, 2nd edn. Springer, Berlin (2005)

Aschwanden, M.J., Nightingale, R.W., Alexander, D.: Evidence for nonuniform heating of coronal loops inferred from multithread modeling of TRACE data. Astrophys. J. **541**, 1059–1077 (2000)

Ashbourn, J.M.A., Woods, L.C.: Energy transport in the solar transition layer. Proc. R. Soc. Lond. Ser. A **457**, 1873 (2001)

Athay, R.G., Holzer, T.E.: The role of spicules in heating the solar atmosphere. Astrophys. J. **255**, 743–752 (1982). https://doi.org/10.1086/159873

Avrett, E.H., Hummer, D.G.: Non-coherent scattering II. Line formation with a frequency independent source function. Mon. Not. R. Astron. Soc. **130**, 295–331 (1965)

Axford, W.I., McKenzie, J.F.: The origin of high speed solar wind streams. In: Marsch, E., Schwenn, R. (eds.) Solar Wind Seven, vol. 3. COSPAR Colloquium Series, pp. 1–5. Pergammon Press, Oxford (1992)

Axford, W.I., McKenzie, J.F.: Coronal heating. Adv. Space Res. **30**, 505 (2002). https://doi.org/10.1016/S0273-1177(02)00328-9

Banks, P.M., Kockarts, G.: Aeronomy. Academic Press, New York (1973)

Barker Bausell, R.: The Problem with Science. The Reproducibility Crisis and What to Do About It. Oxford University Press, Oxford (2021)

Beckers, J.M.: Solar spicules. Sol. Phys. **3**, 367–433 (1968)

Bhattacharyya, R., Low, B.C., Smolarkiewicz, P.K.: On spontaneous formation of current sheets: untwisted magnetic fields. Phys. Plasmas **17**, 112901 (2010)

Biskamp, D.: Magnetohydrodynamic Turbulence. Cambridge University Press, Cambridge (2003)

Bogachev, S.A., et al.: Microflares and nanoflares in the solar corona. Phys. Uspekhi **63**, 783–800 (2020). https://doi.org/10.3367/UFNe.2019.06.038769

Boldyrev, S.: On the spectrum of magnetohydrodynamic turbulence. Astrophys. J. Lett. **626**, L37–L40 (2005)

Boldyrev, S.: Spectrum of magnetohydrodynamic turbulence. Phys. Rev. Lett. **96**, 115002 (2006)

Braginskii, S.I.: Transport processes in a plasma. Rev. Plasma Phys. **1**, 205–311 (1965)

Brandenburg, A., Lazarian, A.: Astrophysical hydromagnetic turbulence. In: Balogh, A., Bykov, A., Cargill, P., Dendy, R., Dudok de Wit, T., Raymond, J. (eds.) Microphysics of Cosmic Plasmas, vol. 87 (2014). https://doi.org/10.1007/978-1-4899-7413-6_5

Brown, J.C., et al.: Interpretation of density sensitive line diagnostics from inhomgeneous plasmas II. Non-isothermal plasmas. Astron. Astrophys. **249**, 277–283 (1991)

Bruls, J.H.M.J., Von der Lühe, O.: Photospheric fine structure: an observational challenge. An analysis of radiative transfer effects on the visibility of small-scale structures. Astron. Astrophys. **366**, 281–293 (2001)

Cally, P.S.: Turbulent thermal conduction in the solar transition region. Astrophys. J. **355**, 693–699 (1990)

Cally, P.S.: Alfvén waves are easy: mode conversion in magnetic regions. In: Astronomical Society of India Conference Series, vol. 2, pp. 221–227 (2011)

Campi, C., et al.: Feature ranking of active region source properties in solar flare forecasting and the uncompromised stochasticity of flare occurrence. Astrophys. J. **883**, 150 (2019)

Carlsson, M., et al.: A publically available simulation of an enhanced network region of the Sun. Astron. Astrophys. **585**, A4 (2016)

Carlsson, M., de Pontieu, B., Hansteen, V.H.: New view of the solar chromosphere. Ann. Rev. Astr. Astrophys. **57**, 189–226 (2019)

Carmichael, H.: A process for flares. NASA Spec. Publ. **50**, 451 (1964)

Casini, R., Landi Degl'Innocenti, E.: Astrophysical plasmas. In: Fujimoto, T., Iwamae, A. (eds.) Plasma Polarization Spectroscopy, vol. 247. Springer, Berlin (2008)

Casini, R., White, S.M., Judge, P.G.: Magnetic diagnostics of the solar corona: synthesizing optical and radio techniques. Space Sci. Rev. **210**, 145–181 (2017)

Cattaneo, F., Hughes, D.W.: Solar dynamo theory: a new look at the origin of small-scale magnetic fields. Astron. Geophys. **42**, 3.18–3.22 (2001)

Chen, F., Rempel, M., Fan, Y.: A comprehensive radiative magnetohydrodynamics simulation of active region scale flux emergence from the convection zone to the corona. Astrophys. J. **937**, 91 (2022). https://doi.org/10.3847/1538-4357/ac8f95

Chitta, L.P., et al.: Solar coronal heating from small-scale magnetic braids. Astron. Astrophys. **667**, A166 (2022). https://doi.org/10.1051/0004-6361/202244170

Cho, J., Lazarian, A.: Compressible magnetohydrodynamic turbulence: mode coupling, scaling relations, anisotropy, viscosity-damped regime and astrophysical implications. Mon. Not. R. Astron. Soc. **345**, 325–339 (2003)

Cho, J., Lazarian, A., Vishniac, E.T.: Simulations of magnetohydrodynamic turbulence in a strongly magnetized medium. Astrophys. J. **564**, 291–301 (2002)

Cooper, J.: Plasma spectroscopy. Rep. Prog. Phys. **29**, 35–130 (1966)

Craig, I.J.D., Brown, J.C.: Fundamental limitations of X-ray spectra as diagnostics of plasma temperature structure. Astron. Astrophys. **49**, 239–250 (1976)

Craig, I.J.D., Brown, J.C.: Inverse Problems in Astronomy. Hilger, Bristol (1986)

Cranmer, S.R., van Ballegooijen, A.A.: On the generation, propagation, and reflection of Alfvén waves from the solar photosphere to the distant heliosphere. Astrophys. J. Suppl. **156**, 265–293 (2005)

Dahlburg, R.B., et al.: Observational signatures of coronal loop heating and cooling driven by footpoint shuffling. Astrophys. J. **817**, 47 (2016). https://doi.org/10.3847/0004-637X/817/1/47

Damé, L., et al.: A solar interferometric mission for ultrahigh resolution imaging and spectroscopy: SIMURIS. Adv. Space Res. **11**, 383–386 (1991). https://doi.org/10.1016/0273-1177(91)90135-7

Damé, L., et al.: The SIMURIS interferometric mission: solar physics objectives and model payload. Adv. Space Res. **14**, 167–170 (1994). https://doi.org/10.1016/0273-1177(94)90177-5

Davila, J.M.: Heating of the solar corona by the resonant absorption of Alfvén waves. Astrophys. J. **317**, 514–521 (1987)

Davila, J.M.: Solar tomography. Astrophys. J. **423**, 871 (1994)

de Pontieu, B., et al.: A tale of two spicules: the impact of spicules on the magnetic chromosphere. Publ. Astron. Soc. Jpn. **59**, 655 (2007)

de Pontieu, B., et al.: The interface region imaging spectrograph (IRIS). Sol. Phys. **289**, 2733–2779 (2014)

de Pontieu, B., et al.: A new view of the solar interface region from the interface region imaging spectrograph (IRIS). Sol. Phys. **296**, 84 (2021)

del Toro Iniesta, J.C., Ruiz Cobo, B.: Inversion of the radiative transfer equation for polarized light. Living Rev. Sol. Phys. **13**, 4 (2016). https://doi.org/10.1007/s41116-016-0005-2

Del Zanna, G., Mason, H.E.: Solar UV and X-ray spectral diagnostics. Living Rev. Sol. Phys. **15**, 5 (2018). https://doi.org/10.1007/s41116-018-0015-3

Démoulin, P., et al.: Can we extrapolate a magnetic field when its topology is complex? Sol. Phys. **174**, 73–89 (1997). https://doi.org/10.1023/A:1004926513260

Dere, K.P.: Turbulent power and dissipation in the solar transition zone. Astrophys. J. **340**, 599 (1989). https://doi.org/10.1086/167420

Dere, K.P.: Explosive events, magnetic reconnection, and coronal heating. Adv. Space Res. **14**, 13–22 (1994)

Dere, K.P., et al.: Discrete subresolution structures in the solar transition zone. Sol. Phys. **114**, 223–237 (1987)

Dere, K.P., Bartoe, J.D.F., Brueckner, G.E.: Explosive events in the solar transition zone. Astrophys. J. **345**, L95 (1989)

Dere, K.P., et al.: Explosive events and magnetic reconnection in the solar atmosphere. J. Geophys. Res. A6 **96**, 9399–9407 (1991)

Domingo, V., Fleck, B., Poland, A.I.: SOHO: the solar and heliospheric observatory. Space Sci. Rev. **72**, 81–84 (1995). https://doi.org/10.1007/BF00768758

Dufton, P.L., Kingston, A.E.: Atomic processes in the Sun. Adv. At. Mol. Phys. **17**, 355–418 (1981)

Dupree, A.K., Reeves, E.M.: The extreme ultraviolet spectrum of the quiet sun. Astrophys. J. **165**, 599 (1971)

Einaudi, G., et al.: Energetics and 3D structure of elementary events in solar coronal heating. Astrophys. J. **910**, 84 (2021)

Fan, Y.: Simulations of prominence eruption preceded by large-amplitude longitudinal oscillations and draining. Astrophys. J. **898**, 34 (2020)

Feldman, U.: On the unresolved fine structures of the solar atmosphere. Astrophys. J. **275**, 367 (1983)

Feldman, U.: On the unresolved fine structures of the solar atmosphere. II - the temperature region 200,000–500,000 K. Astrophys. J. **320**, 426 (1987)

Feldman, U.: Highly ionized atoms: what do they teach us about the solar atmosphere? Phys. Scr. Vol. T **47**, 128–132 (1993). https://doi.org/10.1088/0031-8949/1993/T47/020

Fisher, G.H., et al.: Coronal heating in active regions as a function of global magnetic variables. Astrophys. J. **508**, 885–898 (1998)

Foukal, P.: The temperature structure and pressure balance of magnetic loops in active regions. Sol. Phys. **43**, 327–336 (1975)

Fox, J.L., Kankelborg, C.C., Thomas, R.J.: A transition region explosive event observed in He II with the MOSES sounding rocket. Astrophys. J. **719**, 1132–1143 (2010). https://doi.org/10.1088/0004-637X/719/2/1132

Fox, N.J., et al.: The Solar Probe Plus mission: humanity's first visit to our star. Space Sci. Rev. **204**, 7–48 (2016)

French, R.J., et al.: Spectropolarimetric insight into plasma sheet dynamics of a solar flare. Astrophys. J. Lett. **887**, L34 (2019). https://doi.org/10.3847/2041-8213/ab5d34

Fujimoto, T.: Plasma Spectroscopy. International Series of Monographs in Physics. Clarendon Press/Oxford Science Publications, Oxford (2004)

Gabriel, A.: A magnetic model of the solar transition region. Philos. Trans. R. Soc. Lond. **281**, 339 (1976)

Galsgaard, K., Moreno-Insertis, F.: Explosive events in connection with small scale flux emergence in open field regions. In: 40th COSPAR Scientific Assembly, vol. 40, E2.4–5–14 (2014)

Gandorfer, A., et al.: The Solar Orbiter mission and its Polarimetric and Helioseismic Imager (SO/Phi). In: GONG-SoHO 24: A New Era of Seismology of the Sun and Solar-Like Stars, vol. 271. Journal of Physics Conference Series, 012086 (2011)

Gelfreikh, G.B.: Radio measurements of coronal magnetic fields. In: Rusin, V., Heinzel, P., Vial, J.C. (eds.) IAU Colloq. 144, Solar Coronal Structures, pp. 21–28. VEDA Publishing Company, Bratislava (1994)

Gelfreikh, G.B., Lubyshev, B.L.: Structure of local sources of the s-component of solar radio emission. Soviet Astron. **23**, 316 (1979)

Gibson, E.G.: The Quiet Sun. NASA Sol. Phys. National Aeronautics and Space Administration, Washington (1973)

Gold, T., Hoyle, F.: On the origin of solar flares. Mon. Not. R. Astron. Soc. **120**, 89 (1960)

Goldreich, P., Sridhar, S.: Toward a theory of interstellar turbulence. II. Strong Alfvenic turbulence. Astrophys. J. **438**, 763 (1995)

Golub, L., Pasachoff, J.M.: The Solar Corona, 2nd edn. Cambridge University Press, Cambridge (2009)

Grad, H.: Problems in magnetostatic equilibrium. In: AFSOR-70=1134. New York University, Courant Institute (1970)

Griem, H.: Plasma Spectroscopy. McGraw-Hill, New York (1964)

Gudiksen, B.V., Nordlund, Å.: An *ab initio* approach to the solar coronal heating problem. Astrophys. J. **618**, 1020–1030 (2005)

Halain, J.P. et al., EUV high resolution imager on-board Solar Orbiter: optical design and detector performances. In: Proc. SPIE 10564, International Conference on Space Optics — ICSO 2012, 105643V (20 November 2017). https://doi.org/10.1117/12.2309019

Hanle, W.: Über Magnetische Beeinflussung Der Polarisation Der Resonanzfluoreszenz. Z. Phys. **30**, 93–105 (1924)

Hansteen, V.: A new interpretation of the redshift observed in optically thin transition region lines. Astrophys. J. **402**, 741–755 (1993)

Hansteen, V.: Waves, shocks and non-stationary phenomena in the outer solar atmosphere. In: Wilson, A. (ed.) The Fifth Soho Workshop, the Corona and Solar Wind Near Minimum Activity, pp. 45–54. ESA SP-404, ESTEC, Noordwijk (1997)

Hansteen, V.H., et al.: Dynamic fibrils are driven by magnetoacoustic shocks. Astrophys. J. Lett. **647**, L73–l76 (2006)

Hansteen, V., et al.: The unresolved fine structure resolved: IRIS observations of the solar transition region. Science **346**, 1255757 (2014)

Heap, S.R., et al.: The Goddard high resolution spectrograph: in-orbit performance. Publ. Astron. Soc. Pac. **107**, 871 (1995)

Heggland, L., de Pontieu, B., Hansteen, V.H.: Observational signatures of simulated reconnection events in the solar chromosphere and transition region. Astrophys. J. **702**, 1–18 (2009). https://doi.org/10.1088/0004-637X/702/1/1

Heyvaerts, J., Priest, E.R.: Coronal heating by phase-mixed shear Alfvén waves. Astron. Astrophys. **117**, 220–234 (1983)

Hirayama, T.: Theoretical model of flares and prominences. I: Evaporating flare model. Sol. Phys. **34**, 323–338 (1974)

Holland, J.H.: Complexity: A Very Short Introduction. Oxford University Press, Oxford (2014)

Howson, T.A., de Moortel, I., Reid, J.: Phase mixing and wave heating in a complex coronal plasma. Astron. Astrophys. **636**, A40 (2020)

Huang, Z., et al.: Explosive events in active regions observed by IRIS and SST/CRISP. Mon. Not. R. Astron. Soc. **464**, 1753–1761 (2017). https://doi.org/10.1093/mnras/stw2469

Hubeny, I.: Probabilistic interpretation of radiative transfer I. The $\sqrt{\varepsilon}$ law. Astron. Astrophys. **185**, 332–335 (1987)

Hyder, C.L., Lites, B.W.: Hα Doppler brightening and Lyman-α Doppler dimming in moving Hα prominences. Sol. Phys. **14**, 147–156 (1970)

Iben, I.: Stellar Evolution Physics, Volume 1: Physical Processes in Stellar Interiors. Cambridge University Press, Cambridge University (2012). https://doi.org/10.1017/CBO9781139061223

Innes, D.E.: Transition region dynamics. In: Lacoste, H. (ed.) *SOHO 13 Waves, Oscillations and Small-Scale Transients Events in the Solar Atmosphere: Joint View from SOHO and TRACE*, vol. 547. ESA Special Publication, p. 215 (2004)

Innes, D.E., et al.: Bi-directional plasma jets produced by magnetic reconnection on the Sun. Nature **386**, 811–813 (1997). https://doi.org/10.1038/386811a0

Innes, D.E., et al.: IRIS Si IV line profiles: an indication for the plasmoid instability during small-scale magnetic reconnection on the Sun. Astrophys. J. **813**, 86 (2015)

Ionson, J.A.: Resonant absorption of Alfvénic surface waves and the heating of solar coronal loops. Astrophys. J. **226**, 650–673 (1978)

Ionson, J.A.: A unified theory of electrodynamic coupling in coronal magnetic loops - the coronal heating problem. Astrophys. J. **276**, 357–368 (1984)

Ionson, J.A.: Coronal heating by resonant (A.C.) and nonresonant (D.C.) mechanisms. Astron. Astrophys. **146**, 199–203 (1985a)

Ionson, J.A.: The heating of coronae. Sol. Phys. **100**, 289 (1985b)

Iroshnikov, P.S.: Turbulence of a conducting fluid in a strong magnetic field. Sov. Astron. **7**, 566 (1964)

Jefferies, J.T., Orral, F.Q., Zirker, J.B.: The interpretation of total line intensities from optically thin gases I. General method. Sol. Phys. **22**, 307–316 (1972a)

Jefferies, J.T., Orral, F.Q., Zirker, J.B.: The interpretation of total line intensities from optically thin gases II. The coronal forbidden lines. Sol. Phys. **22**, 317–326 (1972b)

Jordan, C.: The intensities of helium lines in the solar EUV spectrum. Mon. Not. R. Astron. Soc. **170**, 429–440 (1975)

Jordan, C.: The structure and energy balance of solar active regions. Philos. Trans. R. Soc. Lond. **A281**, 391 (1976)

Jordan, C.: The energy balance of the solar transition region. Astron. Astrophys. **86**, 355–363 (1980)

Jordan, C.: Modelling of solar coronal loops. Memorie della Societa Astronomica Italiana **63**, 605–620 (1992)

Judge, P.G.: The monochromatic density diagnostic technique: first detection of multiple density components of the chromosphere of alpha Tauri. Astrophys. J. **430**, 351–359 (1994)

Judge, P.G.: On spectroscopic filling factors and the solar transition region. Astrophys. J. **531**, 585–590 (2000)

Judge, P.G.: Coronal emission lines as thermometers. Astrophys. J. **708**, 1238–1240 (2010)

Judge, P.G.: UV spectra, bombs, and the solar atmosphere. Astrophys. J. **808**, 116 (2015)

Judge, P.G.: Some thoughts on emission-line spectroscopy. Mon. Not. R. Astron. Soc. **491**, 576–579 (2020)

Judge, P.G.: Magnetic connections across the chromosphere-corona transition region. Astrophys. J. **914**, 70 (2021)

Judge, P.G., et al.: On the failure of standard emission measure analysis for solar EUV and UV irradiance spectra. Astrophys. J. Lett. **455**, L85–189 (1995)

Judge, P. G., Hansteen, V., Wikstøl, Ø., Wilhelm, K., Schühle, U., and Moran, T.: Evidence in Support of the "Nanoflare" Picture of Coronal Heating from SUMER Data. Astrophys. J. **502**, 981 (1998)

Judge, P.G., Tritschler, A., Low, B.C.: Thermal fine structure and magnetic fields in the solar atmosphere: spicules and fibrils. Astrophys. J. Lett. **730**, L4 (2011)

Judge, P.G., Reardon, K., Cauzzi, G.: Evidence for sheet-like elementary structures in the Sun's atmosphere? Astrophys. J. Lett. **755**, L11 (2012)

Judge, P.G., et al.: Photon mean free paths, scattering, and ever-increasing telescope resolution. Sol. Phys. **290**, 979–996 (2015)

Judge, P., et al.: Measuring the magnetic origins of solar flares, coronal mass ejections, and space weather. Astrophys. J. **917**, 27 (2021)

Judge, P., et al.: Optimal spectral lines for measuring chromospheric magnetic fields. Astrophys. J. **941**, 159 (2022). https://doi.org/10.3847/1538-4357/aca2a5

Judge, P.G., et al.: Magnetic fields and plasma heating in the Sun's atmosphere. Astrophys. J. **960**, 129 (2024)

Kaiser, M.L., et al.: The STEREO mission: an introduction. Space Sci. Rev. **136**, 5–16 (2008). https://doi.org/10.1007/s11214-007-9277-0

Kinnison, J., et al.: Solar Probe Plus: a mission to touch the Sun. In:Proceedings of the 2013 IEEE Aerospace Conference, 144 (2013)

Kitiashvili, I., Kosovichev, A.G.: Application of data assimilation method for predicting solar cycles. Astrophys. J. Lett. **688**, L49 (2008). https://doi.org/10.1086/594999

Klimchuk, J.A., DeForest, C.E.: Cross sections of coronal loop flux tubes. Astrophys. J. **900**, 167 (2020)

Kobayashi, K., et al.: The high-resolution coronal imager (Hi-C). Sol. Phys. **289**, 4393–4412 (2014). https://doi.org/10.1007/s11207-014-0544-4

Kohl, J.L., et al.: UVCS/SOHO empirical determinations of anisotropic velocity distributions in the solar corona. Astrophys. J. Lett. **501**, L127 (1998)

Kolmogorov, A.: The local structure of turbulence in incompressible viscous fluid for very large Reynolds' numbers. Akad. Nauk SSSR Dokl. **30**, 301–305 (1941)

Kopp, R., Pneuman, G.: Magnetic reconnection in the corona and the loop prominence phenomenon. Sol. Phys. **50**, 85–98 (1976)

Kraichnan, R.H.: Inertial-range spectrum of hydromagnetic turbulence. Phys. Fluids **8**, 1385–1387 (1965)

Kulsrud, R.M.: Plasma Physics for Astrophysics. Princeton University Press, Princeton and Oxford (2005)

Lacatus, D.A., Judge, P.G., Donea, A.: An explanation of remarkable emission-line profiles in post-flare coronal rain. Astrophys. J. **842**, 15 (2017)

Landi, E., Feldman, U.: The thermal structure of an active region observed outside the solar disk. Astrophys. J. **672**, 674–683 (2008)

Landi, E., Feldman, U., Dere, K.P.: CHIANTI-an atomic database for emission lines. V. Comparison with an isothermal spectrum observed with SUMER. Astrophys. J. Suppl. **139**, 281–296 (2002). https://doi.org/10.1086/337949

Landi Degl'Innocenti, E.: Spectropolarimetry with new generation solar telescopes. Memorie della Societa Astronomica Italiana **84**, 391 (2013)

Landi degl'Innocenti, E.L., Landolfi, M.: Polarization in Spectral Lines, vol. 307. Astrophysics and Space Library. Kluwer Academic Publishers, Dordrecht (2004)

Lang, J., Mason, H.E., McWhirter, R.W.P.: The interpretation of the spectral line intensities from the CHASE spectrometer on spacelab 2. Sol. Phys. **129**, 31–81 (1990)

Lee, M.A.: Comments on the dissipation of hydromagnetic surface waves. Astrophys. J. **240**, 693–695 (1980). https://doi.org/10.1086/158276

Leka, K.D., et al.: A comparison of flare forecasting methods. II. Benchmarks, metrics, and performance results for operational solar flare forecasting systems. Astrophys. J. Suppl. **243**, 36 (2019a)

Leka, K.D., et al.: A comparison of flare forecasting methods. III. Systematic behaviors of operational solar flare forecasting systems. Astrophys. J. **881**, 101 (2019b)

Lemen, J., et al.: The interface region imaging spectrograph (IRIS) NASA SMEX. In: AAS/Solar Physics Division Abstracts #42, vol. 289, 1512 (2011)

Levine, R.H.: A new theory of coronal heating. Astrophys. J. **190**, 457–466 (1974)

Lipartito, I., et al.: The solar chromosphere observed at 1 Hz and 0."2 resolution. Astrophys. J. **785**, 109 (2014)

Lites, B.W.: Remote sensing of solar magnetic fields. Rev. Geophys. **38**, 1–36 (2000)

Lites, B.W., et al.: Are Internetwork magnetic fields in the solar photosphere horizontal or vertical? Astrophys. J. **835**, 14 (2017). https://doi.org/10.3847/1538-4357/835/1/14

Litwin, C., Rosner, R.: On the structure of solar and stellar coronae - loops and loop heat transport. Astrophys. J. **412**, 375–385 (1993)

Liu, Y.H., et al.: First-principles theory of the rate of magnetic reconnection in magnetospheric and solar plasmas. Commun. Phys. **5**, 97 (2022). https://doi.org/10.1038/s42005-022-00854-x

Loureiro, N.F., Schekochihin, A.A., Cowley, S.C.: Instability of current sheets and formation of plasmoid chains. Phys. Plasmas **14**, 100703–100703 (2007). https://doi.org/10.1063/1.2783986

Low, B.C.: Topological nature of the Parker magnetostatic theorem. Phys. Plasmas **30**, 012903 (2023). https://doi.org/10.1063/5.0124164

MacKay, D.H., et al.: How accurately can we determine the coronal heating mechanism in the large-scale corona? Sol. Phys. **193**, 93–116 (2000)

Magyar, N., Van Doorsselaere, T., Goossens, M.: Understanding uniturbulence: self-cascade of MHD waves in the presence of inhomogeneities. Astrophys. J. **882**, 50 (2019). https://doi.org/10.3847/1538-4357/ab357c

Malanushenko, A., et al.: The coronal veil. Astrophys. J. **927**, 1 (2022)

Mandrini, C.H., Démoulin, P., Klimchuk, J.A.: Magnetic field and plasma scaling laws: their implications for coronal heating models. Astrophys. J. **530**, 999–1015 (2000)

Mariska, J.T.: The Solar Transition Region. Cambridge University Press, Cambridge (1992)

Marsch, E.: Kinetic physics of the solar wind plasma. In: Schwenn, R., Marsch, E. (eds.) Physics of the Inner Heliosphere Vol II. Particles, Waves and Turbulence, pp. 45–133. Springer-Verlag, Berlin (1990)

Marsch, E., et al.: Solar Orbiter - mission profile, main goals and present status. Adv. Space Res. **36**, 1360–1366 (2005)

Martínez-Sykora, J., Hansteen, V., Moreno-Insertis, F.: On the origin of the type II spicules: dynamic three-dimensional MHD simulations. Astrophys. J. **736**, 9 (2011)

Martínez-Sykora, J., et al.: On the generation of solar spicules and Alfvénic waves. Science **356**, 1269–1272 (2017)

Martínez-Sykora, J., et al.: Impact of type II spicules in the corona: simulations and synthetic observables. Astrophys. J. **860**, 116 (2018). https://doi.org/10.3847/1538-4357/aac2ca

Matthaeus, W.H., et al.: Pathways to dissipation in weakly collisional plasmas. Astrophys. J. **891**, 101 (2020). https://doi.org/10.3847/1538-4357/ab6d6a

Metcalf, T.R., et al.: Is the solar chromospheric magnetic field force-free? Astrophys. J. **439**, 474–481 (1995)

Mihalas, D.: Stellar Atmospheres, 2nd edn. W. H. Freeman and Co., San Francisco (1978)

Mihalas, D., Mihalas, B.W.: Foundations of Radiation Hydrodynamics. Oxford University Press, New York (1984)

Morton, R.J., Tomczyk, S., Pinto, R.: Investigating Alfvénic wave propagation in coronal open-field regions. Nat. Commun. **6**, 7813 (2015). https://doi.org/10.1038/ncomms8813

Mulay, S.M., et al.: Formation and thermodynamic evolution of plasmoids in active region jets. Mon. Not. R. Astron. Soc. **518**(2), 2287–2299 (2023). https://doi.org/10.1093/mnras/stac3035

Nakariakov, V.M., Verwichte, E.: Coronal waves and oscillations. Living Rev. Sol. Phys. **2**, 3 (2005)

Neukirch, T.: Magnetic field extrapolation. In: Innes, D.E., Lagg, A., Solanki, S.A. (eds.) Chromospheric and Coronal Magnetic Fields, vol. 596. ESA Special Publication, 12.1 (2005)

Nóbrega-Siverio, D., et al.: Ambipolar diffusion in the bifrost code. Astron. Astrophys. **638**, A79 (2020)

Pagano, P., de Moortel, I., Morton, R.J.: Effect of coronal loop structure on wave heating through phase mixing. Astron. Astrophys. **643**, A73 (2020). https://doi.org/10.1051/0004-6361/202039209

Parker, E.N.: Nanoflares and the solar X-ray corona. Astrophys. J. **330**, 474–479 (1988)

Parker, E.N.: Heating solar coronal holes. Astrophys. J. **372**, 719 (1991a)
Parker, E.N.: The phase mixing of Alfvén waves, coordinated modes, and coronal heating. Astrophys. J. **376**, 355 (1991b)
Parker, E.N.: Spontaneous Current Sheets in Magnetic Fields with Application to Stellar X-rays. International Series on Astronomy and Astrophyics. Oxford University Press, Oxford (1994)
Parker, E.N.: Conversations on Electric and Magnetic Fields in The Cosmos. Princeton University Press, Princeton (2007)
Pascoe, D.J., Wright, A.N., De Moortel, I.: Propagating coupled Alfvén and kink oscillations in an arbitrary inhomogeneous corona. Astrophys. J. **731**, 73 (2011). https://doi.org/10.1088/0004-637X/731/1/73
Penn, M.J., Livingston, W.: Temporal changes in sunspot umbral magnetic fields and temperatures. Astrophys. J. Lett. **649**, L45–L48 (2006)
Pereira, T.M.D., et al.: An interface region imaging spectrograph first view on solar spicules. Astrophys. J. Lett. **792**, L15 (2014). https://doi.org/10.1088/2041-8205/792/1/L15
Perez, J.C., Chandran, B.D.G.: Direct numerical simulations of reflection-driven, reduced magnetohydrodynamic turbulence from the Sun to the Alfvén critical point. Astrophys. J. **776**, 124 (2013). https://doi.org/10.1088/0004-637X/776/2/124
Pesnell, W.D., Thompson, B.J., Chamberlin, P.C.: The solar dynamics observatory (SDO). Sol. Phys. **275**, 3–15 (2012)
Peter, H.: First high spectral resolution VUV full-sun spectrum compared to cool stars. Astrophys. J. **449**, 759–768 (2006). https://doi.org/10.1051/0004-6361:20054337
Peter, H., et al.: Hot explosions in the cool atmosphere of the sun. Science **346**, 1255726 (2014)
Pietarila, A., Judge, P.G.: On the formation of the resonance lines of helium in the Sun. Astrophys. J. **606**, 1239–1257 (2004)
Pontin, D.I., Peter, H., Chitta, L.P.: Non-thermal line broadening due to braiding-induced turbulence in solar coronal loops. Astron. Astrophys. **639**, A21 (2020). https://doi.org/10.1051/0004-6361/202037582
Pottasch, S.R.: On the interpretation of the solar ultraviolet emission line spectrum. Space Sci. Rev. **3**, 816–855 (1964)
Priest, E.R., et al.: Nature of the heating mechanism for the diffuse solar corona. Nature **393**, 545–547 (1998). https://doi.org/10.1038/31166
Priest, E.R., et al.: A method to determine the heating mechanisms of the solar corona. Astrophys. J. **539**, 1002–1022 (2000). https://doi.org/10.1086/309238
Priest, E.R., Heyvaerts, J.F., Title, A.M.: A flux-tube tectonics model for solar coronal heating driven by the magnetic carpet. Astrophys. J. **576**, 533–551 (2002)
Przybylski, D., et al.: Chromospheric extension of the MURAM code. Astron. Astrophys. **664**, A91 (2022). https://doi.org/10.1051/0004-6361/202141230
Pucci, F., Velli, M.: Reconnection of quasi-singular current sheets: the "ideal" tearing mode. Astrophys. J. Lett. **780**, L19 (2010). https://doi.org/10.1088/2041-8205/780/2/L19
Rabin, D., Moore, R.: Heating the Sun's lower transition region with fine-scale electric currents. Astrophys. J. **285**, 359–367 (1984)
Rachmeler, L.A., et al.: The high-resolution coronal imager, Flight 2.1. Sol. Phys. **294**, 174 (2019)
Raouafi, N.E., Solanki, S.K.: Effect of the electron density stratification on off-limb O VI line profiles: how large is the velocity distribution anisotropy in the solar corona? Astron. Astrophys. **427**, 725–733 (2004). https://doi.org/10.1051/0004-6361:20041203
Rappazzo, A.F., et al.: Coronal heating, weak MHD turbulence, and scaling laws. Astrophys. J. Lett. **657**, L47–L51 (2007). https://doi.org/10.1086/512975
Rappazzo, A.F., et al.: Nonlinear dynamics of the Parker scenario for coronal heating. Astrophys. J. **677**, 1348–1366 (2008). https://doi.org/10.1086/528786
Rempel, M.: Numerical simulations of quiet sun magnetism: on the contribution from a small-scale dynamo. Astrophys. J. **789**, 132 (2014). https://doi.org/10.1088/0004-637X/789/2/132
Rempel, M.: Extension of the MURAM radiative MHD code for coronal simulations. Astrophys. J. **834**, 10 (2017)

Roberts, W.O.: A preliminary report on chromospheric spicules of extremely short lifetime. Astrophys. J. **101**, 136 (1945)
Rosner, R., Tucker, W.H., Vaiana, G.S.: Dynamics of the quiescent solar corona. Astrophys. J. **220**, 643 (1978)
Roumeliotis, G.: Joule heating as an explanation for the differential emission measure structure and systematic Redshifts in the Sun's lower transition region. Astrophys. J. **379**, 392 (1991)
Rust, T., Kankelborg, C.C.: Quiet-Sun explosive events observed in He II λ304 with MOSES-06. Astrophys. J. **877**, 59 (2019). https://doi.org/10.3847/1538-4357/ab12e2
Rutten, R.J.: Radiative Transfer in Stellar Atmospheres, 8th edn. Utrecht University, Utrecht (2003)
Ryutova, M., Frank, Z., Berger, T.: Formation and dynamics of multi-thread arcades of coronal loops. In: Lites, B., Cheung, M., Magara, T., Mariska, J., Reeves, K. (eds.) The Second Hinode Science Meeting: Beyond Discovery-Toward Understanding, vol. 415. Astronomical Society of the Pacific Conference Series, 291 (2009)
Schad, T.A., et al.: Vector magnetic field measurements along a cooled stereo-imaged coronal loop. Astrophys. J. **833**, 5 (2016)
Scharmer, G., et al.: The new Swedish solar telescope. In: Rimmele, T.R., Balasubramaniam, K.S., Radick, R.R. (eds.) High Resolution Solar Physics: Theory, Observations, and Techniques, vol. 183. Astronomical Society of the Pacific Conference Series, 157. Procs. 19th NSO/Sacramento Peak Summer Workshop, ASP Conf. Ser. 183 (1999)
Schekochihin, A.A.: MHD turbulence: a biased review. J. Plasma Phys. **88**, 155880501 (2022). https://doi.org/10.1017/S0022377822000721
Schrijver, C.J., et al.: The nonpotentiality of active-region coronae and the dynamics of the photospheric magnetic field. Astrophys. J. **628**, 501–513 (2005). https://doi.org/10.1086/430733
Seaton, M.J.: The spectrum of the solar corona. Planet. Space Sci. **12**, 55 (1964)
Semel, M.: Boundary conditions and the extrapolation of magnetic fields (invited review). In: Alissandrakis, C.E., Schmieder, B. (eds.) Three-Dimensional Structure of Solar Active Regions, vol. 155. Astronomical Society of the Pacific Conference Series, 423 (1998)
Shibata, K.: Rapidly time variable phenomena: jets, explosive events, and flares. In: Wilson, A. (ed.) The Fifth Soho Workshop, the Corona and Solar Wind Near Minimum Activity, pp. 103–112. ESA SP-404, ESTEC, Noordwijk (1997)
Simnett, G.M.: Energy input for explosive events in the transition zone. Space Sci. Rev. **70**, 69 (1994)
Smith, E.J.: The heliospheric current sheet. J. Geophys. Res. **106**, 15819–15832 (2001). https://doi.org/10.1029/2000JA000120
Smith, L.: Chaos: A Very Short Introduction. Oxford University Press, Oxford (2007)
Socas-Navarro, H.: Are electric currents heating the magnetic chromosphere? Astrophys. J. Lett. **633**, L57–l60 (2005a)
Socas-Navarro, H.: The three-dimensional structure of a sunspot magnetic field. Astrophys. J. Lett. **631**, L167–l170 (2005b)
Squire, J., et al.: High-frequency heating of the solar wind triggered by low-frequency turbulence. Nat. Astron. **6**, 715–723 (2022). https://doi.org/10.1038/s41550-022-01624-z
Stenflo, J.O.: Solar Magnetic Fields, Polarized Radiation Diagnostics. Kluwer, Dordrecht (1994)
Sturrock, P.A.: Model of the high-energy phase of solar flares. Nature **211**, 695–697 (1966)
Sturrock, P.A.: Chromospheric magnetic reconnection and its possible relationship to coronal heating. Astrophys. J. **521**, 451–459 (1999). https://doi.org/10.1086/307544
Sturrock, P.A., Roald, C.B., Wolfson, R.: Chromospheric magnetic reconnection and its implication for coronal heating. Astrophys. J. Lett. **524**, L75–L78 (1999). https://doi.org/10.1086/312301
Suematsu, Y., et al.: High resolution observations of spicules with Hinode/SOT. In: Matthews, S.A., Davis, J.M., Harra, L.K. (eds.) First Results from Hinode, vol. 397. Astronomical Society of the Pacific Conference Series, 27 (2008)

Tavabi, E., Zeighami, S., Heydari, M.: Dynamics of explosive events observed by the interface region imaging spectrograph. Sol. Phys. **297**, 76 (2022). https://doi.org/10.1007/s11207-022-01990-x

Thomas, R.N., Athay, R.G. Physics of the solar hromosphere. New York, Interscience (1961)

Tomczyk, S., McIntosh, S.W.: Time-distance seismology of the solar corona with COMP. Astrophys. J. **697**, 1384–1391 (2009). https://doi.org/10.1088/0004-637X/697/2/1384

Tomczyk, S., et al.: Alfvén waves in the solar corona. Science **317**, 1192 (2007)

Tóth, G.: Comparison of some flux corrected transport and total variation diminishing numerical schemes for hydrodynamic and magnetohydrodynamic problems. J. Comput. Phys. **128**, 82–100 (1996)

Tsiklauri, D.: Missing pieces of the solar jigsaw puzzle. Astron. Geophys. **50**, 5.32–5.38 (2009)

Uzdensky, D.A.: The fast collisionless reconnection condition and the self-organization of coronal heating. Astrophys. J. **671**, 2139–2153 (2007). https://dx.doi.org/10.1086/522915

Uzdensky, D.A., Loureiro, N.F.: Magnetic reconnection onset via disruption of a forming current sheet by the tearing instability. Phys. Rev. Lett. **116**, 105003 (2016). https://doi.org/10.1103/PhysRevLett.116.105003

van Ballegooijen, A.A., Asgari-Targhi, M.: Direct and inverse cascades in the acceleration region of the fast solar wind. Astrophys. J. **835**, 10 (2017). https://doi.org/10.3847/1538-4357/835/1/10

van Ballegooijen, A.A., et al.: Heating of the solar chromosphere and corona by Alfvén wave turbulence. Astrophys. J. **736**, 3 (2011). https://doi.org/10.1088/0004-637X/736/1/3

van Ballegooijen, A.A., Asgari-Targhi, M., Voss, A.: The heating of solar coronal loops by Alfvén wave turbulence. Astrophys. J. **849**, 46 (2017). https://doi.org/10.3847/1538-4357/aa9118

van de Hulst, H.C.: Light Scattering by Small Particles. Wiley, New York (1957)

Vernazza, J.E., Avrett, E.H., Loeser, R.: Structure of the solar chromosphere. I. Basic computations and summary of the results. Astrophys. J. **184**, 605 (1973)

Vernazza, J.E., Avrett, E.H., Loeser, R.: Structure of the solar chromosphere. II. The photosphere and the temperature-minimum region. Astrophys. J. Suppl. **30**, 1 (1976)

Vernazza, J.E., Avrett, E.H., Loeser, R.: Structure of the chromosphere. III. Models of the EUV brightness components of the quiet Sun. Astrophys. J. Suppl. **45**, 635–725 (1981)

Vissers, G.J.M., et al.: Active region chromospheric magnetic fields. Observational inference versus magnetohydrostatic modelling. Astron. Astrophys. **662**, A88 (2022). https://doi.org/10.1051/0004-6361/202142087

Walsh, R., Sarkar, A.: Observed flows in coronal loops generated by multi-thread modelling. In: 38th COSPAR Scientific Assembly, vol. 38, 3. (2010)

Wargnier, Q.M., et al.: Multifluid simulations of upper-chromospheric magnetic reconnection with helium-hydrogen mixture. Astrophys. J. **946**(2), 115 (2023). https://doi.org/10.3847/1538-4357/acbfb1

Warren, H.P.: Multithread hydrodynamic modeling of a solar flare. Astrophys. J. **637**, 522–530 (2006). https://doi.org/10.1086/497904

Wiegelmann, T., et al.: Comparing magnetic field extrapolations with measurements of magnetic loops. Astron. Astrophys. **433**, 701–705 (2005)

Wiegelmann, T., et al.: Can we improve the preprocessing of photospheric vector magnetograms by the inclusion of chromospheric observations? Sol. Phys. **247**, 249–267 (2008). https://doi.org/10.1007/s11207-008-9130-y

Williams, T., et al.: Is the high-resolution coronal imager resolving coronal strands? Results from AR 12712. Astrophys. J. **892**, 134 (2020)

Williams, T., Walsh, R.W., Morgan, H.: The cross-sectional shape and height expansion of coronal loops: high-resolution coronal imager (Hi-C) analysis of AR 12712. Astrophys. J. **919**, 47 (2021)

Wilmot-Smith, A.L.: An overview of flux braiding experiments. R. Soc. Philos. Trans. A **373** (2015)

Woo, R.: Revealing the true solar corona. Am. Sci. **98**, 212–219 (2010)

Wood, B.E., Linsky, J.L., Ayres, T.R.: Evaluating possible heating mechanisms using the transition region line profiles of late-type stars. Astrophys. J. **478**, 745–765 (1997). https://doi.org/10.1086/303821

Xia, C., Keppens, R.: Formation and plasma circulation of solar prominences. Astrophys. J. **823**, 22 (2016)

Yang, Y., et al.: Pressure-strain interaction as the energy dissipation estimate in collisionless plasma. Astrophys. J. **929**, 142 (2022). https://doi.org/10.3847/1538-4357/ac5d3e

Chapter 4
Methodologies

The goal of this chapter is to scrutinize previous work to find out if and why we still have a heating problem, and what the nature of the remaining challenges might be. By necessity some discussion of motivations, rewards and career-building cannot be avoided, science does not operate in a vacuum. It is done by scientists, working within an imperfect system of peer review, funding, politics, social dynamics and specific career choices. We ask readers merely to consider these more subjective arguments that we offer, and to then make their own decisions.

Is there then a weakness in particular in the methods used which might point to more productive approaches? Keeping in mind the formidable challenges of Chap. 3, we extract methodologies from representative articles and suggest new methods and strategies to promote further understanding. In reading the literature we are occasionally reminded of the words of Enrico Fermi recalled by Peierls (1960):

> Quite recently, a friend showed him the paper of a young physicist... Pauli remarked sadly 'It is not even wrong.'

This kind of impression might arise from the observation that much research on coronal heating draws conclusions which are contingent upon multiple *unknown or unknowable premises*. Other influences are likely the innate ill-posedness and non-uniqueness, together with the sticky issue of confirmation bias. Articles with negative or null results are sometimes difficult to get published, even though they can contribute to significant progress. According to philosopher and historian of science Kuhn (1970), natural science progresses in fits and starts. Periods of "normal science" are punctuated by periods of crisis where previous ideas are found to be untenable. Intense activity then follows until a new state of affairs is reached, where scientists begin to adopt the new "paradigm". This is followed by a new episode of "normal science", and these cycles continue. Arguably we have been in a normal science phase since the SKYLAB results fully made their impact (ca. 1980). But unlike the paradigm changes discussed by Kuhn, the existing fundamental paradigm of treating the corona as a magnetized plasma, using classical dynamics and electromagnetic fields, is not in question. Solutions to the heating problem must

P. Judge, J. A. Ionson, *The Problem of Coronal Heating*, Astrophysics and Space Science Library 470, https://doi.org/10.1007/978-3-031-46273-3_4

lie within this paradigm, perhaps even within MHD. Instead, we argue that changes in methodology are called for, in order to clear the landscape of so many competing models.

Throughout this book we advocate for Popper's (1962) suggestion of refutation as a working methodology. This is perhaps easier said than done. For example, in conservative non-linear systems, cause and effect can become difficult to identify, and so testing a hypothesis such as "X causes Y" may not make sense if both phenomena are part of a conservative natural cycle, like Earth's water cycle. What is the origin of rain? Clouds would classify as the proximal cause, but the distal cause in a cycling system might be identified as the rain itself! This is an example of a difficulty: certain questions may not make sense. In solar physics one might ask, for example, do spicules or jets or reconnection heat the corona? One can argue that, energetically at least, on observable scales, coronal heating is not cyclic, there is a definite large-scale source and irreversible sink of energy within the plasma. But in terms of mass supply to the corona, cyclic processes appear to be dominant as spicules (interpreted as jets of plasma) provide far more mass flux than observed in the solar wind at 1 A.U. (Athay and Holzer 1982). All but the small wind mass flux is then returned to the solar surface and observed, according to some interpretations, in transition region emission lines (Mariska 1992). But along the way to energy dissipation, there await many processes able to tap into the free energy, to convert it into different forms, some of which may involve dynamical cycling. These might include various instabilities, wave motions and reconnection. We give voice to these concerns merely as a warning that the problem may have some tricks to play on unsuspecting observers.

Statistically significant correlations should of course be treated with the usual care, and analyses must generally acknowledge tricky possibilities including chaos, attractor solutions, and convergent behavior. Our purpose here is to review methods applied previously to the problem, in order to understand more deeply why results are so provisional, or perhaps "not even wrong". To try to identify profitable new lines of inquiry, we categorize examples of research studies that are closely related to coronal heating, grouping studies together which share a particular overall methodology. Then we dig deeper into representative specific mechanisms, seeking further understanding of strengths and weaknesses of approaches used. Lessons learned from this exercise will be reviewed at the chapter's end.

4.1 Is There Really a Heating Problem?

Up until now we have assumed that there is indeed a problem to be solved, on the basis that energy above that available from thermal processes is required to explain the phenomena. Yet, ignoring our definition on Page 21, we might be tempted to declare that the heating problem has been solved. Mechanisms which generate dissipative scales have been proposed, some, such as current sheets, have been observed in the Sun and in numerical experiments. *If* dissipation is inevitable and independent of details of dissipation processes, then the problem is solved. This is essentially the picture first demonstrated in the 3D "simulation" of Gudiksen and Nordlund (2005, and many subsequent articles, as of the end of 2022 it has been cited over 115 times). But we argue that the real problem here is in the conditionals-the "ifs" and other hidden contingencies, such as assumptions concerning sub-grid-scale physical processes, which we have already discussed. If we adopt our problem definition, this is not satisfactory. At best we would state that the problem's solution is physically incomplete. At worst, the solution has merely identified the source not sink of magnetic free energy, only one third of the problem has been addressed. Further, there is no equivalent solution for regions of the corona open to interplanetary space.

Is our problem definition asking too much? The answer must be an emphatic "no", from both philosophical and pragmatic positions. If we were to declare the problem solved, all future work on heating and energy dissipation through electrodynamic coupling would be a waste of effort. We would deny to ourselves the opportunity to study a large Rm plasma and its evolution, and find out why under some circumstances it might undergo explosive versus quiescent behavior, while experiencing what is measurably similar forcing. In stopping at this point, we would reduce solar physics to an exercise subservient to the mundane and narrow demands of a technologically inquisitive society. We must recall that these are issues deeply related to *all gravitationally bound objects in the universe*, including planetary magnetospheres and ionospheres, and to controlled nuclear fusion (Sect. 1.2).

Calls for science "in service to society" enter scientific discussions and especially those concerning funding for research. Scientists are asked for "deliverables" by agencies purporting to be interested in natural science. Yet even, perhaps especially in the case of coronal heating, physicists can still argue for fundamental research. The argument is simple. Just as flares and CMEs form part of the phenomena of electrodynamic coupling across the solar atmosphere, so does heating. Recall the two branches shown in Fig. 2.4. These are the two main phenomena arising from the same free energy source, and just as dynamic events say something important concerning coronal heating (Page 54), the reverse is logically also true. We already pointed to the need for a source of plasma ionization (heat is a good candidate) in order for the coupling to exist at all (Page 11). We deserve to find out if, then when and why the ARLToR conjecture applies, and if not, what is the more interesting physics underlying its failure? Is the failure related, perhaps causally, to large-scale dynamics?

4.2 Is There a *Coronal* Heating Problem?

Some have proposed a more special idea: instead of a coronal heating problem we have a *chromospheric* heating problem followed by a *loop filling process*. In other words, once some plasma in the upper chromosphere is heated, there is no need for extra heating as unheated plasma radiates, cools, and returns to chromospheric temperatures. Here we refute this picture using physical arguments based upon data. We present the essential arguments from both perspectives, so that the reader may be able to appreciate strengths and weaknesses in methods, presented from both sides.

Following earlier work Mahajan et al. (2001), Aschwanden (2001), and Aschwanden et al. (2007) drew attention to what they call the coronal heating "paradox", asking the question if coronal plasma is heated directly, or if the corona exists merely because the chromosphere (even transition region) plasma is heated, the corona forming as a result of plasma filling the overlying volume. They offer 10 observations in support: (1) the temperature evolution of coronal loops, (2) overdensity of hot coronal loops, (3) upflows in coronal loops, (4) Doppler shifts in coronal loops, (5) upward propagating waves, (6) energy balance in coronal loops, (7) magnetic complexity in the transition region, (8) the altitude of nanoflares and microflares, (9) the cross section of elementary loops, as well as (10) 3D MHD simulations of coronal heating.

A related but quite different concept is that of Scudder (1992a,b) who invoked gravitational "filtering" of chromospheric electron distribution functions, postulating that the corona is formed from the "filtration" of non-Maxwellian high energy tails under almost collisionless conditions. The gravitational force filters charged particles by letting only the fast ones rise, so that the temperature then rises outwards. In a single Maxwellian distribution no filtering occurs because all particles are at the same temperature everywhere. But if one has, say a sum of two Maxwellians, the hotter of the two will survive higher up (e.g. Meyer-Vernet 2001).

These propositions can be dismissed once we accept the fundamental problems of Chap. 3 and the inferences from Chap. 1.

4.2.1 Chromospheric Heating and Loop Filling: A Refutation

From the abstract of Aschwanden et al. (2007) we read:

> "...observations show *no evidence for local heating in the solar corona*, but rather for heating below the corona in the transition region and upper chromosphere, with subsequent chromospheric evaporation as known in flares." (Our emphasis).

The proposition of Aschwanden et al. lies within the ideas encapsulated in the elementary inferences discussed in Sect. 1.6. So, let us assume that the atmosphere shown in Fig. 1.4 is a reasonable starting point for describing typical properties of the Sun's atmosphere. Using these data, the transition region and corona contain just $\lesssim 5\ 10^{-5}$ kg per square meter of material at any given time (Page 13).

Taking the proposition at face value, the chromosphere is to supply all this coronal mass, requiring just the topmost 100 km of the chromosphere. Ignoring gravitational energy, we can compare the steady state energy in a coronal loop pressure scale height (50 Mm) with that in the topmost 100 km of the chromosphere:

$$\frac{(\frac{3}{2}pV)_{corona}}{(\frac{3}{2}pV)_{chromo}} \gtrsim 180$$

Here we have assumed the same area A of a flux bundle in the corona as upper chromosphere, so $V_{corona} = 5\ 10^7 A$ and $V_{chromo} = 10^5 A$, and allowed for the increase of p with depth in the atmosphere. The same number of particles is in each volume so this is also the ratio of the kinetic energies per particle, or temperatures.

Let us now imagine adiabatically compressing the 10^6 K corona back into the chromospheric volume. We would do work according to adiabatic gas laws $p = \rho^\gamma$ and $T = V^{1-\gamma}$, so this plasma would become heated by a logarithmic factor of

$$d \log_{10} T = -(\gamma - 1) d \log_{10} V$$

With $\gamma = 5/3$ and $d \log_{10} V = -2.9$, the temperature of such compressed coronal plasma would increase by a factor of at least 80 to $T > 80$ MK. In the reverse process, the 10^6 K corona could be built from a 100 km thick layer of the upper chromosphere, heated to 80 MK, expanding adiabatically, all energy supplied by internal energy. If we had included gravitational energy, this argument would include the same physics as in Parker's (1958) solar wind model. If we are to believe literally the quotation above (the emphasized text), this amounts to a refutation of their proposal. There are no data to suggest plasma exists at 80 MK in the chromosphere or that it flows into the corona.

But the proposal also entails additional "loop filling processes". So let us permit a mechanical source (such as Alfvén waves) to transport (lift), but not heat, the over-heated chromosphere, such that chromospheric plasma heated to, say, 10^6 K, is merely lifted adiabatically (no heating or cooling) into place in the corona to a final state in which motions are subsonic (Appendix A). The system is thermodynmically the same as before, except work has also been done against gravity. There is no way to avoid the adiabatic cooling of plasma as it expands into the corona. Thus the literal statement of heating the chromosphere followed by adiabatic filling of the loop is manifestly inconsistent with observations. The "evaporating" plasma requires additional heat within the coronal volume if it is to maintain a high temperature.

Let us nevertheless again give Aschwanden et al. (2000) the benefit of the doubt with a looser interpretation of their words. Consider that the unspecified mechanism delivers a vertical force $F > 0$ on a unit volume of plasma of pressure p, against gravity g so that the net upward acceleration of the plasma is A where $\rho A = (F - \rho g - dp/dz)$. To fill the coronal volume, A must be on average > 0. From the proposed initial state, $dp/dz \ll 0$, and $|dp/dz| \gg |\rho g|$, then $A = F + |dp/dz| - \rho g > 0$. Upflows of speed u are generated, each unit volume with a kinetic energy ρu^2 from the work done by A. Initially no matter the (positive) value of F, the pressure gradient will inevitably accelerate u to the sound speed ($c_S \sim 100\,\mathrm{km\,s^{-1}}$ at 10^6 K), with shock formation and viscous heating draining the ordered internal energy. F will only increase any residual kinetic energy of flows or sound waves, which therefore must be dissipated to reach the final subsonic state. Thus, whatever lifts the proposed super-heated chromosphere must also lead to in-situ coronal heating, contradicting the premises of heating + adiabatic coronal loop filling.

Two kinds of models were cited by Aschwanden et al. (2007) to support their proposition: those in which flares dump vast energy fluxes (significant fractions of the entire solar radiative flux, e.g. Fisher et al. 1985) into the denser chromosphere, and others including continual background heating of coronal plasma (Warren et al. 2002, 2003). Their background volumetric heating rate of 10^{-5} W m^{-2} amounts to 1000 W m^{-2} deposited along their 100 Mm half-length of coronal plasma, compared with 600 W m^{-2} needed in Table 1.2, comparable to the impulsive heating introduced by them (2000 W m^{-2}). Neither can be used to support their proposition.

Another refutation of the "paradox" is perhaps already listed in Table 1.2. The average heating rate $\varepsilon = F_M/\rho L$ of the coronal plasma is over one hundred times that of the chromosphere (Anderson and Athay 1989). The heating *per particle* determines the plasma temperature, not per unit volume. Figure 7 of Aschwanden et al. (2007) plots j^2 (the square of the electric current density $\propto$ the heating per unit volume) vs height. Naturally, this peaks in the chromosphere because there are exponentially more particles deeper in the atmosphere. Figure 4.1 shows the heating rates per particle and per unit volume required to balance the radiation losses from the reference model shown in Fig. 1.4. While not from a dynamical calculation, the figure nevertheless reflects the heating requirements in a manner equivalent

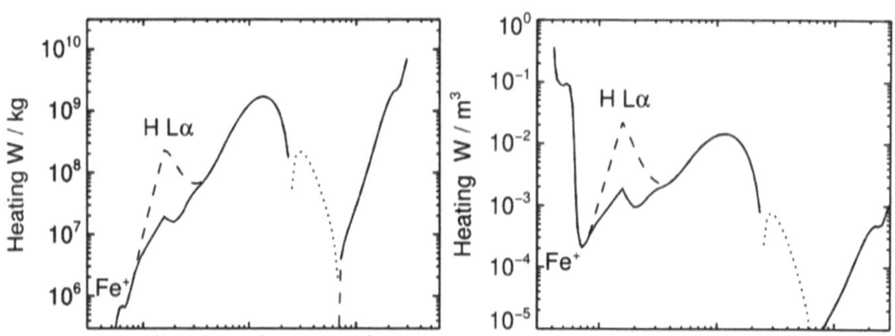

Fig. 4.1 The panels show radiation loss rates computed from the one dimensional model of Fig. 1.4, using radiative loss functions shown in Fig. 3.7, without (solid) and including hydrogen losses (dashed). On the left the heating per kg is shown (proportional to heating per particle). The rightmost panel shows heating per unit volume. The former better represents the heating requirements because temperature is proportional to heat per particle, when the plasma is fully ionized. It shows that the corona requires orders of magnitude more heating per particle than the chromosphere (Anderson and Athay 1989), refuting the idea that coronal heating is merely loop filling following chromospheric heating. Dotted lines show large negative conductive losses (i.e regions heated by conduction). The heating requirements are in broad agreement with those given in Table 1.2

to Figure 7 of Aschwanden and associates. The corona requires more heating per particle than the chromosphere.

Finally, if more evidence were needed, it is provided by young, Sun-like stars, where soft X-ray coronal emission alone can exceed chromospheric emission (Vilhu 1987; Martínez-Arnáiz et al. 2011). Such stars are compatible with a Sun which is almost 100% covered with typical active regions.

We conclude that there is indeed a coronal heating problem. Perhaps as important are the lessons learned. Many lines of circumstantial and contingent arguments including use of numerical calculations cannot survive refutation by one or two critical physical arguments and information-rich data.

4.2.2 "Filtration" of High Energy Particles into the Corona: A Refutation

The Earth possesses an exosphere that sits atop the radiatively heated thermosphere. It consists mostly of hot, upwardly moving hydrogen and helium particles escaping from the upper thermosphere. Scudder's proposal is that the corona is a solar exosphere. The base of the exosphere can be defined as the height where one mean free path equals the local pressure scale height, or where the Knudsen number is 1. The first points to notice in the proposal are

- The energy source in Earth's exosphere is (ironically) solar UV and EUV radiation from the chromosphere and corona.
- The particles in Earth's exosphere are largely neutral. Cross sections are roughly constant with energy ε. The solar particles are ionized, their collisional cross sections vary as ε^{-2}.
- The exosphere's measured composition is dependent on atomic or molecular mass. The Sun's chromosphere-corona is well mixed and almost independent of atomic mass.
- The exospheric temperature is similar to that of the underlying thermosphere. The solar corona is over 100 times hotter than the chromosphere.

Scudder (1992b) proposed that coronal high temperatures arise from the "filtration" effects of the gravitational potential on non-Maxwellian distributions of charged particles generated in the lower atmosphere. The essential difference with Earth's exosphere is the inverse-squared energy dependence of collision cross sections $\sigma \propto \varepsilon^{-2}$ for charged particles, compared with constant cross sections for neutrals in the exosphere. The mean free path of charged particles moving with energies of say 100 eV inside the chromospheric plasma with mean thermal energies of 0.7 eV is thus 2 10^4 larger than the mean. Even though typical chromospheric particles are collisional, relaxing rapidly to Maxwellian distributions, once a faster particle has been created, it can readily escape outwards to large distances (Scudder and Olbert 1979). In the presence of a deep gravitational potential $W(r)$ (r=distance from Sun center), an initially Maxwellian distribution function varies with r as $\exp(-W(r)/k_BT)$: the temperature remains the same but the density decreases. But for the non-equilibrium kappa-distributions, the populations of high energy particles dominates at larger r, and the average speeds increase with r (Scudder 1992a, see the reviews by Meyer-Vernet 2001, 2007).

This picture of heating moves the problem from in-situ heating of coronal plasma to the chromosphere where a mechanism is needed to generate the high energy tails in the distributions of charged particles. The gravitational filtering leads to the average increasing temperatures, defined through the widths of the collisionless particle distribution functions. What are the essential predictions of this picture and can they be refuted by any observations?

Collisionless models of the transport compute the changes in particle distribution functions as the particles move outwards and climb the gravitational potential well. Charge neutrality sets up a gravity-driven electric field, in the solar proton-dominated plasma this serves effectively to halve the gravitational potential for the protons. In heavier trace ion species, such as iron, the net potential modifying the distribution function with height depends on the ion's charge Z_i and mass m_i, according to (equation 17 of Meyer-Vernet 2001)

$$W_i = \langle W \rangle \left[\frac{m_i}{\langle m \rangle} - \frac{Z_i}{1 + \langle Z \rangle} \right], \tag{4.1}$$

where $\langle\rangle$ represents the average over all particles in the plasma ($\langle m \rangle \approx 1.36$, $\langle Z \rangle \approx 1.26$, Table 1.1). For Fe^{8+} $W_i = 37\langle W \rangle$, so these ions are far more tightly bound than protons ($W_p \approx 0.3\langle W \rangle$ in an ionized plasma of solar composition). At heights $z \ll R_\odot$, $W_i(z) \approx m_i g z$. The scale height for an ion at temperature T_i (defined kinetically from the kappa distribution) is $H_i = k_B T_i / m_i g$ for an ion of mass m_i. Equations 17–20 of Meyer-Vernet (2001) yield expressions for temperature and density as a function of height z through the potential $W(z)$:

$$t_i = 1 + \frac{1}{\kappa - 3/2} \frac{z}{H_i} \equiv 1 + x_i \tag{4.2}$$

$$n_i \propto t_i^{1/2} t_i^{-\kappa} \tag{4.3}$$

where $t_i \propto T_i$, the ion kinetic temperature, and n_i is the ion density. These equations allow us to compute the temperatures and densities of any ion i at height z. Equation (4.2) for a given ion has two regimes depending on the magnitude of x_i relative to one.

- When $x_i \ll 1$, then the temperature increases as $1 + x_i$. The density n_i expanded as a Taylor series is

$$n_i \propto 1 - \frac{2\kappa - 1}{2} x_i + \frac{\kappa^2 + 1/4}{2} x_i^2 + O(x^3), \tag{4.4}$$

so that the densities decrease faster than the temperatures rise when, as is almost universally assumed, $\kappa > 3/2$.

- When $x_i \gg 1$, the potential term dominates the temperature, and $T_i \propto m_i z$, and the density varies as $n_i \propto m_i^{1/2-\kappa}$.

These results have two observable consequences for T_i and n_i in this kind of corona. The first is well known, $T_i \propto m_i$ above the first few scale heights $H_i = H/m_i$. More detailed models applied beyond $2R_\odot$ in the solar wind yield ion temperatures T_i varying faster than proportional to m_i (Pierrard et al. 2004). The second is that, depending on the departure of κ from 3/2, the density n_i drops more quickly than temperature below the first few scale heights. Thus the densities of multiply-charged ions should be significantly smaller than, perhaps much smaller than those of protons in the corona, compared with the abundances below.

Points **IX** and **X** of Sect. 1.5, as well as the observations discussed in Appendix A refute these predictions. Neither broad lines nor anomalously low heavy element abundances are measured in the low corona, below the first 100 Mm (two hydrodynamic pressure scale heights) above the photosphere from which the bulk of the UV and X-ray coronal radiation emerges. It is important to recognize that the bulk of the research on exospheric coronae has been applied to plasmas far more tenuous, higher up, and applied mostly to measurements made near 1 AU (e.g. Zurbuchen and Richardson 2006).

Further doubts on the viability of filtration as a cause of coronal heating arise. Values of $\kappa = 2.2$ used by Scudder (1992b) are considered very small in the space science community. What kind of physical processes in the chromosphere might be capable of generating such large wings of distribution functions (Fig. 2.1)? Also Scudder adopted a Knudsen number at the base (where $T = T_0$) of $6\ 10^{-3}$ from Figure 7 of Fontenla et al. (1990), which appears to be too large by a factor 60 (Braginskii 1965). Thus Scudder's estimated energy flux of 3×10^4 W m^{-2} towards the corona must be revised downwards to 500 W m^{-2} owing to the higher collisionality of the plasma. The quiet corona requires 600 W m^{-2}.

The observations simply show that whatever is filling the coronal with plasma, it must do so with little bias towards ion charge or mass. The only statistically significant variation of ion number densities n_i found in coronal plasma is with the first ionization potential of the element (the "FIP" bias, see e.g. Laming 2015). The FIP effect is a relative increase in abundances of easily ionized atoms with low-FIP (e.g., Na, Mg, Si, ...) versus high-FIP atoms (e.g., O, Ne, Ar) in the corona. The effect occurs again because abundant ionized low-FIP chromospheric ions are more strongly coupled by large Coulomb cross sections compared with most gas-kinetic cross sections of neutral particles, and ions experience electrodynamic forces. Being primarily dependent on the difference between ionized and neutral species, this has nothing to do with the physics of the velocity filtration picture *per se*.

Finally, observations with the Parker Solar Probe have revealed an unanticipated level of magnetic complexity in coronal plasma between 13 and $50R_\odot$ (Bale et al. 2019). The magnetic field there exhibits "switchbacks" which contain considerable free energy. While not falsifying the exospheric picture, nevertheless these require an explanation outside of the exospheric model. It may be that other heating mechanisms more naturally and economically account for the measured magnetic complexity or that switchbacks arise from entirely different causes between say 2 and 13 $R_\odot$.

4.3 Methods Addressing the Coronal Heating Problem

Approaches based upon electrodynamic coupling can loosely be categorized. The corona is assumed to form as a result of the generation of magnetic free energy in high-β plasma, which propagates through the photosphere and chromosphere ultimately to be dissipated as heat in the low-β coronal plasma (Fig. 2.4).

- **Blue sky research** in which observations and occasional theoretical ideas are explored without direct reference to existing beliefs. As an example, the ingenious use of novel technology enabled a jump in sensitivity of spectral polarization measurements by two orders of magnitude (Povel 1995). A "second solar spectrum" was detected at the solar limb, opening a new window into solar magnetism based on scattering polarization (Stenflo and Keller 1997; Gandorfer et al. 2004; Kleint and Gandorfer 2017). In another example, multi-layer optical coatings (e.g. Lindblom et al. 1986) enabled the efficient use of normal-incidence optics at EUV wavelengths, which form the basis for modern, high resolution imaging instruments such as the AIA on SDO (Lemen et al. 2012).

  A novel theoretical method for measuring coronal magnetic fields has arisen from a quirk in atomic structure (Beiersdorfer et al. 2003; Grumer et al. 2014; Li et al. 2015). The Fe^{9+} ion, to which the coronal "red line" belongs (Edlén 1943), has two excited levels, $3p^4 3d\,^4D_{5/2}$ and $3p^4 3d\,^4D_{7/2}$, which happen to be separated by < 0.004 eV. Only the $3p^4 3d\,^4D_{5/2}$ level can decay via a dipole transition to the ground $3p^5\,^2P^o_{3/2}$ level, emitting a line at 25.7262 nm in the absence of external influences. However, the Zeeman effect mixes the wavefunctions of the two levels, the strength of the mixing being proportional to $|\mathbf{B}^2|$ divided by the levels' energy difference. This produces a "magnetically induced transition" (MIT) from $3p^4 3d\,^4D_{7/2}$ to $3p^5\,^2P^o_{3/2}$ for modest magnetic field strengths characteristic of the active solar corona, in addition to a magnetic quadrupole transition of comparable strength. This idea has already been applied to measure coronal magnetic fields using several EUV lines of Fe^{9+} observed in active regions on the solar disk (Si et al. 2020; Landi et al. 2020; Brooks et al. 2021). The method is subject to tricky observational issues such as line blends and the effects of absorption in Lyman continuum and helium continuum transitions at EUV wavelengths (Martínez-Sykora et al. 2022). But it deserves further attention.

  A numerical method for studying general force-free states in hydromagnetic structures has been inspired from the notion that magnetic lines of force are vector fields which can be represented by piecewise linear lines in 3D space

(Kankelborg et al. 2001; DeForest et al. 2001; Deforest and Kankelborg 2007). The unique nature of such calculations is that they are free from diffusion, a hybrid Lagrangian/Eulerian scheme solves special equations of motion developed to drive a given force-free system to an equilibrium state. This work has not received wide attention but it extends the "1D" notion further than RTV did to attempt to solve for force-free magnetic configurations in the absence of dissipation. The method has parallels in quantized magnetic systems (Deforest and Kankelborg 2007). The method has been successful in demonstrating the role of ideal instabilities, with no reconnection, to lead to MHD eruptions (Rachmeler 2010), contrary to some commonly-held opinions. Bhattacharyya et al. (2010) developed a minimally diffusive scheme solving not for field lines but for flux surfaces. Magnetic flux surfaces, everywhere tangential to field lines, are fluid surfaces with zero flux across them. In doing so, the evolution is described entirely in terms of the advection of its flux surfaces, integrated using established codes from hydrodynamics (Prusa et al. 2008). Such calculations have strengthened confidence in Parker's fundamental theorem of magnetostatics (Sect. 4.7).

Variations on traditional modes of observation sometimes arise out of the blue. Novel spectra were obtained during the total eclipse of 2017 August 21, of spectra between 310 and 2500 nm with an unprecedented cadence of 50 Hz (Judge et al. 2019). The moon's moving disk covered and uncovered 5.6 km of the chromosphere and corona between each exposure. Images were acquired in the pupil not focal plane to use the detectors as photometers. With limited resources (5 cm telescopes and modest resolution spectrometers) the data mostly demonstrated new technology and novel optical setups, useful perhaps for future experiments.

- **Development and assessment of new hypotheses** prompted directly by observational and theoretical discoveries. Examples include the proposition that unresolved thermal structure can explain departures from observations from otherwise uniform structures (Aschwanden et al. 2000); that spectral line broadening is a signature of turbulence (Billings 1965; Oran et al. 2017), is not (Delone and Makarova 1969); that coronal heating is directly a consequence of magnetic reconnection on observable scales (Longcope and Tarr 2015; Takasao et al. 2015); that there is no coronal heating problem but only a chromospheric one (Aschwanden et al. 2007, refuted in Sect. 4.2); that kinetic processes in the chromosphere naturally lead to a hot corona (Scudder 1992b); that spicules seen at the limb and explosive events result from the same physical process (Wilhelm 2000; Heggland et al. 2009), that sources of kHz frequency MHD waves must exist within the chromosphere, to account for anisotropic distribution functions in the solar wind (Axford and McKenzie 1992; Kohl et al. 1998); that current sheets embedded in a 3D magneto-fluid greatly enhance transport fluxes across them (Judge et al. 2012). The literature is so vast that we are sure to have missed other intriguing ideas.
- **Heuristic observational testing of analytical work** on general aspects of the heating problems. Heuristic strategies seek tractable, approximate solutions to

problems which otherwise would require huge investments of time and resources. An early example is Levine's (1974) proposal, which concerns the dynamic collapse of tangled electric current sheets leading to both particle acceleration and plasma heating. Another is the derivation of thermal scaling laws between observables for uniformly heated coronal plasmas along loops (McWhirter et al. 1975; Rosner et al. 1978); the study of measured deviations from such scalings (Priest et al. 2000); the spontaneous generation of current sheets as magnetic fields relax while maintaining their topology (Parker 1988, 1994); chromospheric magnetic reconnection as a source of wave heating in the corona (Sturrock et al. 1999); in the forced generation of photospheric current sheets extending into the corona (Priest et al. 2002); and phenomenological descriptions of MHD turbulence (Schekochihin 2022 and Appendix C). To some degree, all of these studies have testable predictions, at least in principle.

- **Correlative studies of measurable photospheric magnetic and coronal data.** During the 1980s polarimeters were developed which could measure, modulo a 180° ambiguity, *vector* magnetic fields in the plane-of-the-sky, building upon decades of work measuring just the line-of-sight (longitudinal) magnetic fields (Lites 2000). Thus the transverse components were measured as well as longitudinal components. While the latter produces polarization that is, for small magnetic field strengths, first order in the small parameter of Zeeman splitting divided by Doppler line width, polarized light from the transverse effect are second order (e.g. Jeffries et al. 1989; Landi degl'Innocenti and Landolfi 2004). With such measurements it became possible to estimate magnetic free energies associated with electric currents running within the observed plasmas. Thus, in the 1990s such data were combined with X-ray and EUV data from spacecraft to find empirical relationships between the corona and underlying magnetic fields. First examples include Metcalf et al. (1994) (no correlation with surface electric currents), Moore et al. (1996) and Falconer et al. (1997) (correlations with contours marking changes of sign of vertical magnetic field), Fisher et al. (1998) derived statistical results from over 300 active regions:

> "The X-ray luminosity is highly correlated with all of the global magnetic variables [...and is] best correlated with the total unsigned magnetic flux".

These and many later studies have sought to examine various hypotheses (Wolfson et al. 2000; Mandrini et al. 2000; Yashiro and Shibata 2001; Pevtsov and Acton 2001; Falconer et al. 2003; Schrijver et al. 2005; Tan et al. 2007; Lee et al. 2010; Verbeeck et al. 2013; Hazra et al. 2015; Ko et al. 2016; Kirichenko and Bogachev 2017; Ugarte-Urra et al. 2019). However all are burdened with *ad-hoc* parameters and/or multiple assumptions.

A significant body of work involves the extrapolation of magnetic fields measured in the photosphere into the corona, driven by diverse goals including coronal heating (see the review by Wiegelmann and Sakurai 2021). Owing to particular difficulties presented by the intervening chromosphere, we postpone discussions of this work to Sect. 6.2.

- **Quantitative analytical work on particular MHD processes**, e.g. on the development of waves dissipating in resonant surfaces (Ionson 1978; Hollweg 1984; Davila 1987); phase mixed dissipation of waves in 2D (Heyvaerts and Priest 1983) and 3D (Parker 1991); turbulent MHD cascades (e.g. Einaudi et al. 2021); the role of tearing mode ("plasmoid") instabilities in the dynamics of reconnection and dissipation (Loureiro et al. 2007), even in some collisionless conditions (Huang and Bhattacharjee 2013).
- **Dynamical numerical experiments in 3D non-ideal MHD**. Examples include radiation MHD experiments encompassing the hydromagnetic structure of the core of an active region (Gudiksen and Nordlund 2005; Rempel 2014; Carlsson et al. 2016; Rempel 2017; Przybylski et al. 2022). The algorithms adopt various approximations to enable solutions to be acquired in acceptable computational times. For example, in Rempel (2017) we read

  > "we limit the coronal Alfvén and heat conduction speeds to computationally manageable values using an approximation to semi-relativistic MHD with an artificially reduced speed of light (Boris correction)."

  Less complete physical ingredients are frequently seen in numerical experiments in MHD turbulence (Mallet et al. 2015; Einaudi et al. 2021); propagation of waves in structured media (Similon and Sudan 1989; Goossens et al. 2019; Howson et al. 2020; Pagano et al. 2020); and numerical studies of instabilities, such as Bhattacharjee et al. (2009); Huang and Bhattacharjee (2013, 2016) and reconnection (Daughton and Roytershteyn 2012; Huang, Comisso and Bhattacharjee 2017; Ji et al. 2022). Applied to specific aspects of the heating problem, calculations generally are made using prescriptions for handling scales below the grids (typically tens of km), such as hyper-diffusion (e.g. Rappazzo et al. 2008). The physics below grid scales is missing, but the algorithms are designed to stabilize otherwise unstable integration schemes, while mimicking large Reynolds number conditions. However, they enforce a magnetic Prandtl number of order one, where in reality it is far higher (at least for a collision-dominated plasma, Braginskii 1965). If the plasma develops compressible dynamical modes, ion viscosity can dominate dissipation, and the small scales can evolve dynamically quite differently from the single fluid approaches used in such calculations (Sect. 5.2). These prescriptions are better suited to understanding nonlinear and non-local atmospheric dynamics on larger, resolved scales, not on irreversible heating processes.
- **"Realistic simulations"** (as opposed to numerical experiments) in which all essential physical ingredients are included in 3D time-dependent MHD numerical models, intended to be directly compared with observations, or used to inform researchers how to understand better observations through their synthesis. A remarkable number of studies based upon synthetic data from these calculations have been published. A search of two of the most prominent "radiation-MHD" codes MuRAM (Vögler et al. 2005) and BiFrost (Gudiksen et al. 2011) in the NASA ADS abstract service, both based on finite-difference algorithms, gives 154 and 136 respectively, as of December 2023. Unfortunately almost no obser-

vations of magnetic fields threading the atmosphere have been available, even at the coronal base where residual upward Poynting fluxes might be measured, thereby limiting the ability to refute ideas and learn from comparisons with such models. The researcher is limited to comparing large-scale morphological properties of frozen-in magnetic fields, along with thermal properties of plasmas (see Chap. 3). Examples of this work are reviewed by de Pontieu et al. (2021), but few definitive results related to heating mechanisms can be drawn. Again, the information-poor data available limit the scope of such studies to comparisons between crude observations with synthetic data, both of which are far removed from the scales of heating processes. Such comparisons can bear witness only to the effects, not causes, of coronal heating.

- **Studies of collisionless plasma dynamics** of relevance to the heating problem. Essentially two kinds of approaches have been taken. In the first, it is recognized from the outset that plasma heating deep within a gravitational well can lead to "velocity filtration" in which charged particles in high energy tails of non-Maxwellian distribution functions emerge preferentially from the chromosphere, owing to their long, energy-dependent mean free paths (Scudder 1992b).

  The second kind of study broadly attempts to understand collisionless dynamics on scales of the small diffusion regions (100 m or less, Shay et al. 1998; Huba 2003; Daughton et al. 2006; Huang et al. 2011; Daughton and Roytershteyn 2012; Liu et al. 2017; Matthaeus et al. 2020; Liu et al. 2022). Recent focus has been on their effects on aligned MHD turbulent structures such as sheets and/or plasmoids, enhancing reconnection rates with consequent effects on large MHD-scale dynamics (see Appendix C). Major questions of current interest include whether a theory might be developed to bridge the MHD and kinetic scales, and how particle acceleration occurs during or as a consequence of magnetic reconnection during flares (chapter 9 of Pontin and Priest 2022). Analyses rely heavily on numerical experiments to explore the separate dynamics of protons and electrons. Theory and in-situ observations within the solar wind and Earth's magnetosphere currently drive active research in this area. Even though both concern plasmas of higher β and lower collisionality, at kinetic scales the results should nevertheless be extendible to the low corona. But compared with in-situ particle and field measurements, refutations using remotely sensed data remain challenging. Avenues for advancement are offered later.

  Some (almost) direct evidence of heating at the kinetic level through remote sensing does exist (mentioned already on Page 77). Off-limb observations of line widths in coronal holes, interpreted as signatures of Alfvén waves in the solar wind, have been found to be consistent with free propagation, damping and enhancements of ion temperatures of kinetic origin (e.g. Hahn and Savin 2013). The unique methods reported by Kohl et al. (1998) and many subsequent papers focus on evidence for anisotropy of ion motions at the distribution function level, based upon data from the UVCS instrument on the SoHO spacecraft. Although these measurements concern more tenuous plasma above where most coronal heating occurs, nevertheless they are currently the only solar observational tool enabling studies of the conversion of MHD scale energy into heat. The interpretation is not unique (Raouafi and Solanki 2004), but the techniques exemplify a serious attempt to derive microscopic physics from macroscopic observations. The prize for such work can involve refutation of entire classes of models (for example, fluid dissipation mechanisms), and so we elaborate further on this work.

Specific emission lines measured high in the corona ($0.4R_{\odot}$ above the visible limb) produce macroscopic signals directly reflecting anisotropies in ion distribution functions. Of various kinetic processes, high frequency (> 10 Hz) Alfvén waves are believed to speed-up the ion motions perpendicular to a dominant magnetic field via the cyclotron effect, increasing ion magnetic moments. Reversible work is done on particles by the electric fields of Alfvén waves, creating not heat but anisotropic particle distribution functions. As such it is energy transfer to ion orbits not dissipation *per se*. Instabilities and collisions would be required to make isotropic, i.e. thermalize, the distribution functions. Collisions are rare in the solar wind, so moderately anisotropic distribution functions can remain unchanged in the flow outwards from the resonant regions, to be measured in-situ (Kasper et al. 2013). Kohl et al. (2006) point out that

> "Although the UVCS ion measurements represent strong constraints on the heating and acceleration of the fast solar wind, we do not yet have closure on the exact kinetic microphysics of the heavy ions"

which is a statement prompted perhaps in part by the study of the same spectra by Raouafi and Solanki (2004). These authors showed that the UVCS data were also consistent with *isotropic* ion distribution functions, once stratification and line-of-sight effects were treated. Further doubt on the original (Kohl) interpretation concerns the origin of high frequency waves capable of perpendicular ion heating. Such high frequencies can most readily arise from regions with relatively high Alfvén speeds and small scales (frequencies $\nu \sim c_A/\ell$ exceeding 1 kHz require lengths $\ell < 1$ km for $c_A \sim 1000\,\text{km s}^{-1}$). Suitable reservoirs of energy accordingly include free energy within the uppermost chromospheric regions, and the kinetic energy of the solar wind. Sources of high frequency waves rely on non-linear interactions including turbulence, shocks, and rapid reconnection, generated near the base of the corona or within the wind plasma itself (e.g. Cranmer et al. 1999; Kasper et al. 2013).

- **Empirical studies of observable phenomena** constitute a significant fraction of research on the corona. Examples include polar "plumes" in coronal holes, "X-ray and EUV bright points", "blinkers", "tornadoes", "jets", "campfires", and a host of other phenomena observed since the days of SKYLAB. Well before SKYLAB, observations of the Hα line were used as prime indicators of magnetic activity (e.g. Hale 1908; Kiepenheuer 1953). Researchers have not been shy in assigning entertaining names to observational phenomena, such as "faculae", "flocculi", "filigree", "burning prairie", "red sierra", "moustaches", "whiskers", "bombs", "whirls", "surges", "spicules", "fibrils", "plages", "bushes", "rosettes", "bald patches", "polarity inversion lines" and others. Speculations abound on the possibility of a particular phenomenon of interest causing coronal and or chromospheric heating, for example that flares (an observed phenomenon) are an important component of processes leading to coronal heating. Interpretations of all such phenomena are, however, subject to those fundamental challenges of Chap. 3. These phenomena should probably be seen mostly as manifestations of an electrodynamically coupled atmosphere on a variety of observable scales. Generally speaking, detailed studies of such phenomena have offered few insights into the heating problem as defined (Page 21).

Perhaps this diversity of methods arose because most of the phenomena should ultimately be examined using information-rich observations of both plasmas and magnetic fields. Previous work with photospheric fields and coronal plasma have been productive. But the lack of magnetic measurements within chromospheric and coronal plasmas has produced the embarrassment of riches represented by a large number of unrefuted ideas (Sect. 1.1). Direct approaches which hold promise include measurements of the changing magnetic field across the chromosphere, to set components of the Poynting flux of energy at the coronal base and obtain lower boundary conditions in plasma close to a force-free state. Indirect measurements are of course needed at the smallest scales. Building on decades of research on magnetic reconnection (see, e.g. Sonnerup 1988; Biskamp 2005; Kulsrud 2005; Priest and Forbes 2007; Kulsrud 2011; Dong et al. 2018) Pontin and Priest (2022) noted that,

> "...the dynamics in the vicinity of the current layer itself will not be spatially resolved for the foreseeable future, so of particular interest are secondary effects that can be observed at larger scales."

These words reflect the simple fact that currently available observational tools are *ill suited* to examine the *nature* of the dynamics and energy transport across the decades of scale sizes. The work on UVCS data cited above is an example *par excellence* of the "secondary effects" (Pontin and Priest 2022). It would be difficult to argue for example, that the results of Kasper et al. (2013), combining in-situ measurements with remote sensing, is of no consequence for the heating problem. New approaches might be to identify more of those secondary effects by making predictions of specific observables or designing new experiments to test physical processes which survive spatial and temporal averaging. Such robust, information-rich observations will involve novel instrument development.

We turn next to more specific examples in the hope of revealing more methodological strengths and weaknesses.

4.4 Refutation of Non-magnetic Processes

One important class of mechanism did not clear the hurdle of supplying energy sufficient to power the chromosphere or corona. The 1973 SKYLAB and 1975 OSO-8 missions marked an era of paradigm change. Non-magnetic processes (Schatzman 1949; Osterbrock 1961; Lighthill 1967; Kuperus 1969; Stein 1971), failed to account for the essential observation that the structure, dynamics and heating are so clearly related to emerging magnetism (Rosner et al. 1978; Eddy 1979). It became increasingly difficult to reconcile acoustic heating with the large variation in energy fluxes revealed by SKYLAB observations between quiet and active regions.

The inability of acoustic wave heating was settled by Athay and White (1978), who analyzed observations from the high resolution (typically $\mathcal{R} \approx 50{,}000$) OSO-8 UV spectrometer. They showed that signatures of non-magnetic wave motions, observed in emission lines of Si^+ formed near the top of the chromosphere, fall short of supplying the energy fluxes needed. They estimated the energy flux density associated with upwardly-propagating sound waves:

$$F \approx \rho\, c_S\, \xi_0^2 \lesssim 70 \ \text{W m}^{-2}. \tag{4.5}$$

The quiet Sun radiative losses from all lines formed above the Si^+ lines require 1 kW m^{-2} (Tables 1.3 and B.1). Here, ρ is the density of plasma near where the Si^+ lines are formed, ξ_0 is the speed of oscillations derived from observed line widths, and c_S the sound speed. Both ρ and c_S were taken from a hydrostatic model chromosphere (Vernazza et al. 1973). The factor of 15 shortfall lies far below probable uncertainties. Recently, Molnar et al. (2021) have re-enforced this result using observations and dynamical models, arguing that densities ρ from static models (Vernazza et al. 1973) are likely an overestimate (see Fig. 1.6). But given that spectral lines like those of Si^+ form at a certain mass column density $m = \int \rho ds \sim \rho H$, with H the pressure scale height, whether static or dynamic, Eq. (4.5) shows that ρ will be as accurate as the accuracy for which we know H. Figure 1.6 suggests Athay & White's estimate of ρ is accurate to better than a factor of 2.

4.5 Accelerated Particles Do Not Heat the Quiet Corona

Many have noted the possibility that processes determining coronal heating may involve energetic particles, even at the end of an MHD cascade (e.g. Rappazzo et al. 2008). An interesting contradiction in the literature has arisen recently, which also clearly illustrates differences in methodologies.

Paterson et al. (2023) used EUV, soft X-ray and novel hard X-ray observations from the astronomical NuSTAR mission (Harrison et al. 2013) to set upper limits on the distribution functions of high energy particles in the corona. They note that

> "NuSTAR's use of focusing optics means that it can directly image very faint HXR sources on the quiet Sun, and perform spectroscopy on regions of interest."

Paterson et al. tested the hypothesis that populations of energetic particles, observed during flares, may be a consequence of flaring of all scales, including nano-flares. The origins of such accelerated particles remain debated (reviewed, e.g., by Zharkova et al. 2011; Vlahos et al. 2019; Lysenko et al. 2020, among many others).

In one example of interest to coronal heating (Drake et al. 2006), electrons extract energy from multiple contracting plasmoids forming during magnetic reconnection. (Sect. 4.7). The effective area of NuSTAR $\sim 0.08\,\text{m}^2$ above 8 keV is larger than any previous X-ray astronomical mission, and exceeds that of the solar RHESSI instrument by factors of between 1 at 70 keV to ten at 8 keV (Lin et al. 2002). The NuSTAR measurements were shown to limit the total energy in particles above a few keV to well beneath those needed to sustain the quiet corona, using firm upper limits on hard X-ray intensities. Such limits were made possible owing to the ability of NuSTAR to resolve sources as close as $18''$, as well as the large effective area, a unique combination among many other high energy experiments. For example, the FERMI gamma-ray observatory, with an effective area approaching $1\,\text{m}^2$ between energies 100 MeV–100 GeV, had an angular resolution of 9 arc minutes essentially nullifying the ability to study the quiet Sun (Atwood et al. 2009). At the highest energies of FERMI the Sun emits gamma rays at rates independent of the state of the heliosphere and of the solar magnetic activity cycle. This indicates an origin independent of solar flares, which in hard X-rays follow the Sun's magnetic activity cycle, as measured by the Konus-Wind experiment (0.02–15 MeV) (Lysenko et al. 2020). Closer to 100 MeV energies, flares detected by FERMI are related to fast coronal mass ejections (Ackermann et al. 2014).

The NuSTAR results are therefore unique, they could not have been anticipated from first principles, given that the physics of energy dissipation is probably controlled, at least in part, by collisionless physics (Chap. 2, Sect. 4.2.2). Nevertheless we can conclude that the flow of energy towards dissipation scales does not involve acceleration of significant populations of particles well above $10^6\,\text{K}$ by electric fields, caused by any process. This negative result potentially rules out important scenarios, suggesting that physics of MHD alone might suffice to answer the heating of the quiet corona, as well as significant numbers of active region loops (Duncan et al. 2021, see Sect. 5.2).

The NuSTAR results call into question those based upon far less direct and, consequently, lengthy arguments (de Pontieu et al. 2021, section 4.6) involving "advanced models". While these authors discuss active regions, the nanoflare picture adopted by them (see Sect. 4.7) applies also to the heating of quiet regions (e.g. Chitta et al. 2021). de Pontieu et al. (2021) claim that non-thermal electrons "appear to be present in a significant fraction ($\approx$ half) of the studied small heating events, including the smallest". This claim must be seen with skepticism, even before the hard X-ray NuSTAR results arrived. After all, the authors claim compatibility of a non-thermal model with tools which are insensitive to non-thermal electrons. In related work but applied to a small flare, Judge et al. (2017) demonstrated the existence of a flare ribbon isolated from the overlying coronal plasma. Nevertheless, this same ribbon was satisfactorily modeled using non-thermal electrons by de Pontieu et al. (2021), leading them again to the conclusions that non-thermal electrons from the corona are present and frequently important in coronal heating (Testa et al. 2014).

The simple refutation offered by NuSTAR data offers up two lessons. First, information-rich data, even the absence of a signal, can offer more opportunity for advancement than a multitude of data with tenuous connections to phenomena of interest. Second, simplistic approaches can offer advantages over sophistication (related to Occam's Razor, his principle of parsimony). Neither sophistication nor agreement with observation implies scientific advancement. Nevertheless it seems comparisons of predictions from numerical models encompass a huge volume of the literature today. For example, the review by de Pontieu et al. 2021, "A New View of the Solar Interface Region from the Interface Region Imaging Spectrograph", cites a staggering 574 references 8 years after the instrumental commissioning. All this devoted largely to transition region plasma which contains so little mass, driven from above and below, whose structure is far from being understood (Pietarila and Judge 2004; Judge 2021).

In another interesting development, Duncan et al. (2021) analyzed several active region mini/micro-flares (energies about 10^{21} J, five orders of magnitude larger than nanoflares). Of eleven observed, their NuSTAR observations showed just one with evidence of particle acceleration. Perhaps then only large flares enhance acceleration processes? Or perhaps high energy particles are intermittent and the statistical sample needs increasing? These questions seem worth resolving, given the importance of understanding the origin of accelerated particles during flares (Drake et al. 2006).

4.6 Resonant Absorption: An Ideal Linear Model

One special yet important model for coronal heating exhibits clearly and without ambiguity the phenomenon of spontaneous imperfection and its role in heating active regions of the Sun. Resonant absorption of Alfvén waves is a phenomenon in a natural plasma as apparently "untidy" (inhomogeneous) as the solar corona, resulting from the presence of singularities in the wave dispersion relation. Singularities come and go whenever and wherever the wave speeds happen to equal the dynamically evolving local Alfvén speed. Prompted by active region data reported by Foukal (1976, 1977), Ionson (1978) developed a *linear* theory within ideal MHD, tracing free energy within coronal loop systems, attempting to explain the presence of co-spatial cool (10^5 K) and hot ($\geq$ 2 10^6 K) plasma loops in active region data obtained by the SKYLAB astronauts. Ionson's approach was first to construct the simplest model of (ideal) MHD oscillations which nevertheless captured the inherent inhomogeneities within coronal loops. By introducing a linear radial dependence of density across a circular tube of magnetic flux, he was able to study the effects of inhomogeneity on wave motions to the lowest order. Using linearized, ideal MHD equations subject to analytical boundary conditions, Ionson demonstrated that *ideal MHD leads inevitably to its own demise*. Around the MHD singularities, enormous gradients in wave motions are generated which decay exponentially with distance from them. An MHD surface wave, existing on a resonant surface within the inhomogeneous plasma, generates "kinetic Alfvén waves" propagating across the magnetic field which owe their existence to forces involving finite electron mass and ion gyro-radii. The rate of dissipation balances the rate at which the surface wave pumps energy into the kinetic Alfvén wave, without reference to the ultimate dissipation mechanism. Ionson noted that the ideal magnetic fluctuations may be subject to tearing mode instabilities, a non-ideal effect depending on resistivity, which could modify dissipation (see Sect. 4.9.2).

Evidence for resonant absorption has been reported by Okamoto et al. (2015) in images and slit spectra of prominence plasma. With the complete theoretical picture from Ionson (1978) and many diverse later articles (e.g. Davila 1987), together these studies offer a further opportunity to understand how observation and theory interact, and why refutation can be so difficult. First we note that spatial and spectral excursions over (at most) just one wave cycle were observed by Okamoto et al. (2015), limiting the credibility of their studies based upon wave mechanics. Nevertheless, phase relationships between plane-of-sky motions and line-of-sight motions suggested that a combination of resonant absorption and Kelvin-Helmholtz instabilities might be active. We again encounter weak compatibility statements as conclusions (Okamoto et al. 2015):

> "Our detailed modeling shows that the observations are fully compatible with a state-of-the-art numerical model of transverse MHD oscillations and thus our interpretation of resonant absorption."

The authors valiantly attempt to refute other wave motions. But again the detection limits of observations also limit the information content of the data, denying access to potentially critical signatures of physical processes that might be important, but which are currently unobservable (see Nasreddin's problem described on Page 208). Small amounts of plasma, such as that within individual surface resonance layers intrinsic to the resonance absorption mechanism, would probably be below noise levels. In this sense the data used are not information rich. Again we would need to seek signatures surviving the averages in space and time.

This assessment is not intended to level criticisms at a particular model or article. Instead, we search for reasons why the results are inconclusive, in order to find pointers to new, more penetrating analyses of the right kind of data. It illustrates just why results are almost always dependent upon assumptions, signal-to-noise ratios, and contingencies, and why cause-and-effect continue to evade us.

4.7 Parker's Fundamental Theorem of Magnetostatics

Spontaneous imperfection is epitomized by the remarkable work of Parker (celebrated, elucidated and expanded by Low 2023). Single-fluid magnetostatics involves the balance of the forces of pressure gradients and the Lorentz force, coupled with any external body forces (gravity). In the case of fully ionized plasmas, the dynamics by which static equilibria are approached consists of two mathematically different constraints. First Maxwell's equations describe the fields in terms of sources of electric charge and electric currents. Second, Michael Faraday conceived of the idea of magnetic lines of force, whose local density penetrating a unit area is a measure of field strength $|\mathbf{B}|$, and Alfvén demonstrated that in ideal MHD the plasma and magnetic field enclosed by a given flux surface are two ways of describing the same topological structure. Pragmatically, even though there are technical difficulties with Faraday's picture (Slepian 1951; McDonald 1954), Maxwell's equations give solutions for a local $\mathbf{B}_0(\mathbf{r})$, in terms of charge and current distributions in space. They "know nothing" about field line topologies. The fixed topology associated with the plasma knows nothing locally, but imposes global conditions on the fields. It is the conflict between the local and global constraints which Parker's theorem addresses.

The phenomenon of spontaneous imperfection is perhaps most elegantly made evident in Parker's monograph "Spontaneous Current Sheets in Magnetic Fields with Application to Stellar X-Rays" (Parker 1994). A gentler, more heuristic argument is found in the "conversations" of Parker (2007). The theorem states that

> A magnetic field embedded in a perfectly conducting fluid and rigidly anchored at its boundary has a fixed topology. Subject to that topology, the force-free state of such a field generally requires the presence of tangential discontinuities (TDs) in the magnetic field.

TDs are mathematical discontinuities, surfaces across which the field abruptly changes direction but not magnitude. As the dynamics approaches the formation of a TD, they are ultimately avoided in nature by the inevitable naturally imperfect processes of work done against frictional forces converted by collisions to heat. The TDs thereby dissipate free energy associated with electric currents flowing within the TDs which arise from Ampères law, and any accompanying velocity gradients generated during the process via viscosity. The *electro*-dynamic, dissipative dynamics occurs after the ideal system has reached the limit of its own applicability. Internally-generated electric fields can arise on sub-MHD scales via any of a large number of instabilities available, including tearing modes in MHD and MHD/ kinetic processes like plasmoid disruption (Appendix C).

The theorem originated in the work of Parker (1972, 1988, 1994), and Janse et al. (2010), following some early ideas concerning the tangling of magnetic fields in the corona (e.g. Gold and Hoyle 1960). A formal proof is found in the work of Parker (1972), in considering a uniform magnetic field $\mathbf{B}_0$ threading a perfectly conducting fluid between two infinitely conducting plates. The magnetic field is assumed to evolve from its initial state $\mathbf{B}_0$, ultimately to achieve the condition of hydrostatic equilibrium

$$\mu_0 \nabla p = (\mathbf{curl}\, \mathbf{B}) \times \mathbf{B} \tag{4.6}$$

Along the $\mathbf{B}$ field clearly

$$\mathbf{B} \cdot \nabla p = 0 \tag{4.7}$$

so the pressure is constant along each magnetic line of force. Parker (1972) seeks a new equilibrium of a perturbative expansion of $\mathbf{B}$ in terms of an infinite series of magnetic vectors $\mathbf{B} = \mathbf{B}_0 + \Sigma_\alpha \epsilon^\alpha \, \mathbf{b}_\alpha(\mathbf{r}, t)$ each multiplied by the small parameter $\epsilon^\alpha, \alpha = 1 \ldots \infty, \epsilon = \ell_\perp / L$ where $\ell_\perp$ is the scale of perpendicular stirring motions at one plate, the two plates separated by $L \gg \ell_\perp$. Parker proved that

$$\frac{\partial \mathbf{b}_\alpha}{\partial z} \rightarrow 0 \quad \text{as}\, \epsilon \rightarrow 0 \;\forall\, \alpha. \tag{4.8}$$

This states that there can be no variation of the magnetic pattern along $\mathbf{B}_0$ in a new equilibrium, as a function of position between the two plates, z. The magnetic topology is fixed. This must be true not just for uniform fields but for a general initial state $\mathbf{B}(\mathbf{r}, t = 0)$, complete with its particular innate topology. As a consequence of Eq. (4.8), if there are changes in topology between the plates, then there is no equilibrium possible. The above expansion is an analytic function of space. It follows then that any new physical equilibrium must generally contain non-analytic solutions, those that satisfy an unchanged topology can only be tangential discontinuities. Several more proofs exist (e.g. Janse et al. 2010; Low 2023). Heuristically, the fixed topology and force balance equations together over-constrain the system, no smooth solutions are possible satisfying these two constraints simultaneously except under special, symmetric conditions. Another proof looks at the hyperbolic nature of the partial differential equations in a force-free state, where $\nabla \times \mathbf{B}(\mathbf{r}) = \alpha(\mathbf{r})\mathbf{B}(\mathbf{r})$. Then

$$\mathbf{B}(\mathbf{r}) \cdot \nabla \alpha(\mathbf{r}) = 0, \tag{4.9}$$

so that along each line of force, $\alpha(\mathbf{r})$ is constant. There are no constraints on $\alpha(\mathbf{r})$ from neighboring field lines, each can have its own α, so that the winding of neighboring field lines will generally lead to discontinuities. This is because the characteristics of the magnetic field are real, they are the field lines themselves (Fig. 4.2).

Certain refutations of aspects of the theorem have been shown to be due to misunderstandings (e.g. Low 2023). For example, Van Ballegooijen (1985) developed what is essentially an algorithm for generating smooth solutions (Low 2023). The overly restrictive condition Eq. (4.8) imposed by analyticity implies that analytical force-free fields conceptually belong to a smaller continuum of solutions, than the continuum of solutions with embedded TDs.

The theorem has received observational support in observations of apparently super-Alfvénic motions (Judge et al. 2012; Lipartito et al. 2014, see Fig. 3.9), which require physical motions of sheet-like structures to explain the phase speeds observed. As lace curtains appear with variable striations as they are blown by the wind, the sheets come and go rapidly in 2D images as their projections along the line-of-sight integrate over more or less plasma with time.

Other suggestive but rare observations use language of the (now old-fashioned) view of photospheric magnetic fields as individual tubes of flux (Solanki 2000), before the turbulent nature of photospheric magnetic fields became evident (see Fig. 1.5). Apparently braided magnetic flux bundles (proposed by Glencross 1975,

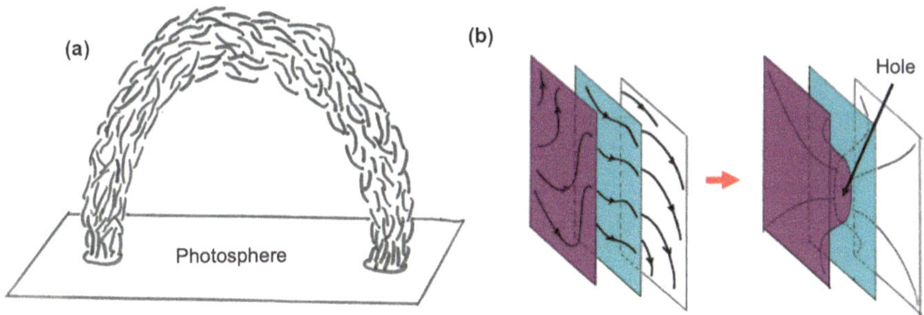

Fig. 4.2 A loop of magnetic fields embedded in perfectly conducting plasmas is shown above an infinitely conducting photosphere. As convective motions twist and inter-weave lines of force to (**a**), the fundamental theorem states that in general the volume must contain tangential discontinuities. Panel (**b**) illustrates how a hole appears in the light blue surface as the dark blue and white surfaces touch to form a TD. Manifested in nature as current sheets, the TDs are the dissipative structures in MHD turbulence (Sect. 4.9). From Low (2023)

1980) appear to be traced out by plasma in coronal images (Cirtain et al. 2013; Chitta et al. 2022). The questions whether TDs would be expected on the scales observed by these authors remains open. But if genuinely braided, such bundles would contain current sheets at their surfaces of inner contact. No heating induced by the morphological simplification of such braids into monolithic structures has yet been seen. Instead people have sought signatures resulting from the *dynamical* reconnection between the bundles. Thus Chitta et al. (2022) emphasize dynamic untangling, not coronal heating *per se*, conflating reconnection (changes in macroscopic fields $\mathbf{B}_0(\mathbf{r}, t)$) with dissipative processes (changes in $\mathbf{b}(\mathbf{r}, t)$, in the language of our $B^\flat$ conjecture, see Eq. (2.25) and accompanying text). The widely-accepted and adopted paradigms of flux tubes, reconnection between them, and the dynamical development of current sheets within otherwise continuous fluid structures and turbulence will be revisited briefly in Chap. 7. Insofar as Parker's picture applies to magnetic structures with a large aspect ratio (loop length/width), Parker's theorem applies to the problem of relaxation of the $\mathbf{b}(\mathbf{r}, t)$ components only. In this sense Parker's arguments share a physical picture with MHD turbulence (Sect. 3.5, Pontin and Hornig 2020). For example, Rappazzo et al. (2008) conclude:

> The time-dependent Parker problem may be seen as an MHD turbulence problem, where the large-scale forcing function is realized by the photospheric motions... This system is genuinely turbulent... a property of the Maxwell stresses developing in the coronal volume. Current sheets... are the result of a nonlinear cascade in a self-organized system.

In choosing to examine MHD equilibria instead of dynamic evolution *per se*, Parker instead showed that the system has no option but to conform to topological constraints by developing internal TDs. To Parker the non-existence of smooth solutions was sufficient to prove scales below the MHD frozen-field scales would spontaneously develop. MHD specialists concerned with explicit time-dependent problems, seek to understand the nature of the nonlinear interactions ultimately leading to irreversible dissipation of current sheets. In both cases current sheets form naturally within the ideal MHD regime, and both lead to the demise of ideal MHD on smaller scales. In the case of turbulence, the development of current sheets is attributed to the internal dynamic evolution of a self-organized non-linear system, not to the quasi-static picture of Parker's fixed topologies.

A distinction should be made between the fundamental theorem and its consequences, and superficially similar work that is less general. As already noted, individual "flux tubes" or "fibrils" were regarded as elementary building blocks of magnetic fields (Solanki 2000). Such ideas prompted the "tectonics model" (Priest et al. 2002), which relies on mixed polarity flux tubes interacting with long-lived network flux. In this it shares features in common with studies of dynamics in the chromosphere (e.g. Heggland et al. 2009). However, Priest et al. (2002) appeal to current sheet formation within multiple separatrix surfaces—surfaces between and separating topologically separate flux systems assumed present already as a result of photospheric dynamics—as the tubes of magnetic flux extend and expand upwards. Just as rims of tectonic plates are seats of earthquake activity, the sheets formed between mis-aligned tubes are supposed to be seats of heating. But modern instruments reveal that individual flux tubes are rare, the photospheric magnetic field instead lies within tortuous eddies of plasma (Fig. 1.5) lacking symmetry. Parker (1988) already stressed the difference between the spontaneous and forced development of current sheets. The work of Priest and associates is a more restricted model, prompted by an observed feature of magnetoconvection (Proctor and Weiss 1982; Thomas and Weiss 2004) from low resolution observations given the fanciful name of the "magnetic carpet" (Schrijver et al. 1998; Schrijver and Zwaan 2000). Parker's picture is therefore more general.

So far, the search for heating in Parker's scenario has almost exclusively focused on nanoflares, a concept and proposal originally put forward by Parker (1988) in reference to the heating of active regions.

4.8 Nanoflares

Parker (1988) proposed that coronal heating is a natural outcome of the fundamental theorem of magnetostatics. The idea was prompted by full-Sun observations by Lin et al. (1984) of *hard X-rays* (HXRs), which are phenomena more generally associated with electrons accelerated during flares. Statistical properties of transient spikes in HXR intensities suggested that the active corona is heated in bursts of energy, with energies extending down to 10^{16} J, compared with the typical large solar flare energy of 10^{25} J. (The relation between hard X-rays observed prior to 1988 and those discussed by NuSTAR was discussed in Sect. 4.5). The concept of a "nanoflare" was born. Now a small industry has built upon Parker's ideas (Bogachev et al. 2020). Parker explicitly treated the case of active regions, not quiet Sun. He estimated energy fluxes injected into the active corona by random stirring of magnetic fields by photospheric motions. Using estimates of magnetic field strengths and photospheric motions from observations, he then supposed that, once the magnetic fields across the forming current sheets reached a certain mis-alignment, reconnection would occur stochastically, converting magnetic free energy into kinetic energy, and perhaps accelerated particles and subsequent heating. By adopting a critical angle at which mis-alignment leads to reconnection, he suggested that packets of energy of $\sim 10^{16}$ J would be released, and if dissipated as heat, would provide sufficient energy to sustain the average corona. The nature of the mechanisms leading to irreversible dissipation were not addressed, basically assuming the ARLToR conjecture.

Ever since, nanoflares have been actively sought and studied theoretically by many research groups (reviewed by Bogachev et al. 2020). As of 2022, Parker's (1988) paper has been cited over forty times per year in diverse papers in astrophysics, reflecting the general nature of the idea. To date there is no agreement on the role of "nanoflares" viewed as an elementary heating event, in the corona, statistical work yields conflicting answers (Aschwanden 1999; Jess et al. 2019; Ulyanov et al. 2019; Aschwanden 2022). At least part of the problem involves the shape of, and extrapolation of power laws of the rate of occurrence of flares with energy (see Table 1 of Bogachev et al. 2020), the treatment of heat conduction to lower temperatures and (unanalyzed, as yet) energy losses in transition region lines. Statistical data analyses including power law studies have struggled to converge on definite answers also because of the small energies and brightness changes of nanoflares, following some difficulties pointed out in Chap. 3. Detailed numerical approaches have presumed the efficacy of a nanoflare before setting up model calculations at various levels of sophistication. All are based on the assumption that reconnection is the engine driving and adopt some level of "steady" coronal heating, a connection which we will argue is as yet unclear (Sect. 4.9), and for which modern hard X-ray data have cast some doubt (Sect. 4.5).

Also, as noted by Bogachev et al. (2020), the very notion that a type of "flare" constitutes a heating mechanism by itself has perhaps been misleading, instead it is a concept prompted by observations (spicules are another example) which has been co-opted as a basis for explaining the problem of transferring energy across scales highlighted in Fig. 3.1. Bogachev et al. (2020) conclude that:

> "Some authors argue that solar flares should be regarded as pico- and even femto-events... According to an alternative opinion, the energy of low-energy flares constitutes but a minor addition to the known large-scale activity and has no appreciable influence on its total balance. None of these concepts can be ruled out... we are of the opinion that both the nature and the role of low-energy flares still await clarification, despite the progress achieved over the past 50 years."

This opinion is unsurprising, revealing yet again methodological weaknesses highlighted in Chap. 3. Bogachev et al. (2020) also highlight further problems, concerning the slope α of probability distributions of the number of flares with energy E: $dN \sim E^{-\alpha}$ measured in multiple analyses of EUV and X-ray observations focusing on coronal radiation. Derived values of α are spread around the value of 2 which therefore includes infinite energies if integrated to smaller and smaller values of nanoflare energy E, when $\alpha > 2$, and places all the energy in larger flares if $\alpha < 2$. This explicitly demonstrates the effect of ill-posedness in the methods used. To remedy this a low energy cutoff is introduced, but it too is poorly constrained by observations. Nanoflares are generally too faint to be clearly seen, therefore instrument performance at low count rates becomes perhaps an interesting challenge. The statistical data used consist of X-ray or EUV brightness measurements, often with no account taken of conductive transport to cooler plasmas which radiate at different wavelengths, and no direct assessment of the kinetic component of the energy through, for example. line-of-sight Doppler motions.

"Avalanches" of nano-flares, leading to full scale flares, have been proposed based upon ideas of self-organized criticality prompted by the existence of power law distributions (Bak et al. 1987; Charbonneau et al. 2001). The idea is that, consistent with the ideas of Gold and Hoyle (Page 82), non-linear systems evolve slowly to a state of high free energy which becomes unstable to small perturbations, in this case nanoflares can trigger large flares in a dramatic inverse cascade, through reconnections on all scales. Wheatland et al. (1998) and Wheatland (2000) examined observations of waiting time distributions (intervals between flares) in full-disk soft X-ray data. The distributions were found to be consistent with Poisson processes, and so also with avalanches, for larger flares.

In the latest development, evidence from the EUI instrument on the Solar Orbiter mission (Marsch 2005; Marsden 2013; Rochus et al. 2020, 2022) has recently been found to support the idea that flares, if responsible for heating stable active region plasma loops, must be pico-scale, 10^{13} J or less (Judge 2023b).

4.9 From MHD to Dissipation

The transition from MHD scales to dissipation scales is an active research area. An excellent introduction can be found in the book by Kulsrud (2005). Formally one might expect MHD only to be valid on scales larger than particle mean free paths, $> 200\,\text{km}$ in the corona (Table 1.2). But MHD can be applied to some aspects of problems at smaller scales, with care (Parker 2007), because MHD is built upon the same conservation equations for the plasma, without reference to the mean free path of particles. Depending on the particular problem, with appropriate modifications to the pressure tensor in the equation of motion, we might expect to be able to treat the plasma as a single-fluid to scales an order of magnitude or two lower. Below these scales kinetic equations must be solved, or the consequences of long-range free particle transport at least examined (Scudder 1992a,b). In the journey from driving scales ($\sim$ Mm) to dissipation, Scudder emphasized that we must be able to assess if and when kinetic scales will influence ideas based upon MHD.

Such theories cannot be built from first principles. Instead, *phenomenological* theories using numerical experiments as if real experiments have been constructed, under the umbrella of "MHD turbulence". Thus we can take advantage of developments in MHD and plasma physics reviewed by Schekochihin (2022). Schekochihin reviews current sheet formation, tearing instabilities, plasmoids, reconnection, collisional and collisionless regimes and of course turbulence and dissipation. He argues that, under conditions of strong turbulence in a strong guide field $\mathbf{B}_0$, then a reasonable picture of events witnessed in numerical experiments and the solar wind can be described with a few provisionally reliable concepts. We will briefly examine the meaning of this viewpoint noting the caveat that (Schekochihin 2022):

> "the 'logical completeness' of our current picture of MHD turbulence...only applies to forced, balanced, inertial-range turbulence, with collisional dissipation and in the presence of a strong mean field."

Aside perhaps from the weak collisionality of the plasma, these are precisely the conditions which apply to the coronal heating problem in closed-field regions. A bare-bones summary of this area is given in Appendix C. The phenomenology emerging in MHD contains the following elements, the dynamics are assumed to be incompressible.

- The cascade originates from the non-linear interactions between two oppositely-directed Alfvén waves.
- Turbulence occurs asymmetrically. Motions occur predominantly in planes perpendicular to the mean local magnetic field $\mathbf{B}_0$.
- A state of *critical balance* between parallel wave crossing time $\lambda_{\parallel}/c_A$ and non-linear interaction time $\lambda_{\perp}/\delta u_{\perp}$ is reached (Goldreich and Sridhar 1995).
- The motions within the perpendicular plane also become asymmetric and intermittent, forming "linguini-like" structures themselves resembling current sheets.
- Such sheets can be disrupted by the plasmoid instability, breaking (through reconnection on small scales) the sheet into a series of topologically separate plasmoids (Fig. 4.3).

The above sequence of events potentially applies across structures with large enough magnetic Reynolds numbers (numerical experiments suggest Rm$\gtrsim 10^4$ suffices, Loureiro et al. 2007). What happens next is again an active research area. Is this remarkably intricate sequence of events fields and motions quenched by collisions? Or does the sequence of events survive down to kinetic scales intact, ready for fast rapid dynamical processes to complicate things further? Might we probe these processes by seeking observational evidence for or against associated signatures surviving averages in space and time? Or are in-situ measurements required? These questions are important, for if frictional dissipation occurs before kinetic scales are reached, then this would explain naturally the recorded absence of accelerated particles in the quiet Sun and steady active region loops (Sect. 4.5).

Lastly, recent work has indicated that MHD turbulence can be generated even if initially only uni-directional waves are introduced (Magyar et al. 2019). These authors showed that by relaxing assumptions of incompressibility and homogeneity, counter-propagating Alfvén waves, as well as magnetosonic waves and surface waves, can be generated. Thus richer turbulent phenomena can be anticipated, and some MHD turbulent power should be anticipated even in locally "open" magnetic fields such as those within coronal holes. Named "uniturbulence", this picture

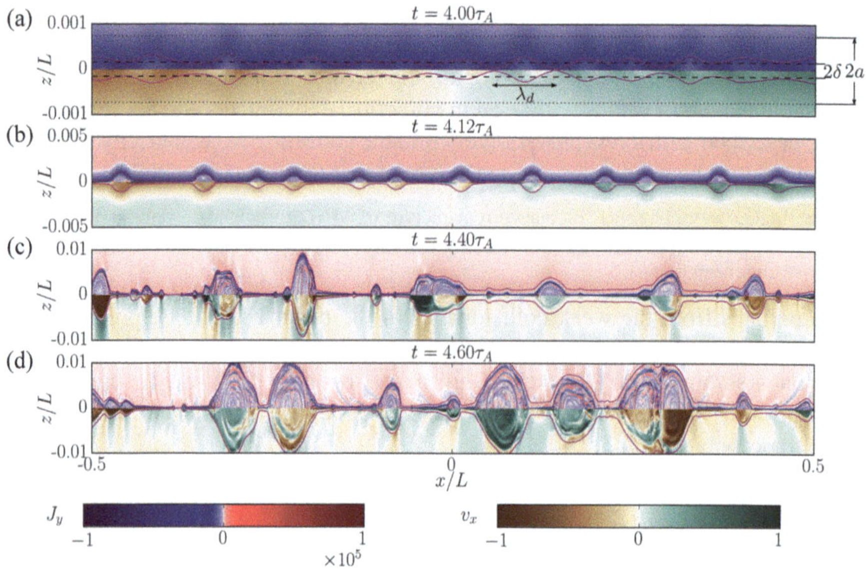

Fig. 4.3 The development and disruption of islands resulting from the tearing instability is shown as a function of time expressed in units of Alfvén crossing time. In each panel, the upper half shows the out-of-plane current density profile J_y and the bottom half shows the outflow profile v_x. At $t = 4.12$ Alfvén times, extended secondary current sheets have developed between plasmoids. Secondary current sheets are susceptible to the plasmoid instability and further fragmentation. From Huang et al. (2017)

clearly has parallels with phase-mixing and dynamics leading to surface waves, such as resonant absorption.

4.9.1 Resistive and Viscous Dissipation of Turbulence

So according to current understanding, MHD turbulence is highly anisotropic, with different dynamics in directions perpendicular and parallel to guide fields $\mathbf{B}_0$ (Appendix C). MHD turbulence also tends to form small-scale aligned structures (linguini-like current sheets) *in the plane perpendicular to* $\mathbf{B}_0$. Such structures are themselves subject to the plasmoid instability which lead to changes in topology. Thus, in the absence of mitigating effects, *small-scale reconnection within the linguini should be expected to partake in the conversion of MHD turbulence into heat*. The formation of plasmoids in a current sheet is illustrated in Fig. 4.3, displaying clearly the changing topology of the magnetic field across the computed domain.

The time needed to halt a dynamical cascade with characteristic scale $\lambda_\perp$ (ignoring the additional alignment noted above) can be estimated as $t \approx \lambda_\perp^2 \eta_\perp^{-1}$, which must be less than $\lambda_\perp u_\perp^{-1}$, $u_\perp$ being the amplitude of the turbulent motions, and $\eta_\perp$ the resistive or viscous diffusivity ($m^2\,s^{-1}$), also perpendicular to $\mathbf{B}_0$. Using estimates of the scales $\lambda_\perp$ and speeds $u_\perp$ of eddies from Goldreich and Sridhar (1995, henceforth "GS95", see Appendix C),

$$u_\perp \approx (\varepsilon\lambda_\perp)^{1/3}, \tag{4.10}$$

we see that

$$\lambda_\perp < \eta^{3/4}\varepsilon^{-1/4}, \tag{4.11}$$

where the average heating rate is ε. If we set this rate to the observed value $\varepsilon = F_M/\rho L \approx 10^7$ W kg^{-1} (Table 1.2, see also Fig. 4.1), we can find the perpendicular scale at which turbulence-driven resistive and viscous dissipation heats the corona. Using a magnetic diffusivity of $1.8\,m^2\,s^{-1}$, a perpendicular viscous diffusivity of $\nu_0/\omega_p^2\tau_p^2 \approx 0.3\,m^2\,s^{-1}$, we find length scales $\lambda_\perp \approx 0.03$ and 0.007 m respectively. In both cases, viscous or current dissipation would only occur on ridiculously small scale lengths, well below the ion inertial length ($\approx$ 10 m), so that kinetic effects would instead control dissipation.

However, the scale at which viscous dissipation of compressions *parallel* to **B** is potentially far larger, leading to dissipation on scales 10^5 times larger (3 and 0.7 km respectively). Only if the derivative $\frac{\partial u_\parallel}{\partial x_\parallel}$ is identically zero is there no dissipation. Generally then the stress tensor components $\frac{\partial u_j}{\partial x_j}$ must be evaluated using non-linear fluid calculations.

In the particular two-scale case of MHD turbulence (GS95's scales $\lambda_\perp$ and $\lambda_\parallel$) driven by the interaction of Alfvén waves (Appendix C), theory delivers a robust estimate of $k_\parallel u_\parallel$ in terms of c_A and ε, with a relation between $k_\parallel$ and $k_\perp$. The model yields (equations 2.9 and 2.11 of Schekochihin 2022):

$$k_\parallel \approx \frac{\varepsilon^{1/3}}{c_A} k_\perp^{2/3}, \tag{4.12}$$

$$u_\parallel^2 \approx \frac{\varepsilon}{c_A k_\parallel} \tag{4.13}$$

where c_A is the Alfvén speed. Then the rate of work against viscous forces is

$$Q_0 \approx \eta_0 \frac{\varepsilon}{c_A} k_\parallel \approx \eta_0 \frac{\varepsilon^{4/3}}{c_A^2} k_\perp^{2/3} \quad \text{W m}^{-3}. \tag{4.14}$$

If we assume that the observed heating is supplied by Q_0, we can set $Q_0 = Q = F_M/L$ from Table 1.2 and solve for the value of $k_\perp$ needed to provide the heating. Then

$$k_\perp \approx \left(\frac{Qc_A^2}{\eta_0}\right)^{3/2} \varepsilon^{-2}. \tag{4.15}$$

Using the values from Table 1.2 we therefore find $k_\perp \approx 1/35\,\text{m}$, i.e. we should expect structures of $< 35\,\text{m}$ scale to be irreversibly removed by viscous dissipation. This scale is just 3.5 times the ion inertial length, suggesting that if kinetic processes are important in energy dissipation in MHD turbulence, they would have to occur in the sub-viscous range. Eq. (4.15) will be useful in Chap. 6.

In conclusion, without measurements or first principles estimates of the power in compressible ($\frac{\partial u_\parallel}{\partial x_\parallel}$) versus incompressible modes, *we cannot assume that viscosity dissipates turbulent motions before kinetic scales are reached*, in both quiet and active coronal conditions. Given that the whole business of turbulence concerns non-linear interactions, we should expect that compressive modes will not, in general, be absent. But first we examine the stability of current sheets including the linguini-like structures, assuming that viscous dissipation is ineffective. Additional dependencies of growth rates on magnetic Prandtl number and wavenumber are to be found in section D.1 of Schekochihin (2022).

4.9.2 *Plasmoid Interruption of Turbulent Cascades*

The MHD tearing mode is a resistive MHD instability occurring within current sheets of finite thickness (Furth et al. 1963). As originally formulated, growth rates were found to scale as

$$\gamma \sim S^{-3/5}\frac{c_A}{L} \tag{4.16}$$

for Lundqvist number $S = c_A L/\eta$. This theory was modified to include lower wavenumber modes by Coppi et al. (1976), yielding two kinds of behavior ("FKR" and "Coppi" modes), depending on a parameter Λ measuring the ratio of the growth rate of the mode to the rate of resistive diffusion across a layer of the width of the resistive boundary layer δ_{in} (Schekochihin 2022).

The realization that sheets with especially large aspect ratios are subject to tearing became clear (e.g. Loureiro et al. 2007; Bhattacharjee et al. 2009; Uzdensky and Loureiro 2016). In the particular case of the Sweet-Parker configuration (aspect ratio $S^{1/2}$ with $S > 10^4$), the instability meant that such sheets would be disrupted before they could form. This "plasmoid" instability generated plasmoids, multiple islands of plasma separated themselves by additional Sweet-Parker type layers (see Fig. 4.3 Huang et al. 2017). Unlike the tearing mode, the instability generates many small plasmoids and is consequently far faster (Loureiro et al. 2007):

$$\gamma \approx S^{1/4}\frac{c_A}{L}, \tag{4.17}$$

exceeding the inverse dynamical time by the factor $S_\lambda^{1/4}$. The maximum growth rate arises at a "sweet spot" k^* in a normalized wavenumber κ (= wavenumber $\times L$)

between the Coppi and FKR limits (equations 19 and 20 of Loureiro et al. 2007), $\kappa^* \approx S_\lambda^{3/8}$, which is clearly just the number of plasmoids generated along length λ. A little care is needed in applying these scalings to coronal loops with $|\mathbf{B}_0| \gg \mathbf{b}$. Then L is the transverse scale λ associated with the fluctuating $\mathbf{b}$. The dynamical time $\tau_\lambda = \lambda/c_A$ seconds, c_A is the Alfvén speed $|\mathbf{b}|/\sqrt{\rho\mu_0}$, and $S_\lambda = c_A\lambda/\eta$. Numerical calculations of current sheets with aspect ratios $\lambda/\delta \gtrsim 100$ appear to be subject to this very rapid instability. This result carries over to collisionless conditions (e.g. Huang and Bhattacharjee 2013). Thus, if in the development of aligned MHD turbulence such aspect ratios are achieved, the growth rate can be very large, rapidly destroying the sheets and causing a hierarchy of plasmoids, perhaps with observable consequences for kinetic acceleration of particles and spectroscopy and spectropolarimetry. With λ set to the current observational limit of $\sim 10^5$ m, $c_A \sim 10^4$ m s^{-1}, we find $\gamma \sim 15\,\mathrm{s}^{-1}$. For $\lambda = 1000$ m, $\gamma \sim 470\,\mathrm{s}^{-1}$. Thus, MHD turbulence can in principle be disrupted by plasmoids across a wide range of scales. If such a sheet were to be forming along the length of a loop $L \approx 150$ Mm), as envisioned by Parker (1988) then γ would become growth rate would be $\gamma \sim 20$ per second. Given the times of several minutes over which such structures are buffeted by dense plasma below, such long and narrow sheets cannot exist in the corona, they are rapidly disrupted by plasmoids.

Plasmoids accelerate larger-scale reconnection, by chopping up the "outer scale" ℓ of a current sheet effectively into smaller scales, the essential breakthrough in generating fast reconnection made early by Petschek (1964) through slow shocks. Drake et al. (2006) showed how collisionless plasmoids can also accelerate charged particles. Loureiro et al. (2007) stressed that growth rates are larger by $S^{1/4} > 10^3$ than c_A/L (Eq. (4.17)), thus plasmoid widths should become comparable to the inner thickness (boundary layer) of the current sheet before plasmoids can be advected outwards by reconnection outflows.

We arrive at a picture of non-linear evolution in such driven MHD systems in which anisotropic, aligned and intermittent turbulence is probably interrupted by rapid plasmoid formation and reconnection. With Parker's picture together with numerical simulations of evolving current sheets developing as part of an MHD cascade, this (almost!) completes a new (post-2017) picture of turbulence and the journey of energy at large scales to scales of small plasmoids. Table 4.1 lists estimates of scales at which these phenomena arise, applied to the quiet and active conditions listed in Table 1.2.

An entirely different concept of MHD turbulence has developed in parallel to these developments. At the heart of the theory is a rejection of determinism in the trajectories—and flux-freezing condition—of magnetic field lines (Lazarian and Vishniac 1999). This work is elaborated on in the review of Brandenburg and Lazarian (2013). In part these ideas arose from the investigation of stochastic acceleration of energetic particles in flares (Blackman 1997) in a turbulent magnetic field, as well as in macroscopic shocks. Selkowitz and Blackman (2004) explored the observable consequences of stochastically varying field-line trajectories which can, through Fermi acceleration, generate local sites for particle acceleration. As

Table 4.1 Critical scales in MHD turbulence and kinetics

Region	L m	Viscous regime	λ_{CB} m	λ_D m	λ_η m	Ion inertial length m	R*
Quiet	1.5 10^8	η_0	6 10^6	2 10^3	9 10^5	13	19
Quiet		$\eta_0/\omega^2\tau^2$	6 10^6	2.5	0.2	13	2 10^{11}
Active	1 10^8	η_0	2 10^5	600	7 10^5	3	0.1
Active		$\eta_0/\omega^2\tau^2$	2 10^5	0.6	0.07	3	5 10^9

λ_{CB} = critical balance scale below which turbulence is strong. λ_D = scale below which tearing modes and plasmoids dominate. λ_η = kinetic dissipation scale. R* is the Reynolds number at the critical balance scale λ_{CB}, computed using the larger of viscous and magnetic diffusion coefficients (section 6.4.1 Schekochihin 2022)

particles exhibit Brownian motion owing to random excursions around the mean thermal behavior, the authors argue that the magnetic field has random components due to the roughness of the advecting velocity field. This picture makes its impact known in the nature and rate at which reconnection occurs, and perhaps in the aligned cascade/plasmoid destruction model. The model leads to rapidly enhanced reconnection over the laminar case as illustrated in Fig. 1.9.

Akin to a turbulent diffusivity in fluids, in this "stochastic turbulence" MHD picture there arises a turbulent resistivity, and so a turbulent reconnection with an effective rate independent of diffusivity arises, compatible with numerical experiments and solar observations (Brandenburg and Lazarian 2013). The model seems to deny the existence of structures aligned in the plane transverse to the magnetic field, there being no preferred direction in that plane if turbulence is genuinely random. Schekochihin (2022) (sections 8.3.3 and D7.4) concludes that the stochastic fields are not inconsistent with aligned cascades, because the magnetic fields in the aligned cascade reconnect just as fast as the turbulent structures decorrelate. Again this represents just the current understanding of an active research area.

But the final part of this journey surely involves kinetic scales of tens of meters. Several numerical studies already exist of hybrid calculations in which kinetic and MHD calculations are coupled in order to patch across the entire k-space (Tóth et al. 2017; Shou et al. 2021; Wang et al. 2022). Such work is in its infancy, and given the uncertainties in the journey already accumulated during the interrupted cascade picture, it seems premature to look for specific predictions of this regime to be tested against observations (Schekochihin 2022). But at least one more thing is known: even in the collisionless regime, plasmoid formation cannot be stopped in magnetized plasmas with large enough Reynolds numbers.

Returning to the goal of following energy from large-scale injection to thermalization, this (absurdly) rapid overview of an unsolved fundamental problem in plasma physics does suggest qualitative tests that might be dreamed up. The immature state of understanding suggests that such observational tests, if possible, might be better suited to understanding more about the detailed processes as opposed to identifying or rejecting specific heating mechanisms. But earlier in this chapter, one observational test has refuted a role for accelerated electrons in the coronal heating story. Encouraged by this, we can suggest future work which attempts to find anisotropic turbulent motions, push deeper into signatures of non-thermal electrons, to seek radiative signatures of plasmoids in polarized coronal spectra (French et al. 2019), and build on ways to probe different ion and electron temperatures and other mean properties of distribution functions. Some new ideas are presented in the final chapters.

4.10 Philosophical Considerations

Several issues have emerged from the review in this chapter, the most important of which are given in the chapter summary below. Before proceeding, we must first consider two methodologies because they have become so frequently encountered in modern research. First we look to numerical experiments, and contrast these with the increasingly common practice of assessing numerical simulations against real data.

4.10.1 Numerical Experiments

Many acknowledge that computers are capable of executing simulated experiments in natural science. In setting up a calculation to represent physical systems beyond the reach of laboratory experiments, numerical experiments have become regarded as a "third pillar of science" (e.g. Skuse 2019). Results are interpreted as if they were real experiments with equal potential to assess hypotheses as laboratory measurements. No actual measurements or observations are involved in the arguments, the "data" used come only from computers. Of course, in the coronal heating problem, a complete treatment of processes across the huge range of scales, even using the fluid approximation, is far beyond current computer capabilities. Instead, calculations are made on a subset of the problem, with boundary conditions and sub-grid scale physics replaced with mathematical algorithms representing the rest. The qualitative differences encountered in "open" versus "closed" computations of coronal heating arise entirely through just one different boundary condition. Recent examples relevant to coronal heating include the study of kinetic processes in magnetic reconnection (Comisso and Sironi 2022; Yao et al. 2022), seeking to understand particle acceleration mechanisms at an elementary level, at scales far below particle mean free paths. At the other extreme, numerical experiments are used to test arguments based upon ideal MHD, such as numerical experiments set up to test Parker's theorem on current sheet formation using "reduced MHD" (Rappazzo et al. 2007, 2008; Rappazzo and Parker 2013; Rappazzo 2015). MHD turbulence phenomenology is built almost entirely on results from numerical experimentation (Schekochihin 2022).

There is no prior reason to question the methodology associated with numerical experiments, there is a clean separation between data and any hypotheses being tested, or conclusions drawn. One might question assumptions concerning boundary conditions and sub-grid scales, or perhaps the applicability of say a non-relativistic formulation of equations of motion. But provided the equations and boundary and initial conditions are accurately specified, then there is in principle no difference between calculated and measured data.

The same cannot be said for calculations considered to be simulations, which are compared with observations, for better or worse.

4.10.2 Numerical "Simulations"

For the sake of argument, we consider a ficticious sophisticated numerical calculation targeted at chromospheric or coronal heating. It contains sufficient ingredients, with appropriate initial and boundary conditions, and a long enough (multiple wave-crossing) time to run. Recent impressive examples might include those of Martínez-Sykora et al. (2020) and Przybylski et al. (2022). Also consider that a single run indicates that, after the decay of initial transients, simulated observations have similarities with observations. Not only is a complete 3D atmospheric model accessible to the researcher, but post-processing has been done to mimic real observations, in order to compare properties of observations such as those of Figs. 1.5 and 1.8.

Assume further that the models use lower time-dependent boundary conditions (which drive the calculation) which have been credibly assimilated, or which match statistical properties of observations, using precise observations of the photosphere. The calculation, subject to boundary conditions imposed also on the sides and top of the computational domain, represents a credible realization of the atmosphere well above the photosphere. These calculations are considered "simulations", like local weather forecasts or global climate models, an attempt to mimic reality as accurately as possible.

Like a weather forecast, the system is modeled as a set of advection-diffusion partial differential equations (Chap. 2), for which there is in principle a fundamental sensitivity to initial conditions (Smith 2007). They are also dependent on the algorithmic suppression of the smallest scales. Unlike a weather forecast, the principal energy source for the plasmas of interest is unknown- it *is* the principal goal of the coronal heating problem. Yet, the most detailed information available to us about the coronal plasma comes from emission lines from ions which are most highly populated in the regime where energies of ionization and excitation significantly exceed the mean thermal energy, $k_B T_e$ of the plasma electron population. For example, the well-known "green line" of Fe^{13+} originates from thermal plasmas close to $\log_{10} T_e = 6.3$, yet to ionize Fe^{12+} requires energies above $I = 331\,\text{eV}$, so that

$$I \sim 2k_B T_e.$$

Under these conditions just a fraction $\lesssim \exp(-2^2) \sim 0.02$ of the thermal electrons can produce emission in the Fe^{13+} spectrum. Modest changes in temperature lead to large changes in this emission, a change in T_e of 10% changes the emission by a factor of 2. Yet T_e in the synthetic calculation is determined by the imbalance of larger terms in the energy equation. For example, when fully ionized, changes in kT_e follow from the dynamical update of the equations for conservation of energy. This involves differences between (unknown) heating and cooling terms as well as changes in internal energy and enthalpy due to dynamics. In other words, what we measure from highly ionized ions in the coronal plasma is the end result of the history of the plasma's evolution including some of the unknown terms we aim to find. Even if we can eliminate algebraically, say by taking ratios, the absolute brightness of these coronal emissions, our derivative quantities such as Doppler line widths and shifts, they are naturally skewed, being heavily weighted by brightness and its exponential dependence on T_e. In short, we are comparing synthetic data with observations knowing that we are looking at tails of distributions, and weighted averages of structures in the plasma which depend on the unknown conditions within the plasma. A crude analogy might be that our observations are akin to trying to understand the structure of a Christmas tree being blown around merely by observing the few moving lights strung loosely around it.

The question thus arises: *How should the researchers draw conclusions about chromospheric and coronal physics from such a comparison?* Here is a representative conclusion from a study published in a peer reviewed journal during the last decade:

> "...other physical processes, such as ..., could drive [dynamics] that might produce similar observables, but this needs to be investigated...A satisfactory model ...must be able to reproduce these observables."

The first statement reflects a mix of observational data limited by availability and technology, and by necessary restrictions in the physical model's ingredients and/or boundary conditions. The second reveals a natural urge to reproduce certain observations with calculations. The particular study did not concern itself explicitly with coronal heating, but it does fairly represent an increasingly dominant methodology reported in articles in solar and coronal physics. We should also recall the words of warning from Tsiklauri (2009) quoted on Page 90.

The difference with a numerical experiment seems clear. The latter is an experiment under the control of the programmer, who can be entirely disconnected from the scientist analyzing the output "data" and an accompanying solution "vector" of physical quantities at every point or cell in the calculation. The development of emergent or other non-linear behavior is studied in order to understand better what leads to unanticipated or anticipated behavior (equivalent to a formal hypothesis).

In contrast, the comparison of synthetic with actual observations is not clear cut. The simulation has included observational data for some part of the computational domain by assimilation, or something equivalent. The comparison of observed and computed data elsewhere in the computation is limited to those observables acquired in the real data, which are integrals in time and space of the physical properties sought. To date, no measurements of magnetic field evolution have been made in the corona, only a few have attempted the same in the chromosphere, the work of Kleint (2017) was the first to find changes in chromospheric magnetic fields before and after a major (X-class) flare. Changes were observed only in line-of-sight fields, thereby containing no information on the free magnetic energy. Kleint compared photospheric and chromospheric magnetic fields, their changes in time and space, and in relation to changes in regard to a sun quake and the radiative emissions at EUV wavelengths. She argued that the observations opened more questions instead of providing answers, pointing to the need for more observations including all three components of the changing magnetic field. Only the morphology of coronal plasma loops is routinely reported from EUV and X-ray "measurements" of magnetic properties from spacecraft. Even for the extensively studied 'Masuda flare" of 1992 January 13 observed at the solar limb (e.g. Liu et al. 2011), the lack of chromospheric and coronal magnetic field measurements and their changes remains a major weakness in flare studies. In the coronal heating problem, our measurements therefore are limited to revealing only the *effects* of unseen yet overwhelming magnetic forces.

The researcher comparing simulations with observations also does not know how physics missing on sub-grid scales affects the plasma dynamics and observables. The common trick of shrinking model grid sizes and checking for signs of grid-dependent behavior is a tool to build confidence that results are insensitive to physical processes on those grid scales within the particular model's framework (e.g. fluid mechanics). It is only through the ARLToR conjecture, essentially an extrapolation from numerical to physical dissipation scales, that full independence from grid scales has meaning. A clear example of known problems with this approach is the identification of a "helicity barrier", a predicted phenomenon waiting to halt an otherwise downward cascade in scale of energy of Alfvénic turbulence towards electron heat. Instead the barrier leads to ion cyclotron waves which heat the ions (Squire et al. 2022, see Page 55). It is unclear if either the fluid or hybrid kinetic models would lead to distinguishable observables (refer to the debate highlighted on Page 191). This example illustrates just how potentially damaging the convenient ARLToR conjecture might be. We are left to ponder what the comparison of data and necessarily simplified simulations, both undersampled, actually means.

These physical worries are accompanied by concerns over methodology. What might a student of scientific method conclude? The best and most exciting outcome, scientifically, is refutation, meaning that *in some critical fashion*, there is a definite disagreement between simulated and information-rich observed data. These might be qualitative, such as a phenomenon is not present in observations or model, or quantitative, where the magnitude of some observation differs outside uncertainties from the modeled phenomenon. In the "toy" simulation work imagined above, just four outcomes are possible: (1) The observations are insufficient for discrimination; this might entail cases of agreement and disagreement in the comparisons with diverse data; (2) the observation refutes the model; (3) the observations support the model; (4) the observation partly supports some aspect of the model. Of these outcomes, (1) and (4) are the most commonly encountered, but are arguably the least scientifically interesting or successful. Outcome (1) points to the need for better data, perhaps new instrumentation looking at the solar corona in an uncharted fashion rich in critical information. But outcomes (4) and especially (3) present us with a difficulty. A first question one might ask is whether agreement with observation is, in fact, an *unreasonable* expectation, given the regimes of validity the model which will tend to be outside solar regimes. A second is how to understand what is meant by partial support? Is it in the eye of the beholder? Or are we to reject such an idea outright à la Karl Popper?

Are we to accept that the model represents processes in the Sun, do we declare the problem solved, even though we are well aware of limited model ingredients within physical regimes (such as Reynolds and magnetic Prandtl numbers) far from those on the Sun? Speaking technically, general purpose PDE algorithms usually seek solutions which are *continuous* to avoid de-stabilizing discontinuities, even though thought experiments reveal the central importance of discontinuous solutions for the problem at hand (Parker 1994). Special approaches have been developed for solving idealized MHD problems (e.g. Deforest and Kankelborg 2007), including the unique nature of tangential discontinuities (Bhattacharyya et al. 2010). Related are the notions (Einaudi et al. 2021) that the corona is a "self-regulating forced system", more specifically that the system adjusts itself to dissipate the energy input without knowing the dissipation mechanisms themselves (ARLToR). If so, the question then arises: why study the dissipation at all? In our framing of the problem (Page 21), the answer to the question at the beginning of this paragraph is unequivocally "'no" simply because the physics is incomplete, and we already have examples refuting ARLToR under *some* conditions. Not only this, but theoretical understanding has arguably advanced enough to make particular predictions on the basis of, for instance, the signature scales of processes listed in Table 4.1 made specifically for recent models of MHD turbulence. But these tricky questions and debates have, as yet, no answers.

Perhaps there is a way forward which avoids such difficulties.

4.10.3 A Road Forward Based Upon Bayesian Inference

There is no agreed methodology in science, but there are diverse methods. Two extreme views were represented by philosophers Feyerabend (1975) and Popper (1962). In an absurd reduction of decades of debate, we might say that Feyerabend argued, on the basis of historical precedent, against rationalism, i.e. that there is no fixed methodology, scientists choose methods as they suit specific problems. In contrast, Popper argued for refutation as a means to identify and avoid "pseudo science". Most scientific research falls between these extremes.

The coronal heating problem is, by any measure, mature. But arguably, in recent years, while we have witnessed progress in many specific but related issues (e.g., in our understanding of chromospheric jets, signatures of unresolved "threads", MHD turbulence. . .), the progress in understanding the full problem has been lacking. An over-abundance of diverse solutions, without quantifiable probabilities of success or failure, only add to a perceived lack of progress. Perhaps research on problem has reached a phase of "normal science", between epochs of unusual "paradigm-changing" breakthroughs (Kuhn 1970). Seen from this perspective, an obvious way forward is to promote research work which emphasizes attempts to refute hypotheses and models, assigning quantitative probabilities to different outcomes. After all, there is no prior reason to believe that just one mechanism dominates all of chromospheric and coronal heating, indeed, the different large-scale environments of locally closed and "open" field regions might be used to argue otherwise. If in the process of eliminating successfully various models, we might then be on the cusp of Kuhn's changes in paradigm.

Such a change in approach will require adaptation by a community which currently is highly rewarded for producing positive outcomes, both in proposals and in research reports. Naturally there will be resistance to such a change, but perhaps this suggestion is more a question of attitude than actual change requiring significant attention. Nevertheless, success must surely be measured less in terms of "agreement with observations" (Tsiklauri 2009), but in firm disagreement. In appreciating that "disagreement" is not synonymous with "failure", that "dead ends" are just as useful to find in complex systems as "open ends", perhaps we find common ground with words colloquially attributed to John Archibald Wheeler:

> "Knowledge only progresses by making mistakes as fast as possible"

where we interpret "making mistakes" brazenly to include "taking the wrong paths". Currently we travel many paths, but have yet to assess their likelihood of compatibility against critical data. But, how do we know that we are actually wrong? Consider the opening lines of a new review in solar physics (Asensio Ramos et al. 2023):

> "Astrophysics, and solar physics in particular, is an observational science in which we cannot change the experimental conditions, we simply observe. Therefore, the only way of learning is by confronting observations with state-of-the-art theoretical modeling. The models are then tuned until the observations are explained and conclusions are drawn from this comparison."

With such an approach the mistakes are hopefully found during the "tuning" process. But this view of method in modern astrophysics does not emphasize refutation. Perhaps it describes why we have so many research articles with conclusions depending so prominently on contingencies, (Sect. 3.3). The possibility of tuning models towards a false solution is concerning, given ill-posedness (Sect. 3.7), and especially the prescient observation of Tsiklauri (2009) on Page 90. A worrisome parallel might exist in the global re-generation of magnetism in Sun-like stars. Some popular models include a "tachocline", a region of steep radial rotational velocity gradients between radiative cores and convection zones. Yet, a tachocline must barely exist in cooler stars which are almost fully convective (Judge 2020, 2023a). Such stars frequently show similar "sunspot" activity cycles to the Sun.

In the coronal heating problem, a balanced approach might be as simple as adopting Bayesian methods from the outset (Jeffreys 1961; Jaynes 2003). In requiring a researcher to adopt prior hypotheses H and/or parameterized models, the approach offers benefits for the coronal heating problem. First it can avoid the problem of being "not even wrong". All outcomes (that the hypothesis is rejected, not rejected or neither) are advancements, not merely statements about compatibility. Second, it permits a definition for "information rich" data. Given data d, Bayes' theorem gives the probability $P(H \mid d)$ that H is compatible with d,

$$P(H \mid d) = \frac{P(d \mid H)\, P(H)}{P(d)}, \tag{4.18}$$

a quantity called the *posterior* probability. $P(H)$ is the (subjective) *prior* probability that the hypothesis or model is correct, in the absence of d. The prior $P(H)$ could, for example, be drawn from a uniform, binomial or a Poisson distribution, reflecting what we believe we know in the absence of data d. $P(d \mid H)$ is the probability of observing d given H, it is a function of d for a given hypothesis and is called the likelihood. $P(d)$ is the *marginal likelihood*, the same for all H and it does not affect the relative probabilities of different hypotheses, H. It is precisely these relative probabilities which are desired, not absolute values. Equation (4.18) can be viewed as an updating procedure where the prior probability $P(H)$, when challenged with d, yields the new posterior probability, showing how the probability of the hypothesis changes in the light of d. Further data might then be used to obtain another posterior probability, using the original posterior as a new prior. The benefits of Bayesian methods are clear:

- Only in the case where new data d do not influence $P(H \mid d)$ (i.e. $P(H \mid d) =$ the new prior $P(H)$) is the result "not even wrong". An example might be to confront a model of viscous heating of ions with the brightness of EUV line emission. Among many difficulties, the latter depends sensitively on the electron population which can take some time to exchange energy with ions, and existing data may be unable to resolve critical ion motions on scales approaching dissipation. The researcher then must acknowledge that better hypotheses or data are needed for genuine advancement.
- In being obliged to pose a prior hypothesis or model, this approach should be used to identify precisely *which* data are information-rich.
- In turn this should prompt development of new experiments, numerical and physical, to try to produce clear differences in posterior probabilities. Some ideas along these lines are presented in Chap. 6.

UV, EUV and X-ray data have informed us of large-scale phenomena in the corona, the morphology of such magnetic structures using stereoscopy. But they have only informed us of plasma properties *after* irreversible coronal heating has occurred. Many of these properties have been well-studied and typical values are listed in Table 1.2. But as in the above example, such data contain little new information on heating mechanisms. In the next chapters some new approaches are proposed, placing the coronal heating problem first and foremost, by identifying information-rich data types and possible future experiments, numerical and observational.

Chapter Summary

We have categorized research on coronal heating (arbitrarily) into several methods (Sect. 4.3). None has convincingly followed energy from injection scales to dissipation as heat. *We still have a coronal heating problem.*

Failure to find definitive results is a result not only of the practical problems (non-uniqueness, ill-posedness, separation of scales, non-local transport, etc.), but also weaknesses in *methodology*.

In perennially observing thermal and morphological properties of plasmas, we are reduced to analyzing the *aftermath* of heating process. Novel, information-rich chromospheric and coronal instruments are needed to target open theoretical questions.

Seeking contingent agreement of models with information-poor data has become a dominant methodology, an exhaustingly successful enterprise. *Information-rich* data are needed to reduce incipient non-uniqueness and ill-posedness.

Refutation is key to solving the problem. The community would be well served to value negative outcomes, and to reject mere compatible outcomes, endowed heavily with contingencies. *Confirmation bias* should be more widely recognized.

The *absence* of a predicted phenomenon can be more important than its presence. An upper limit on hard X-ray intensities (Paterson et al. 2023; Duncan et al. 2021) denies a role for accelerated particles in heating the quiet and active corona.

Until coupling across scales is achieved numerically, one must be skeptical of comparisons between sophisticated "simulations" and observations. Numerical work is *best seen as an experiment to understand coronal physics.* On scales below current grid scales, a veritable zoo of theoretical non-linear phenomena await before kinetic scales of 10 m are reached.

A clean separation of observations from models is often not possible, and refutation impossible. A given model may depend on observations for an initial state, for example. Some articles forgo such a separation in order to derive conclusions. Bayesian methods might offer an alternative methodology, delivering probabilities of model agreement and informing researchers of areas in need of attention.

(continued)

In conclusion, after this broad look at methodology in the light of challenges involved, a more focused approach recognizing the intrinsic and practical difficulties seems appropriate. The problem deserves to be treated with a synthesis of a broad range of theories which can be challenged by the identification and acquisition of new, information-rich data.

References

Ackermann, M., et al.: High-energy gamma-ray emission from solar flares: summary of fermi large area telescope detections and analysis of two M-class flares. Astrophys. J. **787**, 15 (2014). https://doi.org/10.1088/0004-637X/787/1/15

Anderson, L.S., Athay, R.G.: Chromospheric and coronal heating. Astrophys. J. **336**, 1089–1091 (1989)

Aschwanden, M.J.: Do EUV nanoflares account for coronal heating? Sol. Phys. **190**, 233–247 (1999). https://doi.org/10.1023/A:1005288725034

Aschwanden, M.J.: An evaluation of coronal heating models for active regions based on Yohkoh, SOHO, and TRACE observations. Astrophys. J. **560**, 1035–1044 (2001). https://doi.org/10.1086/323064

Aschwanden, M.J.: Reconciling power-law slopes in solar flare and nanoflare size distributions. Astrophys. J. **934**, L3 (2022). https://doi.org/10.3847/2041-8213/ac7b8d

Aschwanden, M.J., Nightingale, R.W., Alexander, D.: Evidence for nonuniform heating of coronal loops inferred from multithread modeling of TRACE data. Astrophys. J. **541**, 1059–1077 (2000)

Aschwanden, M.J., et al.: The coronal heating paradox. Astrophys. J. **659**, 1673–1681 (2007)

Asensio Ramos, A., et al.: Machine learning in solar physics (2023). arXiv e-prints, arXiv:2306.15308. https://doi.org/10.48550/arXiv.2306.15308

Athay, R.G., Holzer, T.E.: The role of spicules in heating the solar atmosphere. Astrophys. J. **255**, 743–752 (1982). https://doi.org/10.1086/159873

Athay, R.G., White, O.R.: Chromospheric and coronal heating by sound waves. Astrophys. J. **226**, 1135 (1978)

Atwood, W.B., et al.: The large area telescope on the Fermi gamma-ray space telescope mission. Astrophys. J. **697**, 1071–1102 (2009). https://doi.org/10.1088/0004-637X/697/2/1071

Axford, W.I., McKenzie, J.F.: The origin of high speed solar wind streams. In: Marsch, E., Schwenn, R. (eds.) Solar Wind Seven. COSPAR Colloquium Series, vol. 3, pp. 1–5. Pergamon Press, Oxford (1992)

Bak, P., Tang, C., Wiesenfeld, K.: Self-organized criticality: an explanation of the 1/F noise. Phys. Rev. Lett., **59**, 381–384 (1987). https://doi.org/10.1103/PhysRevLett.59.381

Bale, S.D., et al.: Highly structured slow solar wind emerging from an equatorial coronal hole. Nature **576**, 237–242 (2019). https://doi.org/10.1038/s41586-019-1818-7

Beiersdorfer, P., Scofield, J.H., Osterheld, A.L.: X-ray line diagnostic of magnetic field strength for high-temperature plasmas. Phys. Rev. Lett., **90**, 235003 (2003). https://doi.org/10.1103/PhysRevLett.90.235003

Bhattacharjee, A., et al.: Fast reconnection in high-Lundquist-number plasmas due to the plasmoid instability. Phys. Plasmas **16**, 112102 (2009)

Bhattacharyya, R., Low, B.C., Smolarkiewicz, P.K.: On spontaneous formation of current sheets: untwisted magnetic fields. Phys. Plasmas **17**, 112901 (2010)

Billings, D.: The Solar Spectrum: The Coronal Spectrum. D. Reidel Publishing Co., Dordrecht (1965)

Biskamp, D.: Magnetic Reconnection in Plasmas. Cambridge University Press, Cambridge (2005)

Blackman, E.G.: Distinguishing solar flare types by differences in reconnection regions. Astrophys. J. Lett. **484**, L79–L82 (1997). https://doi.org/10.1086/310763

Bogachev, S.A., et al.: Microflares and nanoflares in the solar corona. Phys. Uspekhi, **63**, 783–800 (2020). https://doi.org/10.3367/UFNe.2019.06.038769

Braginskii, S.I.: Transport processes in a plasma. Rev. Plasma Phys. **1**, 205–311 (1965)

Brandenburg, A., Lazarian, A.: Astrophysical hydromagnetic turbulence. In: Balogh, A., Bykov, A., Cargill, P., Dendy, R., Dudok de Wit, T., Raymond, J. (eds.) Microphysics of Cosmic Plasmas, Space Science Reviews, vol. 178, pp. 163–200, Academic Press, London (2013). https://doi.org/10.1007/978-1-4899-7413-6_5

Brooks, D.H., Warren, H.P., Landi, E.: Measurements of coronal magnetic field strengths in solar active region loops. Astrophys. J. Lett. **915**, L24 (2021). https://doi.org/10.3847/2041-8213/ac0c84

Carlsson, M., et al.: A publically available simulation of an enhanced network region of the sun. Astron. Astrophys. **585**, A4 (2016)

Cassak, P.A., Liu, Y.H., Shay, M.A.: A review of the 0.1 reconnection rate problem. J. Plasma Phys. **83**, 715830501 (2017). https://doi.org/10.1017/S0022377817000666

Charbonneau, P., et al.: Avalanche models for solar flares (invited review). Sol. Phys. **203**, 321–353 (2001). https://doi.org/10.1023/A:1013301521745

Chitta, L.P., Peter, H., Young, P.R.: Extreme-ultraviolet bursts and nanoflares in the quiet-sun transition region and corona. Astron. Astrophys. **647**, A159 (2021). https://doi.org/10.1051/0004-6361/202039969

Chitta, L.P., et al.: Solar coronal heating from small-scale magnetic braids. Astron. Astrophys. **667**, A166 (2022). https://doi.org/10.1051/0004-6361/202244170

Cirtain, J.W., et al.: Energy release in the solar corona from spatially resolved magnetic braids. Nature **493**, 501–503 (2013). https://doi.org/10.1038/nature11772

Comisso, L., Sironi, L.: Ion and electron acceleration in fully kinetic plasma turbulence. Astrophys. J. Lett. **936**, L27 (2022). https://doi.org/10.3847/2041-8213/ac8422

Coppi, B., et al.: Resistive internal kink modes. Sov. J. Plasma Phys. **2**, 533–535 (1976)

Cranmer, S.R., Field, G.B., Kohl, J.L.: Spectroscopic constraints on models of ion cyclotron resonance heating in the polar solar corona and high-speed solar wind. Astrophys. J. **518**, 937–947 (1999). https://doi.org/10.1086/307330

Daughton, W., Roytershteyn, V.: Emerging parameter space map of magnetic reconnection in collisional and kinetic regimes. Soc. Sci. Res. **172**, 271–282 (2012). https://doi.org/10.1007/s11214-011-9766-z

Daughton, W., Scudder, J., Karimabadi, H.: Fully kinetic simulations of undriven magnetic reconnection with open boundary conditions. Phys. Plasmas, **13**, 072101 (2006). https://doi.org/10.1063/1.2218817

Davila, J.M.: Heating of the solar corona by the resonant absorption of alfvén waves. Astrophys. J. **317**, 514–521 (1987)

Deforest, C.E., Kankelborg, C.C.: Fluxon modeling of low-beta plasmas. J. Atmos. Sol. Terr. Phys. **69**, 116–128 (2007). https://doi.org/10.1016/j.jastp.2006.06.011

DeForest, C.E., Kankelborg, C.C., Longcope, D.W.: Fluxon modeling of force free magnetic fields: Voronoi method. In: AGU Spring Meeting Abstracts, SH41A–18 (2001)

Delone, A.B., Makarova, E.A.: Interferometric investigation of the red and green coronal lines during the total solar eclipse of May 30, 1965. Sol. Phys. **9**, 116–130 (1969). https://doi.org/10.1007/BF00145733

de Pontieu, B., et al.: A new view of the solar interface region from the interface region imaging spectrograph (IRIS). Sol. Phys. **296**, 84 (2021)

Dong, C., et al.: Role of the plasmoid instability in magnetohydrodynamic turbulence. Phys. Rev. Lett.**121**, 165101 (2018). https://doi.org/10.1103/PhysRevLett.121.165101

Drake, J.F., et al.: Electron acceleration from contracting magnetic islands during reconnection. Nature **443**, 553–556 (2006). https://doi.org/10.1038/nature05116

Duncan, J., et al.: NUSTAR observation of energy release in 11 solar microflares. Astrophys. J. **908**, 29 (2021). https://doi.org/10.3847/1538-4357/abca3d

Eddy, J.A.: New Sun: The Solar Results from Skylab. Nasa Sp-402. NASA, Washington (1979)

Edlén, B.: Die Deutung Der Emissionslinien Im Spektrum Der Sonnenkorona. Zeitschrift für Astrofysik **22**, 30–64 (1943)

Einaudi, G., et al.: Energetics and 3D structure of elementary events in solar coronal heating. Astrophys. J. **910**, 84 (2021)

Falconer, D.A., et al.: Neutral-line magnetic shear and enhanced coronal heating in solar active regions. Astrophys. J. **482**, 519–534 (1997). https://doi.org/10.1086/304114

Falconer, D.A., et al.: Solar coronal heating and the magnetic flux content of the network. Astrophys. J. **593**, 549–563 (2003). https://doi.org/10.1086/376359

Feyerabend, P.: Against Method. New Left Books, London/New York City (1975)

Fisher, G.H., Canfield, R.C., McClymont, A.N.: Flare loop radiative hydrodynamics. V - response to thick-target heating. Astrophys. J. **289**, 414–441 (1985). https://doi.org/10.1086/162901

Fisher, G.H., et al.: Coronal heating in active regions as a function of global magnetic variables. Astrophys. J. **508**, 885–898 (1998)

Fontenla, J.M., Avrett, E.H., Loeser, R.: Energy balance in the solar transition region. I. Hydrostatic thermal models with ambipolar diffusion. Astrophys. J. **355**, 700–718 (1990)

Foukal, P.: The pressure and energy balance of the cool corona over sunspots. Astrophys. J. **210**, 575 (1976)

Foukal, P.: Magnetic loops, downflows and convection in the solar corona. Astrophys. J. **223**, 1046 (1977)

French, R.J., et al.: Spectropolarimetric insight into plasma sheet dynamics of a solar flare. Astrophys. J. Lett. **887**, L34 (2019). https://doi.org/10.3847/2041-8213/ab5d34

Furth, H.P., Killeen, J., Rosenbluth, M.N.: Finite-resistivity instabilities of a sheet pinch. Phys. Fluids, **6**, 459–484 (1963). https://doi.org/10.1063/1.1706761

Gandorfer, A.M., et al.: Solar polarimetry in the near UV with the Zurich imaging polarimeter ZIMPOL II. Astron. Astrophys. **422**, 703–708 (2004). https://doi.org/10.1051/0004-6361:20040254

Glencross, W.M.: Heating of coronal material at X-ray bright points. Astrophys. J. Lett. **199**, L53–L56 (1975). https://doi.org/10.1086/181847

Glencross, W.M.: Plasma flow along sheared magnetic arches within the solar corona. Astron. Astrophys. **83**, 65–72 (1980)

Gold, T., Hoyle, F.: On the origin of solar flares. Mon. Not. R. Astron. Soc. **120**, 89 (1960)

Goldreich, P., Sridhar, S.: Toward a theory of interstellar turbulence. II. Strong Alfvenic turbulence. Astrophys. J. **438**, 763 (1995)

Goossens, M.L., Arregui, I., Van Doorsselaere, T.: Mixed properties of MHD waves in non-uniform plasmas. Front. Astron. Space Sci. **6**, 20 (2019). https://doi.org/10.3389/fspas.2019.00020

Grumer, J., et al.: Unexpected transitions induced by spin-dependent, hyperfine and external magnetic-field interactions. Phys. Scripta **89**, 114002 (2014)

Gudiksen, B.V., Nordlund, Å.: An *ab initio* approach to the solar coronal heating problem. Astrophys. J. **618**, 1020–1030 (2005)

Gudiksen, B.V., et al.: The stellar atmosphere simulation code BIFROST. Code description and validation. Astron. Astrophys. **531**, A154 (2011)

Hahn, M., Savin, D.W.: Observational quantification of the energy dissipated by Alfvén waves in a polar coronal hole: evidence that waves drive the fast solar wind. Astrophys. J. **776**, 78 (2013). https://doi.org/10.1088/0004-637X/776/2/78

Hale, G.E.: Solar vortices. Astrophys. J. **28**, 100–116 (1908)

Harrison, F.A., et al.: The nuclear spectroscopic telescope array (NUSTAR) high-energy X-ray mission. Astrophys. J. **770**, 103 (2013). https://doi.org/10.1088/0004-637X/770/2/103

Hazra, S., Nandy, D., Ravindra, B.: The relationship between solar coronal X-ray brightness and active region magnetic fields: a study using high-resolution hinode observations. Sol. Phys. **290**, 771–785 (2015). https://doi.org/10.1007/s11207-015-0652-9

Heggland, L., de Pontieu, B., Hansteen, V.H.: Observational signatures of simulated reconnection events in the solar chromosphere and transition region. Astrophys. J. **702**, 1–18 (2009). https://doi.org/10.1088/0004-637X/702/1/1

Heyvaerts, J., Priest, E.R.: Coronal heating by phase-mixed shear Alfven waves. Astron. Astrophys. **117**, 220–234 (1983)

Hollweg, J.V.: Resonances of coronal loops. Astrophys. J. **277**, 392–403 (1984). https://doi.org/10.1086/161706

Howson, T.A., de Moortel, I., Reid, J.: Phase mixing and wave heating in a complex coronal plasma. Astron. Astrophys. **636**, A40 (2020)

Huang, Y.M., Bhattacharjee, A.: Scaling laws of resistive magnetohydrodynamic reconnection in the high-Lundquist-number, plasmoid-unstable regime. Phys. Plasmas **17**, 062104–062104 (2010). https://doi.org/10.1063/1.3420208

Huang, Y.M., Bhattacharjee, A.: Plasmoid instability in high-Lundquist-number magnetic reconnectiona) . Phys. Plasmas **20**, 055702 (2013). https://doi.org/10.1063/1.4802941

Huang, Y.M., Bhattacharjee, A.: Turbulent magnetohydrodynamic reconnection mediated by the plasmoid instability. Astrophys. J. **818**, 20 (2016). https://doi.org/10.3847/0004-637X/818/1/20

Huang, Y.M., Comisso, L., Bhattacharjee, A.: Plasmoid instability in evolving current sheets and onset of fast reconnection, Astrophys. J. **849**, 75 (2017). https://doi.org/10.3847/1538-4357/aa906d

Huang, Y.M., Bhattacharjee, A., Sullivan, B.P.: Onset of fast reconnection in hall magnetohydrodynamics mediated by the plasmoid instability. Phys. Plasmas **18**, 072109–072109 (2011). https://doi.org/10.1063/1.3606363

Huba, J.D.: Hall magnetohydrodynamics - a tutorial. In: Büchner, J., Dum, C., Scholer, M. (eds.) Space Plasma Simulation, vol. 615, pp. 166–192. Springer-Verlag, Berlin (2003)

Ionson, J.A.: Resonant absorption of Alfvénic surface waves and the heating of solar coronal loops. Astrophys. J. **226**, 650–673 (1978)

Janse, A.M., Low, B.C., Parker, E.N.: Topological complexity and tangential discontinuity in magnetic fields. Phys. Plasmas **17**, 092901 (2010)

Jaynes, E.: Probability theory: the logic of science. Cambridge University Press, Cambridge (2003)

Jeffreys, H.: Theory of Probability, 3rd edn. Clarendon Press, Oxford (1961)

Jeffries, J., Lites, B., Skumanich, A.: Transfer of line radiation in a magnetic field. Astrophys. J. **343**, 920–935 (1989)

Jess, D.B., et al.: Statistical signatures of nanoflare activity. I. Monte Carlo simulations and parameter-space exploration. Astrophys. J. **871**, 133 (2019). https://doi.org/10.3847/1538-4357/aaf8ae

Ji, H., et al.: Magnetic reconnection in the era of exascale computing and multiscale experiments. Nat. Rev. Phys. **4**, 263–282 (2022)

Judge, P.G.: The Sun: A Very Short Introduction. Oxford University Press, Oxford (2020)

Judge, P.G.: Magnetic connections across the chromosphere-corona transition region. Astrophys. J. **914**, 70 (2021)

Judge, P.G.: The mercurial sun at the heart of our solar system. In: Lara, L.M., Jewitt, D. (eds.) Planetary Systems Now, pp. 49–70. World Scientific, London (2023a). https://doi.org/10.1142/q0388

Judge, P.G.: Steadiness of coronal heating, Astrophys. J. **957**, 25 (2023b)

Judge, P.G., Reardon, K., Cauzzi, G.: Evidence for sheet-like elementary structures in the sun's atmosphere? Astrophys. J. Lett. **755**, L11 (2012)

Judge, P.G., et al.: Are all flare ribbons simply connected to the corona? Astrophys. J. **838**, 138 (2017). https://doi.org/10.3847/1538-4357/aa656c

Judge, P., et al.: High-cadence visible and infrared spectra of the sun during eclipse. Astrophys. J. **877**, 10 (2019). https://doi.org/10.3847/1538-4357/ab0e04

Kankelborg, C.C., Longcope, D.W., DeForest, C.E.: Lagrangian modeling of force free fields and current sheets: fluxon representation and the kernel method. In: AGU Spring Meeting Abstracts, vol. 2001, SH41A–17 (2001)

Kasper, J.C., et al.: Sensitive test for ion-cyclotron resonant heating in the solar wind. Phys. Rev. Let, **110**, 091102 (2013). https://doi.org/10.1103/PhysRevLett.110.091102

Kiepenheuer, K.O.: Solar activity. In: Kuiper, G.P. (ed.) The Sun, pp. 322–465. Chicago University Press, Chicago (1953)

Kirichenko, A.S., Bogachev, S.A.: The relation between magnetic fields and x-ray emission for solar microflares and active regions. Sol. Phys. **292**, 120 (2017). https://doi.org/10.1007/s11207-017-1146-8

Kleint, L.: First detection of chromospheric magnetic field changes during an X1-flare. Astrophys. J. **834**, 26 (2017)

Kleint, L., Gandorfer, A.: Prospects of solar magnetometry – from ground and in space. Space Sci. Rev. **210**, 397–426 (2017)

Ko, Y.K., et al.: Correlation of coronal plasma properties and solar magnetic field in a decaying active region. Astrophys. J. **826**, 126 (2016). https://doi.org/10.3847/0004-637X/826/2/126

Kohl, J.L., et al.: UVCS/SOHO empirical determinations of anisotropic velocity distributions in the solar corona. Astrophys. J. Lett. **501**, L127 (1998)

Kohl, J.L., et al.: Ultraviolet spectroscopy of the extended solar corona. Astron. Astrophys. Rev. **13**, 31–157 (2006). https://doi.org/10.1007/s00159-005-0026-7

Kuhn, T.S.: The Structure of Scientific Revolutions, 2nd edn. University of Chicago Press, Chicago (1970)

Kulsrud, R.M.: Plasma Physics for Astrophysics. Princeton University Press, Princeton/Oxford (2005)

Kulsrud, R.M.: Intuitive approach to magnetic reconnection. Phys. Plasmas **18**, 111201 (2011)

Kuperus, M.: The heating of the solar corona. Space Sci. Rev. **9**, 713–739 (1969)

Laming, J.M.: The FIP and inverse FIP effects in solar and stellar coronae. Living Rev. Solar Phys. **12**, 2 (2015). https://doi.org/10.1007/lrsp-2015-2

Landi, E., et al.: Hinode/EIS measurements of active-region magnetic fields. Astrophys. J. **904**, 87 (2020). https://doi.org/10.3847/1538-4357/abbf54

Landi degl'Innocenti, E.L., Landolfi, M.: Polarization in Spectral Lines. Astrophysics and Space Library, vol. 307. Kluwer Academic Publishers, Dordrecht (2004)

Lazarian, A., Vishniac, E.T.: Reconnection in a weakly stochastic field. Astrophys. J. **517**, 700–718 (1999). https://doi.org/10.1086/307233

Lee, J.Y., et al.: The role of magnetic topology in the heating of active region coronal loops. Astrophys. J. **723**, 1493–1506 (2010). https://doi.org/10.1088/0004-637X/723/2/1493

Lemen, J.R., et al.: the atmospheric imaging assembly (AIA) on the solar dynamics observatory (SDO). Sol. Phys. **275**, 17–40 (2012). https://doi.org/10.1007/s11207-011-9776-8

Levine, R.H.: A new theory of coronal heating. Astrophys. J. **190**, 457–466 (1974)

Li, W., et al.: A novel method to determine magnetic fields in low-density plasma facilitated through accidental degeneracy of quantum states in Fe^{9+}. Astrophys. J. **807**, 69 (2015)

Lighthill, M.J.: Predictions on the velocity field coming from acoustic noise and a generalized turbulence in a layer overlying a convectively unstable atmospheric region. In: Thomas, R.N. (ed.) Aerodynamic Phenomena in Stellar Atmospheres, vol. 28, p. 429, Academic Press, London (1967)

Lin, R.P., et al.: Solar hard X-ray microflares. Astrophys. J. **283**, 421–425 (1984). https://doi.org/10.1086/162321

Lin, R.P., et al.: The Reuven Ramaty high-energy solar spectroscopic imager (RHESSI). Sol. Phys. **210**, 3–32 (2002)

Lindblom, J.F., Walker, A.B.C., Barbee, T.W.J.: Normal incidence reflection multilayer optics for solar soft X-ray/extreme ultraviolet (XUV) observations. In: X-ray Imaging II. Society of Photo-Optical Instrumentation Engineers (SPIE) Conference Series, vol. 691, pp. 11–19 (1986). https://doi.org/10.1117/12.936615

Lipartito, I., et al.: The solar chromosphere observed at 1 Hz and 0."2 resolution. Astrophys. J. **785**, 109 (2014)
Lites, B.W.: Remote sensing of solar magnetic fields. Rev. Geophys. **38**, 1–36 (2000)
Liu, R., Xu, Y., Wang, H.: A revisit of the Masuda flare. Sol. Phys. **269**, 67–82 (2011). https://doi.org/10.1007/s11207-010-9688-z
Liu, Y.H., et al.: Why does steady-state magnetic reconnection have a maximum local rate of order 0.1? Phys. Rev. Lett. **118**, 085101 (2017). https://doi.org/10.1103/PhysRevLett.118.085101
Liu, Y.H., et al.: First-principles theory of the rate of magnetic reconnection in magnetospheric and solar plasmas. Commun. Phys. **5**, 97 (2022). https://doi.org/10.1038/s42005-022-00854-x
Longcope, D.W., Tarr, L.A.: Relating magnetic reconnection to coronal heating. Philos. Trans. R. Soc. Lond. Series A **373**, 20140263–20140263 (2015). https://doi.org/10.1098/rsta.2014.0263
Loureiro, N.F., Schekochihin, A.A., Cowley, S.C.: Instability of current sheets and formation of plasmoid chains. Phys. Plasmas **14**, 100703–100703 (2007). https://doi.org/10.1063/1.2783986
Low, B.C.: Topological nature of the parker magnetostatic theorem. Phys. Plasmas **30**, 012903 (2023). https://doi.org/10.1063/5.0124164
Lysenko, A.L., et al.: X-ray and γ-ray emission from solar flares. Phys. Uspekhi **63**, 818–832 (2020). https://doi.org/10.3367/UFNe.2019.06.038757
Magyar, N., Van Doorsselaere, T., Goossens, M.: Understanding uniturbulence: self-cascade of MHD waves in the presence of inhomogeneities. Astrophys. J. **882**, 50 (2019). https://doi.org/10.3847/1538-4357/ab357c
Mahajan, S.M., et al.: Formation and primary heating of the solar corona: theory and simulation. Phys. Plasmas **8**, 1340–1357 (2001). https://doi.org/10.1063/1.1350670
Mallet, A., Schekochihin, A.A., Chandran, B.D.G.: Refined critical balance in strong Alfvenic turbulence. Mon. Not. R. Astron. Soc. **449**, L77–L81 (2015)
Mandrini, C.H., Démoulin, P., Klimchuk, J.A.: Magnetic field and plasma scaling laws: their implications for coronal heating models. Astrophys. J. **530**, 999–1015 (2000)
Mariska, J.T.: The Solar Transition Region. Cambridge University Press, Cambridge (1992)
Marsch, E. et al.: Solar orbiter - mission profile, main goals and present status. Adv. Space Res. **36**, 1360–1366 (2005)
Marsden, R.G., Müller, D., St. Cyr, O.C.: Solar orbiter - close-up view of the sun. In: Zank, G.P. (ed.), Solar Wind 13, American Institute of Physics Conference Series, vol. 1539, 448–453 (2013)
Martínez-Arnáiz, R., et al.: Effect of magnetic activity saturation in chromospheric flux-flux relationships. Mon. Not. R. Astron. Soc. **414**, 2629–2641 (2011)
Martínez-Sykora, J., et al.: Ion-neutral interactions and nonequilibrium ionization in the solar chromosphere. Astrophys. J. **889**, 95 (2020). https://doi.org/10.3847/1538-4357/ab643f
Martínez-Sykora, J., et al.: A novel inversion method to determine the coronal magnetic field including the impact of bound-free absorption. Astrophys. J. **938**, 60 (2022). https://doi.org/10.3847/1538-4357/ac8d5b
Matthaeus, W.H., et al.: Pathways to dissipation in weakly collisional plasmas. Astrophys. J. **891**, 101 (2020). https://doi.org/10.3847/1538-4357/ab6d6a
McDonald, K.L.: Topology of steady current magnetic fields. Am. J. Phys. **22**, 586–596 (1954). https://doi.org/10.1119/1.1933854
McWhirter, R.W.P., Thoneman, P.C., Wilson, R.: Heating of the solar corona II. A model based on energy balance. Astron. Astrophys. **40**, 63–73 (1975)
Metcalf, T.R., et al.: Electric currents and coronal heating in NOAA active region 6952. Astrophys. J. **428**, 860 (1994)
Meyer-Vernet, N.: Large scale structure of planetary environments: the importance of not being Maxwellian. Plan. Space. Sci. **49**, 247–260 (2001). https://doi.org/10.1016/S0032-0633(00)00146-X
Meyer-Vernet, N.: Basics of the Solar Wind. Cambridge University Press, Cambridge/New York (2007)
Molnar, M.E., et al.: High-frequency wave power observed in the solar chromosphere with IBIS and ALMA. Astrophys. J. **920**, 125 (2021). https://doi.org/10.3847/1538-4357/ac1515

Moore, R.L., et al.: Evidence that strong coronal heating results from photospheric magnetic flux cancellation. In: American Astronomical Society Meeting Abstracts. American Astronomical Society Meeting Abstracts, vol. 188, 86.04 (1996)

Okamoto, T.J., et al.: Resonant absorption of transverse oscillations and associated heating in a solar prominence. I. Observational aspects. Astrophys. J. **809**, 71 (2015). https://doi.org/10.1088/0004-637X/809/1/71

Oran, R., et al.: Alfvén wave turbulence as a coronal heating mechanism: simultaneously predicting the heating rate and the wave-induced emission line broadening. Astrophys. J. **845**, 98 (2017). https://doi.org/10.3847/1538-4357/aa7fec

Osterbrock, D.E.: The heating of the solar chromosphere, plages, and corona by magnetohydrodynamic waves. Astrophys. J. **134**, 347 (1961)

Pagano, P., de Moortel, I., Morton, R.J.: Effect of coronal loop structure on wave heating through phase mixing. Astron. Astrophys. **643**, A73 (2020) https://doi.org/10.1051/0004-6361/202039209

Parker, E.N.: Dynamics of the interplanetary gas and magnetic fields. Astrophys. J. **128**, 664–676 (1958)

Parker, E.N.: Topological dissipation and the small-scale fields in turbulent gases. Astrophys. J. **174**, 499 (1972)

Parker, E.N.: Nanoflares and the solar X-ray corona. Astrophys. J. **330**, 474–479 (1988)

Parker, E.N.: The phase mixing of Alfvén waves, coordinated modes, and coronal heating. Astrophys. J. **376**, 355 (1991)

Parker, E.N.: Spontaneous Current Sheets in Magnetic Fields with Application to Stellar X-Rays. International Series on Astronomy and Astrophyics. Oxford University Press, Oxford (1994)

Parker, E.N.: Conversations on Electric and Magnetic Fields in The Cosmos. Princeton University Press, Princeton (2007)

Paterson, S., et al.: The first survey of quiet sun features observed in hard X-rays with NuSTAR. Sol. Phys. **298**(3), 47 (2023). https://doi.org/10.1007/s11207-023-02135-4

Peierls, R.: Wolfgang Ernst Pauli. 1900-1958. Mem. Fellows R. Soc. **5**, 174–192 (1960)

Petschek, H.E.: Magnetic field annihilation. NASA Special Publication **50**, 425 (1964)

Pevtsov, A.A., Acton, L.W.: Soft X-ray luminosity and photospheric magnetic field in quiet sun. Astrophys. J. **554**, 416–423 (2001). https://doi.org/10.1086/321342

Pierrard, V., Lamy, H., Lemaire, J.: Exospheric distributions of minor ions in the solar wind. J. Geophys. Res. Space Phys. **109**, 118 (2004). doi:https://doi.org/10.1029/2003JA010069

Pietarila, A., Judge, P.G.: On the formation of the resonance lines of helium in the sun. Astrophys. J. **606**, 1239–1257 (2004)

Pontin, D.I., Hornig, G.: The Parker problem: existence of smooth force-free fields and coronal heating. Living Rev. Sol. Phys. **17**, 5, (2020). https://doi.org/10.1007/s41116-020-00026-5

Pontin, D.I., Priest, E.R.: Magnetic reconnection: MHD theory and modelling. Living Rev. Sol. Phys. **19**, 1 (2022)

Popper, K.R.: Conjectures and Refutations. The Growth of Scientific Knowledge. Routledge, London (1962)

Povel, H.P.: Imaging Stokes polarimetry with piezoelastic modulators and charge-coupled-device image sensors. Opt. Eng. **34**, 1870–1878 (1995). https://doi.org/10.1117/12.200596

Priest, E., Forbes, T.: Magnetic Reconnection. Cambridge University Press, Cambridge (2007)

Priest, E.R., et al.: A method to determine the heating mechanisms of the solar corona. Astrophys. J. **539**, 1002–1022 (2000). https://doi.org/10.1086/309238

Priest, E.R., Heyvaerts, J.F., Title, A.M.: A flux-tube tectonics model for solar coronal heating driven by the magnetic carpet. Astrophys. J. **576**, 533–551 (2002)

Proctor, M.R.E., Weiss, N.O.: Magnetoconvection. Rep. Progress Phys. **45**, 1317–1379 (1982). https://doi.org/10.1088/0034-4885/45/11/003

Prusa, J.M., Smolarkiewicz, P.K., Wyszogrodzki, A.A.: EULAG, a Computational model for multiscale flows. Comput. Fluids **37**(9), 1193–1207 (2008)

Przybylski, D., et al.: Chromospheric extension of the MURAM code. Astron. Astrophys. **664**, A91 (2022). https://doi.org/10.1051/0004-6361/202141230

Rachmeler, L.A.: Modeling eruptive coronal magnetohydrodynamic systems with FLUX. Ph.D. Thesis, University of Colorado, Boulder, 2010

Raouafi, N.E., Solanki, S.K.: Effect of the electron density stratification on off-limb O VI line profiles: how large is the velocity distribution anisotropy in the solar corona? Astron. Astrophys. **427**, 725–733 (2004). https://doi.org/10.1051/0004-6361:20041203

Rappazzo, A.F.: Equilibria, dynamics, and current sheet formation in magnetically confined coronae. Astrophys. J. **815**, 8 (2015). https://doi.org/10.1088/0004-637X/815/1/8

Rappazzo, A.F., Parker, E.N.: Current sheets formation in tangled coronal magnetic fields. Astrophys. J. Lett. **773**, L2 (2013). https://doi.org/10.1088/2041-8205/773/1/L2

Rappazzo, A.F., et al.: Coronal heating, weak MHD turbulence, and scaling laws. Astrophys. J. Lett. **657**, L47–L51 (2007). https://doi.org/10.1086/512975

Rappazzo, A.F., et al.: Nonlinear dynamics of the Parker scenario for coronal heating. Astrophys. J. **677**, 1348–1366 (2008). https://doi.org/10.1086/528786

Rempel, M.: Numerical simulations of quiet sun magnetism: on the contribution from a small-scale dynamo. Astrophys. J. **789**, 132 (2014). https://doi.org/10.1088/0004-637X/789/2/132

Rempel, M.: Extension of the MURAM radiative MHD code for coronal simulations. Astrophys. J. **834**, 10 (2017)

Rochus, P. et al.: The solar orbiter EUI instrument: the extreme ultraviolet imager. Astron. Astrophys. **642**, A8 (2020). https://doi.org/10.1051/0004-6361/201936663

Rochus, P. et al.: The solar orbiter EUI instrument: the extreme ultraviolet imager (Corrigendum). Astron. Astrophys. **665**, C1 (2022). https://doi.org/10.1051/0004-6361/201936663e

Rosner, R., Tucker, W.H., Vaiana, G.S.: Dynamics of the quiescent solar corona. Astrophys. J. **220**, 643 (1978)

Rosner, R., et al.: Heating of coronal plasma by anomalous current dissipation. Astrophys. J. **222**, 317–332 (1978)

Schatzman, E.: The heating of the solar corona and chromosphere. Annales d'Astrophysique **12**, 203 (1949)

Schekochihin, A.A.: MHD turbulence: a biased review. J. Plasma Phys. **88**, 155880501 (2022). https://doi.org/10.1017/S0022377822000721

Schrijver, C.J., Zwaan, C.: Solar and Stellar Magnetic Activity. Cambridge University Press, New York (2000)

Schrijver, C.J., et al.: Large-scale coronal heating by the small-scale magnetic field of the sun. Nature **394**, 152–154 (1998). https://doi.org/10.1038/28108

Schrijver, C.J., et al.: the Nonpotentiality of active-region coronae and the dynamics of the photospheric magnetic field. Astrophys. J. **628**, 501–513 (2005). https://doi.org/10.1086/430733

Scudder, J.D.: On the causes of temperature change in inhogeneous low-density astrophysical plasmas. Astrophys. J. **398**, 299–318 (1992a)

Scudder, J.D.: Why all stars should possess circumstellar temperature inversions. Astrophys. J. **398**, 319–349 (1992b)

Scudder, J.D., Olbert, S.: A theory of local and global processes which affect solar wind electrons 1. The origin of typical 1 AU velocity distribution functions-steady state theory. J. Geophys. Res. **84**, 2755–2772 (1979). https://doi.org/10.1029/JA084iA06p02755

Selkowitz, R., Blackman, E.G.: Stochastic Fermi acceleration of subrelativistic electrons and its role in impulsive solar flares. Mon. Not. R. Astron. Soc. **354**, 870–882 (2004). https://doi.org/10.1111/j.1365-2966.2004.08252.x

Shay, M.A., et al.: Structure of the dissipation region during collisionless magnetic reconnection. J. Geophys. Res. **103**, 9165–9176, 1998. https://doi.org/10.1029/97JA03528

Shou, Y., et al.: Magnetohydrodynamic with adaptively embedded particle-in-cell model: MHD-AEPIC. J. Comput. Phys. **446**, 110656 (2021). https://doi.org/10.1016/j.jcp.2021.110656

Si, R., et al.: A first spectroscopic measurement of the magnetic-field strength for an active region of the solar corona. Astrophys. J. Lett. **898**, L34 (2020). https://doi.org/10.3847/2041-8213/aba18c

Similon, P.L., Sudan, R.N.: Energy dissipation of Alfvén wave packets deformed by irregular magnetic fields in solar-coronal arches. Astrophys. J. **336**, 442 (1989)

Skuse, B.: The third pillar. Phys. World **32**, 40–43 (2019)

Slepian, J.: Lines of force in electric and magnetic fields. Am. J. Phys. **19**, 87–90 (1951). https://doi.org/10.1119/1.1932718

Smith, L.: Chaos: A Very Short Introduction. Oxford University Press, Oxford (2007)

Solanki, S.: Solar photospheric magnetic flux tubes: observations. In: Murdin, P. (ed.) Encyclopedia of Astronomy and Astrophysics, 2013. Institute of Physics Publishing, Bristol (2000). https://doi.org/10.1888/0333750888/2013

Sonnerup, B.U.Ö.: On the theory of steady state reconnection. Comput. Phys. Commun. **49**, 143–159 (1988). https://doi.org/10.1016/0010-4655(88)90222-6

Squire, J., et al.: High-frequency heating of the solar wind triggered by low-frequency turbulence. Nature Astron. **6**, 715–723 (2022). https://doi.org/10.1038/s41550-022-01624-z

Stein, R.F.: Reflection, refraction, and coupling of MHD waves at a density step. Astrophys. J., Suppl. Ser. **22**, 419 (1971). https://doi.org/10.1086/190231

Stenflo, J.O., Keller, C.U.: The second solar spectrum. A new window for diagnostics of the sun. Astron. Astrophys. **321**, 927–934 (1997)

Sturrock, P.A., Roald, C.B., Wolfson, R.: Chromospheric magnetic reconnection and its implication for coronal heating. Astrophys. J. Lett. **524**, L75–L78 (1999). https://doi.org/10.1086/312301

Takasao, S., et al.: Magnetohydrodynamic shocks in and above post-flare loops: two-dimensional simulation and a simplified model. Astrophys. J. **805**, 135 (2015). https://doi.org/10.1088/0004-637X/805/2/135

Tan, C., et al.: Statistical correlations between parameters of photospheric magnetic fields and coronal soft X-ray brightness. Astrophys. J. **665**, 1460–1468 (2007). https://doi.org/10.1086/519304

Testa, P., et al.: Evidence of nonthermal particles in coronal loops heated impulsively by nanoflares. Science **346**, 1255724 (2014)

Thomas, J.H., Weiss, N.O.: Fine structure in sunspots. Ann. Rev. Astron. Astrophys. **42**, 517–548 (2004). https://doi.org/10.1146/annurev.astro.42.010803.115226

Tóth, G., et al.: Scaling the ion inertial length and its implications for modeling reconnection in global simulations. J. Geophys. Res. (Space Physics) **122**, 10,336–10,355 (2017). https://doi.org/10.1002/2017JA024189

Tsiklauri, D.: Missing pieces of the solar jigsaw puzzle. Astron. Geophys. **50**, 5.32–5.38 (2009)

Ugarte-Urra, I., et al.: The magnetic properties of heating events on high-temperature active-region loops. Astrophys. J. **877**, 129 (2019). https://doi.org/10.3847/1538-4357/ab1d4d

Ulyanov, A.S., et al.: The energy distribution of nanoflares at the minimum and rising phase of solar cycle 24. Astron. Lett. **45**, 248–257 (2019). https://doi.org/10.1134/S1063773719040078

Uzdensky, D.A., Loureiro, N.F.: Magnetic reconnection onset via disruption of a forming current sheet by the tearing instability. Phys. Rev. Lett. **116**, 105003 (2016). https://doi.org/10.1103/PhysRevLett.116.105003

Van Ballegooijen, A.A.: Electric currents in the solar corona and the existence of magnetostatic equilibrium. Astrophys. J. **298**, 421–430 (1985)

Verbeeck, C., et al.: A multi-wavelength analysis of active regions and sunspots by comparison of automatic detection algorithms. Sol. Phys. **283**, 67–95 (2013). https://doi.org/10.1007/s11207-011-9859-6

Vernazza, J.E., Avrett, E.H., Loeser, R.: Structure of the solar chromosphere. I. Basic computations and summary of the results. Astrophys. J. **184**, 605 (1973)

Vilhu, O.: Observational constraints and evidence for saturation. In: Linsky, J.L., Stencel, R.E. (eds.) Proceedings of the Fifth Cambridge Workshop on Cool Stars, Stellar Systems, and the Sun, pp. 110–122. Springer, Berlin (1987)

Vlahos, L., et al.: Sources of solar energetic particles. Philos. Trans. R. Soc. Lond. Series A **377**, 20180095 (2019). https://doi.org/10.1098/rsta.2018.0095

Vögler, A., et al.: Simulations of magneto-convection in the solar photosphere. equations, methods, and results of the MURAM code. Astron. Astrophys. **429**, 335–351 (2005)

Wang, X., Chen, Y., Tóth, G.: Global magnetohydrodynamic magnetosphere simulation with an adaptively embedded particle-in-cell model. J. Geophys. Res. (Space Physics) **127**, e30091 (2022). https://doi.org/10.1029/2021JA030091

Warren, H.P., Winebarger, A.R., Hamilton, P.S.: Hydrodynamic modeling of active region loops. Astrophys. J. Lett. **579**, L41–L44 (2002). https://doi.org/10.1086/344921

Warren, H.P., Winebarger, A.R., Mariska, J.T.: Evolving active region loops observed with the transition region and coronal explorer. II. Time-dependent hydrodynamic simulations. Astrophys. J. **593**, 1174–1186 (2003). https://doi.org/10.1086/376678

Wheatland, M.S.: The origin of the solar flare waiting-time distribution. Astrophys. J. Lett. **536**, L109–L112 (2000). https://doi.org/10.1086/312739

Wheatland, M.S., Sturrock, P.A., McTiernan, J.M.: The waiting-time distribution of solar flare hard X-ray bursts. Astrophys. J. **509**, 448–455 (1998). https://doi.org/10.1086/306492

Wiegelmann, T., Sakurai, T.: Solar force-free magnetic fields. Living Rev. Sol. Phys. **18**, 1 (2021). https://doi.org/10.1007/s41116-020-00027-4

Wilhelm, K.: Solar spicules and macrospicules observed by SUMER. Astron. Astrophys. **360**, 351–362 (2000)

Wolfson, R., et al.: Coronal X-ray brightness and photospheric magnetic field: a study in correlations. Astrophys. J. **539**, 995–1001 (2000). https://doi.org/10.1086/309245

Yao, X., Muñoz, P.A., Büchner, J.: Non-thermal electron velocity distribution functions due to 3D kinetic magnetic reconnection for solar coronal plasma conditions. Phys. Plasmas **29**, 022104 (2022). https://doi.org/10.1063/5.0061151

Yashiro, S., Shibata, K.: Relation between thermal and magnetic properties of active regions as a probe of coronal heating mechanisms. Astrophys. J. Lett. **550**, L113–L116 (2001). https://doi.org/10.1086/319486

Zharkova, V.V., et al.: Recent advances in understanding particle acceleration processes in solar flares. Space Sci. Rev. **159**, 357–420 (2011). https://doi.org/10.1007/s11214-011-9803-y

Zurbuchen, T.H., Richardson, I.G.: In-situ solar wind and magnetic field signatures of interplanetary coronal mass ejections. Space Sci. Rev. **123**, 31–43 (2006). https://doi.org/10.1007/s11214-006-9010-4

Chapter 5
Seeking the Sun's Achilles' Heels

Homer's Achilles was endowed with seeming immortality, his mother Thetis dipping her baby son into the River Styx in an attempt to make him invulnerable, for it had been foretold that he was to die young. Although the greatest warrior-king, he was not quite perfection—the legends tell of Thesis' finger and thumb covering Achilles' heel from the protective power of the water. While not "spontaneous", his death was nevertheless inevitable owing to just this small imperfection. We might consider the solar corona's imperfection in the same light, its very invulnerability leading to its demise, it cannot escape heating once magnetism is a component of the physics of the sun-as-a-star. Unlike Homer's story, our Trojan war is still incomplete. The purpose of the current chapter is to find some unprotected weakness, exploiting not an arrow, but both our understanding of plasma physics and observational capabilities to find the Sun's own weaknesses leading us along that road to imperfection, and heating. So, out of the rich, chaotic and still-to-be-refuted ideas reviewed thus far, we embark on a brief odyssey. This will naturally entail subjective detective work. Subjectivity in a short book devoted to a long-unsolved problem is unavoidable. But it will surely be an interesting odyssey given the many challenges discussed in Chap. 3.

5.1 Power Spectra and MHD Turbulence

Does evidence exist to support or refute the proposition that ordered energy on large scales is a cause of heating in the corona, via a turbulent cascade towards dissipation? In the higher-β regions of the solar wind sampled in situ, analyses of particle velocity and magnetic field fluctuations form a backbone for current understanding of dynamics (Horbury et al., 2008; Wicks et al., 2010). However, quantitative remote studies of turbulence in the low$-\beta$ corona are rare, other than

P. Judge, J. A. Ionson, *The Problem of Coronal Heating*, Astrophysics and Space Science Library 470, https://doi.org/10.1007/978-3-031-46273-3_5

frequently encountering a passing reference to non-thermal line profiles in relation to a generalized concept of non-thermal motions.

Serendipity provides opportunities. In 2016, the Akatsuki Spacecraft passed behind the solar corona at heliocentric distances between 1.5 and 8 $R_\odot$, after failing to achieve an orbit around Venus. The scintillation of known radio signals from the spacecraft enabled Chiba et al. (2022) to identify inner dissipation scales from fits to measured scintillation power spectra. They could identify inner (dissipative) scales which increase with heliocentric distance, and measure changing amplitudes of compressive (magneto-acoustic) waves, assumed to be driven by non-linear effects of upward propagating Alfvén waves. The scintillation phase spectra could not be fit by a single power law, suggesting an excess power other than turbulence at lower frequencies. Beyond $6R_\odot$ the flux of compressive waves drops, indicating a transfer of wave power to heat in the absence of mode conversion. In spite of the large path lengths through plasmas along each line of sight, this impressive work clearly deserves further attention.

One of the few studies addressing the K.E. power spectra $E(k)$ remotely, as a function of physical scale ($\lambda = 1/k$) was made on transition region lines of C^{3+}, observed using the HRTS instrument and analyzed by Dere (1989). There are potential problems relating these observations of a transition region line to the corona, but for now we suspend disbelief and follow Dere's arguments. $E(k)$ is the same quantity as discussed in MHD turbulence. HRTS was a remarkable instrument, registering photographic spectra with a spectral resolution $\mathcal{R} \approx 30{,}000$ at each of over 900 positions along the focal plane slit, covering a solar radius with an angular resolution of about $1''$. By obtaining centroids and widths of line profiles, Dere explored both resolved power spectra $E(\hbar)$ from autocorrelations of spectral line shifts along the slit (wavenumbers $0.5 < \hbar < 8$ Mm^{-1}, Dere used the more usual definition of wavenumber, $\hbar = 2\pi k$), complemented with typical line widths which are used to constrain the integral of $E(\hbar)$ for all wavenumbers $\hbar > 8$ Mm^{-1}. At least for the C IV resonance lines (155 nm), he concluded that the power spectrum on wavenumbers > 8 Mm^{-1} was not as steep as turbulence theories predicted. On the large scales ($\hbar < 8$ Mm^{-1}) the spectral index was $-3/2$, but more power was needed to account for line broadening than available from a 3/2 spectrum. Dere concluded that some process other than a turbulent cascade *more directly* drives small scales required to heat the plasma (schematically shown by dot-dashed lines in the sketch in Fig. 3.1). This important result has received scant attention with just 21 citations before the beginning of 2023. Of these, just the papers by Doyle et al. (1997) and Marsch and Tu (1997) recognized the significance of Dere's work for heating and MHD processes, and briefly investigated resonant absorption as a possible mechanism, recognizing also the perpendicular nature of the resulting motions.

Dere studied both Joule and viscous dissipation in compressive and incompressive regimes, noting that the former, if assigned the full observed power, would exceed the ability of the plasma to radiate it. Thus, given the modest, subsonic Doppler shifts, at face value this would imply that whatever drives the small scales cannot be predominantly compressive.[1] But, as noted above, the results of analyzing transition region lines are not directly linked to turbulence in the corona. First we are obliged to understand what structures cause the bulk of the transition region emission in lines of C^{3+}- is it or is it not energetically and dynamically connected to coronal plasma (Sects. 1.5 and 5.12)? The transition region plasma is subject to influences from two large and variable reservoirs of energy above and below (Sect. 3.8). Thus, until this "transition region problem" is resolved, the meaning of these analyses are open to debate (Sect. 5.12).

Can the role of MHD turbulence be addressed with new, remote coronal observations? Technical issues and observational selection effects make similar work on coronal lines challenging. Ground-based observations of coronal lines require us to occult the entire solar disk, which itself is unfortunately precisely the target needed to obtain data with high enough spatial and spectral resolution to build a power spectrum in k, avoiding line-of-sight issues as much as possible. The vast bulk of lines from the corona which can be measured against the solar disk lie in the X-ray and EUV wavelength regions. Of these by far the strongest lie below $\sim$110 nm, where uncoated magnesium fluoride optics cannot be effectively used. Spectroscopy below 110 nm is possible with normal-incidence coated optics (the SUMER instrument on SoHO made measurements from between 49 and 155 nm, Wilhelm et al., 1995), and also grazing-incidence optics which intrinsically have a lower resolution and owing to lower photon fluxes also require large optical components than typical normal-incidence spectrographs, like HRTS. Table 5.1 lists some parameters of instruments flown and under consideration. The SUMER spectrometer could measure a few lines including Ne^{7+} (77 and 78 nm) in second order, with a spectral resolution of 38,000, but its reflectivity was reduced compared with first order transition region lines. More recent EUV instruments have lower spectral and angular resolutions, see Table 5.1. This table reveals that *no instrument past, present or future will readily enable measurements of the power spectra of coronal emission lines sufficient to try to refute different models producing different spectra.* In examining the important proposition that turbulence is critical to energy transport, and assuming that the current understanding is represented by the review of Schekochihin (2022), then we must seek angular *asymmetries* in spectral line profiles (largely in the line widths) which are robustly predicted to exist across large-scale magnetic fields (see Sect. 4.9.1). A prime target for such observations might be relatively quiescent active region loops as they traverse the solar disk (Figs. 2.2 and 1.11), or if they present a favorable aspect to the observatory so that both parallel and perpendicular widths can be measured unambiguously (Sects. 5.8 and 5.9).

Dere's approach seems worth pursuing using data from IRIS which has a significantly higher angular resolution. However, the value of these data will remain contingent upon properly understanding the nature of the transition region

[1] Dere erroneously used the non-dissipative part of the rate-of-strain tensor in assessing incompressible viscous heating. However there are bigger questions concerning the use of transition region lines for coronal implications, discussed in the text.

Table 5.1 Spectrometers capable of measuring coronal emission lines against the solar disk

Instrument	Platform	Range λ nm	$\overline{\mathcal{R}}$	Resolution arcsec	FOV arcsec	Epoch
Past instruments						
HRTS[a]	Rockets, Spacelab	117.5–170	30,000	1	1×900	1976–1992
SUMER[b]	SoHO	46.9–161*	29,000	2	1×300	1996–2014
CDS[c]-NIS	SoHO	30.8–63.3†	1290	≈ 6	2 or 4×240	1996–
CDS-GIS	SoHO	15.1–78.5†	920	≈ 6	$2 \times 2, 4 \times 4$	1996–
Present and future instruments						
EIS[d]	Hinode	17–21 and 25–29	3500	3–4	$\geq 1 \times 512$	2006–
IRIS[e]	SMEX	133–136, 139–141 278–283	50,000	0.33	0.33×175	2013–
SPICE[f]	Solar Orbiter	70.4–79.0 97.3–104.9	1,040 1300	> 1.2	$> 2 \times 840$	2020–
MUSE[g]	MIDEX	10.4–11.4, 16.2–18.3 26–30.5	2000	0.4	170×170	2026?–
EUVST[h]	SOLAR-C	17–21.5	5000	0.4–0.8	280×280	2025?–

* Combination of first and second order grating reflections
† Range has some spectral gaps in coverage
[a] Bartoe and Brueckner (1975)
[b] Wilhelm et al. (1995)
[c] Harrison et al. (1996)
[d] Culhane et al. (2007)
[e] de Pontieu et al. (2014)
[f] SPICE Consortium et al. (2020); Fludra et al. (2021)
[g] de Pontieu et al. (2020)
[h] Shimizu et al. (2020), short wavelength channel

(Sect. 5.12) and the non-local influences of heat transport in the corona (Sect. 3.8), respectively.

The solution to this problem probably resides in developing new instrumentation to observe coronal spectra with $\mathcal{R} \geq 20{,}000$, along with $0.1''$ resolution EUV filter images over relatively narrow, carefully selected wavelength ranges.

5.2 Scales of Dissipation and the Narrowest Coronal Structures

Eventually, one might hope that observations will reveal an "elementary scale" below which the cross-field thermal structure of the corona can be considered uniform, or "resolved". If and when this is found, it will reveal the scale at which dissipation or thermalization of ordered free energy occurs. The lack of evidence for high energy electrons outside of flares (Sect. 4.5) might already suggest that a process on scales much larger than kinetic scales ($\approx$10 m) might thermalize particle populations before kinetic processes can occur. Other hints come from the imaging capabilities of the HiC instrument and analysis of HiC data (Rachmeler et al., 2019), and similar work from the EUI instrument (Chitta et al., 2022, see Sect. 3.3.3), suggesting that scales of 200 km or so may be the smallest observable elementary structures in the corona.

These analyses plus estimates of dissipation scales given in Sect. 4.9.1 prompt us to ask: are the smallest scales set by field-aligned compressible viscosity? Or must we deal with a possible plasmoid-truncated cascade (Appendix C) and then with kinetic processes on scales of meters? Depending on the diagonal components of the rate-of-strain viscous tensor (Appendix D), we argued that scales of 100 km might set the smallest structure scales, or it could be determined by the ion inertial lengths which are four orders of magnitude smaller. The diffraction limit of the 4 m DKIST operating at a wavelength λ nm corresponds to $\sim$ 18 $\left(\frac{\lambda}{500}\right)$ km on the solar disk. But DKIST will observe the corona only above the limb and preferentially at longer, infrared wavelengths. If coronal observations with DKIST operate successfully near this limit, useful information on viscous heating should nevertheless be expected from coronal imaging alone. At 63 nm (the EUV wavelengths of strong lines of Mg^{9+} and O^{4+}), a 0.25 m aperture telescope in Earth orbit would achieve the same angular resolution as DKIST observing at 1 μm. Normal incidence telescopes such as AIA on SDO, coated with interference coatings, have routinely observed wavelengths at wavelengths of 9–30 nm with modest apertures. But they have always observed using pixels far larger than diffraction limits. The coronal heating problem would benefit from diffraction-limited EUV observations (see Sect. 6.3.2).

These estimates suggest that measurements of EUV coronal emission should be attempted at angular resolutions of order 50 km. The fact that this has never been attempted attests in part to the conservative attitudes of review panels responsible for selecting missions, whether through suborbital rocket flights of a few minutes, or satellite missions. But it does not recognize the salient points regarding coronal heating discussed above. Consider the number of counts per second acquired at the focal plane of an imaging telescope of aperture D which performs ideally, producing simple Airy disks of each point in the sky with angular diameter $\theta = 1.22\lambda/D$. With a system focal length of fD with f the usual focal ratio, each disk will subtend a circle of diameter $d = 1.22\lambda f$ meters at the detector. For example, for a compact Gregorian telescope with $f = 30$, $D = 0.25$ m operating at $\lambda = 63$ nm, $\theta = 3\ 10^{-7}$ radians, $\equiv 0.063''$, and $d = 2.3\,\mu$m. Such small pixels are not beyond current technological limits. At the distance of the Sun (150,000 km) the Airy disk diameter corresponds to 46 km at the Sun's surface.

What other reasons might explain why solar EUV images been acquired so far from the diffraction limit? A desire for a large field-of-view is one factor. To approach this limit, there are certainly new technical challenges, but perhaps the reason is that there is a (false) belief that nothing will be seen by spreading light over so many pixels. The fallacy of this argument is simply that it implicitly assumes that all structure has been resolved in existing data. But these optical parameters are not (necessarily) good reasons to reject a proposal. In the minds of some, the Sun's corona is not considered bright enough to acquire sufficient counts above noise levels, and the instrument is rejected on the grounds that nothing might be observed. But there is a simple fact, when sampling intensity at the critically-sampled diffraction limit of any imaging telescope (i.e. using pixels of size $d/2 \times d/2$ m^2 on the detector), the number of counts per pixel per unit time acquired on an ideal detector is independent of aperture. Also, there is much to be learned from pixels with zero counts, providing there are at least some pixels with detected photons.

So as a familiar example, let us assume that HiC, with a 200 km spatial resolution, has resolved all thermal structure in the corona (Rachmeler et al., 2019). A new coronal instrument is constructed with, say, a resolution of 50 km (25×25 km^2 pixels). Two outcomes are possible. In the first, the field-of-view appears structureless on, say, a 200 km scale. The images are either noise, or signal and noise, and all pixels are consistent with an equal brightness. But in the second, the same 200 km-scale area has a distribution of brightnesses, the sum of which is the same as in the first case (conservation of radiant energy). Some pixels are below average, many may well be consistent with zero photons, but others are non-zero and reveal some structure. When many pixels have counts consistent with zero, others must be brighter to conserve the average. The point is simple, *the absence of a signal*[2] *is every bit as crucial as the presence, and every pixel is telling us something about the elementary structures in the heated corona.* So, no matter which of the two outcomes is found, this would advance our understanding significantly, without all the contingencies which are too-often present in work to date. In confirming the hypothesis, we would reject models of coronal heating involving plasmoid formation, magnetic reconnection and kinetic physics, and it would deny many empirical proposals concerning myriads of unresolved threads. In refuting the hypothesis, we would then be obliged to pursue these concepts further.

A technical study to build and fly such an instrument, at least on a sub-orbital mission, seems justified.

5.3 "Fuzziness" of Hot Plasma Loops

A familiar observation of hot plasma ($T_e \geq 2\ 10^6$ K) in the solar corona has long resisted explanation: hotter loop images appear "fuzzier" than their cooler counterparts in images of active regions (Bray et al., 1991; Tripathi et al., 2009).

Data acquired with electronic detectors during eclipse reveal clearly the two types of structure (Habbal et al., 2011). Two images from Figure 6 of their work are

[2] As signified by counts which are below the noise level.

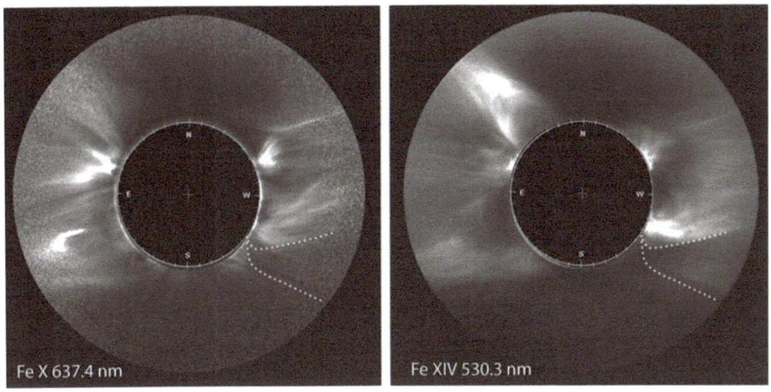

Fig. 5.1 The solar corona as seen in background-subtracted images of the red and green coronal lines. From Habbal et al. (2011)

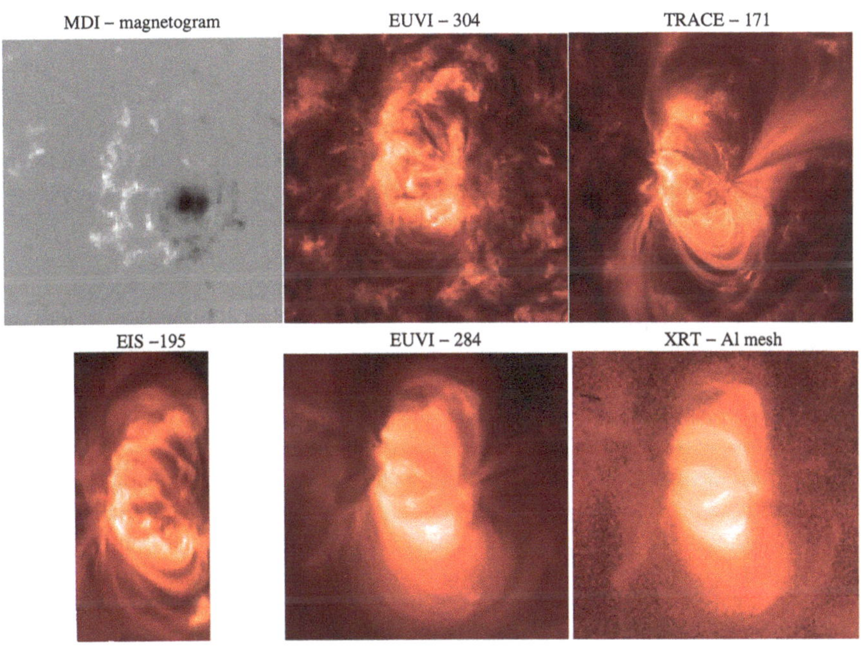

Fig. 5.2 Active region images taken on 2007 July 3 using five space-borne instruments (adapted from Noglik et al., 2009). The corona gets more diffuse in appearance with increasing electron temperature. These images of NOAA 108561 have a N-S extent of roughly 150"

reproduced in Fig. 5.1. The hotter spectral lines of Fe^{13+} (and Ca^{14+}, not shown) observed from the ground happen to be at shorter wavelengths, where atmospheric seeing makes images more diffuse. But seeing cannot be the explanation, because images obtained from space at X-ray and EUV wavelengths show the same effect (Fig. 5.2). Therefore fuzziness is associated with the coronal plasmas.

Given the frozen-in plasma condition, and the small gyroradii of electrons and protons in Table 1.2, this is an intriguing observation for the heating problem. The gyroradii of iron ions vary as mass/charge ratio, thus the largest gyroradius encountered by the prominent ions contributing most to radiative emission would be around seven times that of protons, or about 7 m. Clearly the plasma must be frozen to the magnetic field on all observable scales (>200 km). So is the magnetic field itself suffering from a kind of dynamical diffusion in space? This is precisely the idea of Lazarian and Vishniac (1999) already encountered. But can this picture account for the observed fuzziness which seems to demand a scale of several seconds of arc, i.e. a few Mm? How might we refute such an interpretation?

First reconsider the results of Malanushenko et al. (2022), based upon synthetic images from coronal structure calculated using the MuRAM code (Rempel, 2017). The eye is always attracted to coherent structures, which mostly have a loop-like morphology. But when viewed in the direction along these structures as in Fig. 3.4 we see a more turbulent morphology, consistent with the ideas that the heating of the corona occurs mostly in these planes, qualitatively similar to the $B^\flat$ conjecture, and the ideas behind reduced MHD and MHD turbulence as applied to coronal loops (Appendix C). Yet ironically these structures, potentially central to the heating problem, do not attract our attention as much as the loops! The latter are usually close to a potential state, merely setting the Alfvén speed along the loops, with (perhaps) little more influence on heating mechanisms. The cross-field intensity images of Fig. 3.4 are superficially reminiscent of the turbulence observed so clearly in cool prominence plasmas (Berger et al., 2008).

The deterministic MuRAM calculations produce images which are not at all fuzzier at higher temperatures. This is just as expected, based upon the validity of frozen-plasma assumption at the grid scales accessible to the calculations, and the physical processes included in the computer code. Whatever is responsible for this fuzziness is clearly not present in MuRAM or other modern MHD codes.

Therefore we must appeal to a different, perhaps more interesting, explanation. Other explanations include ad-hoc spatial distributions of multi-threads (Guarrasi et al., 2010), of inhomogeneous cross-field density profiles (Goddard et al., 2017), and numerical MHD studies of the instability and dynamics of multi-threads, presuming a non-potential initial configuration (Tam et al., 2015). Again we remind the reader of the potential contradiction between the multi-thread proposition and the claim that observations have perhaps resolved the corona at far larger scales of 200 km (Sect. 3.3.3).

In global MHD numerical experiments of coronal loops (Gudiksen and Nordlund, 2005; Rempel, 2014; Carlsson et al., 2016; Rempel, 2017; Einaudi et al., 2021; Przybylski et al., 2022; Breu et al., 2022), viscous and current dissipation are determined numerically, and the ARLToR conjecture is enforced by using tricks such as hyper-viscosity, solutions are made stable, and can mimic higher magnetic Reynolds numbers (see page 127). However, physical viscosity is strongly temperature-dependent, so assessing the meaning of a comparison of such models with coronal images as a function of electron, or more challenging, ion temperature becomes difficult. Nevertheless, cross-field viscous diffusion is small, and these models have no physical mechanism for diffusing particles or magnetic fields across their grids at coronal temperatures. Such processes are entirely determined by the numerical grids used. Thus they have not (and should not) be applied to this problem.

It seems therefore that the stochastic model of Lazarian and Vishniac (1999) might be invoked as being the only picture that has not yet been refuted by these remarkable observations. But can the model be quantitatively defined? Might it be consistent with the observed multi-Mm dispersion of magnetic lines of force seemingly required by the fuzziness? Only future research can answer this question. The point here is that here is potentially an important macroscopic clue to mechanisms of coronal heating that should be exploited.

5.4 Is Coronal Thermal Structure Unreasonably Well-Organized?

This question has rarely, if ever been seriously raised (Judge, 2018). The focus of the literature has been on inhomogeneous thermal fine structure related to solar magnetism, emphasizing the "multiple 1D atmospheres" picture of Rosner et al. (1978), as well as variations in time, particularly as seen in transition region lines. The non-symmetric corona was regularly observed during eclipses, along with what came to be known as "spicules", in the late nineteenth century by Jesuit priest Angelo Secchi. Fine structure in the chromosphere was discovered by Hale and Deslandres shortly after the contemporaneous invention of the spectro-heliograph in the late 1890s. These inhomogeneities have drawn the vast bulk of the attention of researchers ever since (e.g. Kiepenheuer, 1953; Nikolsky, 1970; Athay, 1974, 1986; Grossmann-Doerth and Schmidt, 1992; Koza, 2014; Henriques et al., 2020; de Pontieu et al., 2021). The ability to trace the morphology of magnetic flux bundles via fine structures has formed a basis for subsequent analyses of the physics of the chromosphere and corona. Broadband observations of polarized coronal light during eclipse and/or using coronagraphs have been made beginning as early as 1868. Later such work revealed the morphology of magnetic structures in the plane of the sky, integrated along each line of sight around the solar limb (e.g. Van de Hulst, 1953). From the 1960s, theory evolved (Charvin, 1965; House, 1972, 1974; Sahal-Bréchot, 1977) to describe the imprint of coronal magnetic fields on linearly polarized coronal forbidden lines. Instruments were developed and again, the morphology of large magnetic loops revealed itself with the aid of Occam's razor to resolve a 90° ambiguity, in the direction of polarization vectors (Querfeld, 1977, 1982; Arnaud, 1982; Querfeld and Smartt, 1984; Arnaud and Newkirk, 1987). Coronal polarimeters with higher angular resolution and cadence have revealed morphology on smaller scales ($\geq 5''$) through the same techniques (Tomczyk et al., 2008), as well as through following wave packets on the sky (Tomczyk et al., 2007). Thus, the majority of solar studies today are devoted to understanding such detailed small-scale structures and morphology of coronal plasmas. In terms of coronal heating, morphological studies reveal only *large-scale* structures far above scales of dissipation (see Fig. 1.5). Ironically, these morphological studies of coronal magnetic fields may have drawn attention away from some remarkable facts concerning the plasma itself, with potentially deep implications for coronal heating. When viewed against this body of literature, the corona seems to be especially *ordered* on large scales, beyond the simple identification of coronal holes, quiet and active corona.

There are two relevant aspects to this viewpoint, which should not be confused: why are only some hydromagnetic structures in the active corona filled with bright plasma? This is discussed in Sect. 5.6. Then, how, within such structures, does the corona somehow know how to be roughly the same brightness and temperature? It is the latter concerns us here. One aspect of this point was emphasized by Parker (1988) at the beginning of his nanoflare paper:

> ...the surface brightness of the active X-ray corona is essentially independent of the dimensions of the confining bipole.

Another aspect of order at large scales was noted earlier, when ground-based coronagraphic observations revealed that images in the red and green lines (Fe^{9+} and Fe^{13+}) showed structures which complemented one another (Fort and Martres, 1974, see also Billings, 1965):

> ...contrary to what might have been expected if the associated fine structures were randomly distributed on the surface of the Sun, there is little superposition between red and green line emission.

The authors anticipated the separation between large-scale structures observed far later with EUV images. An example is Fig. 1.3, illustrating more clearly how the corona appears to be rather well organized in EUV light. The red-green-blue colors reflect decreasing electron temperatures, because electrons collide with ions of Fe^{14+}, Fe^{11+} and Fe^{8+} and Fe^{9+}, to generate the observed radiation depicted in red, green and blue respectively. The figure shows examples of the three types of structure: the dark coronal holes, the cool quiet regions and the hotter active regions.

The differences between the three types of large-scale structures have well-known heuristic explanations. Firstly, coronal holes (CHs) are cool, dim coronal plasmas lying above regions of accumulation of mostly unipolar magnetic flux. Figure 1.3 shows coronal holes as dark patches at 12 and 4 o'clock close to the limb on the Sun's disk. Quiet regions of the corona (QRs) are revealed as blue emission (dominated by emission from Fe^{8+} and Fe^{9+} ions) seen across most of the solar disk. Both CH and QR regions are punctuated with small (mostly white) "coronal bright points", some of the smallest active X-ray structures pointed to by Parker (1988). The macroscopic difference between CHs and QRs is usually interpreted in terms of the locally open vs. closed magnetic configurations. In CHs the magnetic fields open to the heliosphere and the plasma expands into interplanetary space. QRs are assumed to be closed "loops" although this closed morphology is relatively difficult to glean from the characteristic diffuse EUV images of quiet coronal plasmas. For a given rate of heating, CH plasma must do extra work against gravity and to accelerate the wind, energy that is otherwise available to increase temperature and ionization in quiet regions.

Secondly, active regions (ARs) emerge from the convection zone under otherwise quiet regions, as a result of a large-scale global dynamo (e.g. Charbonneau, 2020). The corona overlying ARs is powered by magnetic flux fighting for dominance with pre-existing quiet magnetic fields, perhaps also pre-loaded with magnetic free energy generated by helical and other motions deep below by the dynamo. Active region magnetic fields are far stronger than in either QS or CH areas. The net Poynting fluxes entering the corona must almost certainly be larger than in QR or CH areas, owing to the larger estimates of their energy losses (Table 1.3). For a given plasma mass density, the Alfvén speeds are also higher. All other things being equal, this difference in Alfvén speed alone might explain much of the increased Poynting flux in active regions. Observationally, the higher states of ionization of iron ions indicate higher electron temperatures (Figs. 1.3 and 3.5).

Here we point to a more subtle property of these three types of corona with potentially important consequences. Not only do they have individual energy requirements and typical temperatures, but the colored images show that *each has its own narrow distribution of electron temperatures, each centered at temperatures quite different from each another*. Each region lives a long time (months or longer) and each occupies its own, slowly changing, vast area, with linear scales $\gtrsim 0.05 - 1R_\odot$. What then organizes the temperatures to make images like Fig. 1.3 colorfully organized on large scales, and not just a random distribution of colors and brightnesses on small scales, a pattern that might appear just shades of gray at a lower resolution? Is this telling us something important about coronal heating?

A partial answer has already been given in Eq. 3.7 in Sect. 3.3.3. For a given loop length L, $d\log_{10} T = \frac{2}{7}\, d\log_{10} \mathcal{F}$. A distribution of $d\log_{10} \mathcal{F}$ entering an ensemble of loops (resolved or unresolved) of length L therefore translates into a narrower distribution of $\log_{10} T$ because of electron heat conduction, which acts as a thermostat. Landi and Feldman (2008) have introduced the notion that the coronal temperature is discretized:

> [we] look at the Sun as a superposition of structures characterized by different values of temperatures that are fixed like eigenvalues: coronal hole ($\log_{10}T \approx 5.90$, [...] quiet Sun $\log_{10}T \approx 6.15$, [...] active regions $\log_{10}T \approx 6.5$).

Their analysis is based upon spectroscopy of many emission lines between 49 and 155 nm observed with the SUMER instrument from above the solar limb. The spectra were shown to be consistent with narrow temperature distributions centered at the above three, and only three, temperatures. They also emphasized (their table 6) the remarkably similar coronal temperatures of 4 different active regions. Their results do not conflict with the statements of Sect. 3.3.3 concerning the accuracy of temperatures derived from spectra, because the important aspect of their results here is that the measurements are sufficiently *precise*, if not accurate, to separate these temperature distributions. Although perhaps not universally accepted, the idea

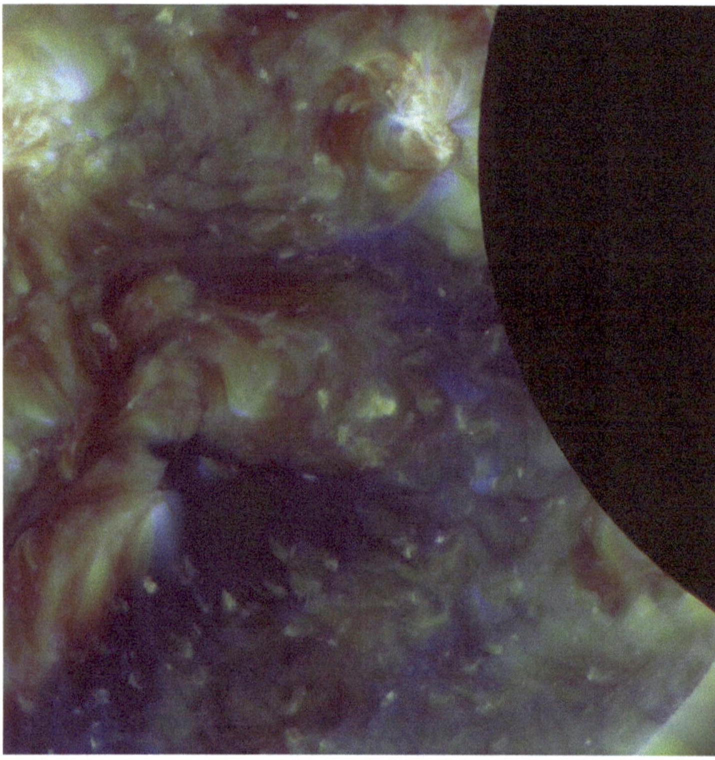

Fig. 5.3 A three-color image obtained during an eclipse from the AIA instrument on SDO, illustrating the relatively uniform brightness of quiet Sun (blue, green) and coronal hole areas (dark). Images obtained on 2012 February 21 at 17.1 nm (blue), 19.5 nm (green) and 21.1 nm (red). Several active regions are present on the solar disk (white and a mix of colors)

has not been refuted, and further supporting observational evidence can be found in Landi and Klimchuk (2010).

However, the distribution of *brightnesses* within the ensemble of course is given by the distribution of $\log_{10} \mathcal{F}/L$. So we ask the different question, based in part on the appearance of images such as Figs. 1.3 and 5.3:

> How wide are the distributions in brightness of QS, CH and even some AR plasmas which appear almost uniformly bright?

While the data of Fig. 1.3 have a 5" resolution, the conclusions survive when the corona is observed also at 1" resolution (Fig. 5.3). To address this question over large areas with adequate statistical samples, Fig. 5.4 shows data from a period when significant numbers of active regions, quiet Sun and coronal hole plasmas were present on the solar disk. Remarkably, the quiet and coronal hole plasmas possess narrow distributions with full widths at half peak of 0.5 in the logarithm of brightness, showing very little overlap and separated by 0.8. The active coronal

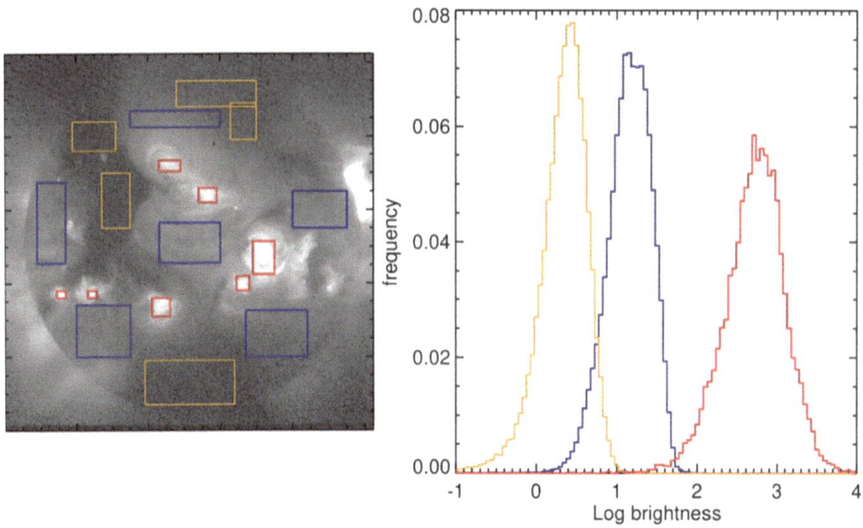

Fig. 5.4 The left panel shows a level-1 processed image from March 29 2015 obtained at 06:29:04 UT, with the XRT telescope on the Hinode satellite. The image is in soft X-rays using the thin beryllium filter. Superposed are boxes in blue, orange and red (quiet Sun, coronal hole and active region plasmas respectively). The right panel shows normalized histograms of the frequency of the logarithm of the brightness, for each feature, taken from well exposed images obtained at 06:04:25 and 06:04:35 UT

brightness is from a less homogeneous set of areas, but nevertheless is 1.6 in the logarithm brighter still. This distribution is naturally broader, and not well defined because of the crude rectangular selection of small areas. But the differences between quiet and coronal hole areas is robust, as are the narrow distributions of these areas: 67% of each plasma has a brightness within a factor of 1.8 of the median value. There is no prior reason that the energy deposited across each region should lead to plasma brightnesses so uniform in X-rays.

In fact, given the turbulent nature of magnetic fields in the photosphere in both quiet and coronal hole regions, each region of the corona will be threaded by intermittent magnetic fields, at least in the chromosphere, which expands to fill the bulk of the coronal volume above. One might therefore expect that, as QR and CH magnetic fields driven by granular motions vary across the solar photosphere, a variety of 1D atmospheres (in the RTV sense) will be realized with a broad variety of brightnesses, if not temperatures. But this is evidently not seen (Fig. 5.4). If further evidence is necessary, Fig. 5.5 shows the average of images acquired with HiC of an active region at 0.3" resolution (Rachmeler et al., 2019). The right panel of the figure shows the normalized distribution of brightnesses measured in the two boxes shown on the image, chosen to avoid serious line-of-sight additions to intensity close to the active region's magnetically strong footpoints. Even at the resolution of close to 200 km, these data confirm the modest changes in brightness between different regions of the active corona. The standard deviations are about 1/2 of the median

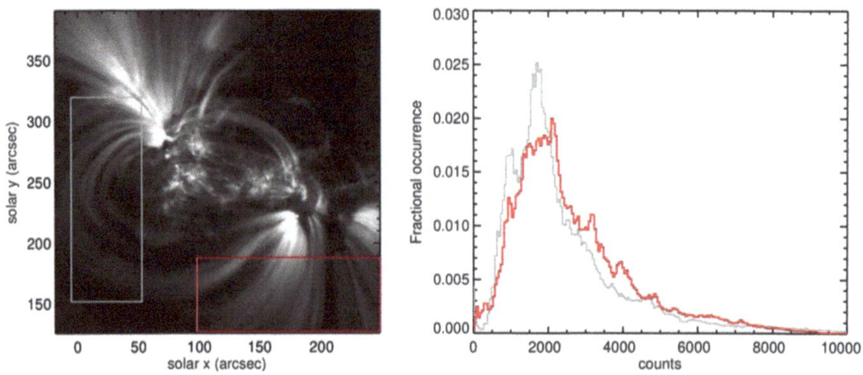

Fig. 5.5 An average image of the corona is shown from HiC 2.1 (Rachmeler et al., 2019). The two boxes are areas mostly away from line-of-sight addition of the coronal emission near the active region footpoints. The plot on the right shows histograms of the two areas encompassed by boxes of the same color in the image. The brightness of the areas of these loop systems has a mean of 3800 and a standard deviation of 2000 counts, but a better measure of the spread of intensities of the "loops" is obtained on inspection of the line plot- these structures have a median of around 2000 and a spread of ± 1000 counts, at most

brightnesses. This is even more remarkable when we consider that the corona's energy requirements are about 20 times smaller than those of the chromosphere, begging the question, what processes let precisely $5 \pm 2.5\%$ of the chromospheric Poynting flux emerge and dissipate in the corona? It is as if the flow via a waterfall out of a lake remains always proportional to the flow into the lake, even if the lake naturally fills non-uniformly as sources and sinks of water change with water depth.

Although we have used rule-of-thumb RTV scalings, we can reasonably conclude that *the uniform brightness of the quiet corona is wanting of an explanation.* While active regions are all different, the work of Landi and Feldman (2008) does suggest that the bulk of the coronal plasma in active regions *when observed far above footpoints* (between 34 and 49 Mm above the limb) prefers temperatures near $\log_{10} T \approx 6.5$ K. At the highest angular resolution, HiC data of an active region at 17.1 nm indicate a relatively uniform brightness (Fig. 5.5), with an rms scatter of about 1/2 of the median value. Thus, the tentative implication is that *the Poynting fluxes must be relatively uniform within QR and AR areas, while being quite different between them.* These observational results should be explored further to understand why the inferred Poynting fluxes into the corona appear so uniform, at the current angular resolutions of 200 km over vast areas of the surface.

5.5 Coronal Oscillations

Oscillations of coronal plasmas have found application primarily as indicators of Alfvén speeds in loops which, combined with a variety of estimates of plasma data, lead to estimates of the coronal magnetic field strength $|\mathbf{B}|$. But other applications, perhaps of more interest to the heating problem, can be conceived. Interesting MHD transverse wave interactions in numerical experiments of mildly inhomogeneous plasmas were first investigated by Pascoe et al. (2011). Their results revealed that "simple" modes of oscillation (e.g. kink modes) become well mixed with Alfvén modes owing to the rather small variations in densities (factor 2 variations), so that one might not expect to observe such clear signals revealed by Tomczyk et al. (2007). So questions again arise. What does the existence of a propagating wave imply concerning small-scales in coronal loops? Under what conditions are waves observed by Tomczyk et al. (2007) and Morton et al. (2015) so clear? How might such observations be related to the turbulent and wave driven pictures of coronal heating? The oscillation velocity amplitudes are far smaller than line widths, and questions arise as to whether this is a line-of-sight averaging effect over very different bundles of magnetic flux, or if it is intrinsically a local effect in the plasmas.

All of these questions may be addressed using the coronagraphic, infrared capability of DKIST, which has the additional capability of measuring properties of the plasma's magnetic field through polarimetry, even on unresolved scales that might be associated with plasmoid formation (French et al., 2019). The VELC coronagraph instrument on the Aditya L1 spacecraft may also provide novel information, observing the corona above the occulted solar disk (Singh et al., 2019). Although such work initially will test no definite hypothesis, the methodology falls under the umbrella of "empirical studies of observable phenomena" of Chap. 4 (page 129).

5.6 Why Are Only Some Magnetic Loops Loaded with Dense, Bright Plasma?

This question has concerned researchers over many years (e.g. Rosner, 1990; Litwin and Rosner, 1993; Falconer et al., 1997; Chitta et al., 2017; Tiwari et al., 2017). It concerns the thermal structure of active regions for which imaged loops are readily separated from one another (a particularly clear example is shown in Fig. 2.2), in spite of the line-of-sight confusion which is so clear in numerical models (Malanushenko et al., 2022). The coronal magnetic field effectively pervades all neighboring space. Yet only selected subsets of the volume-filling magnetic system are loaded with bright plasma. For example, a potential field extrapolation of the photospheric sources in Fig. 2.2 would not single out any flux bundles for hosting bright observed coronal structures, the volume would be filled with a smooth magnetic field. One flux bundle has (without specifying the Poynting flux) almost the same macroscopic magnetic properties as another. More generally, large-scale, steady current sheets may hold some of the answer, although their potential instability in their thinnest form questions whether they can exist (page 102). But if such sheets are so unstable, we would then be left seeking other sources of free energy for coronal dynamics, such as in flares and CMEs, all the energy being otherwise dissipated on small scales in narrow sheets (pages 54 and 82).

Let us continue by first assuming continuous magnetic fields. In this case it seems that, again, the guiding fields $\mathbf{B}_0$ have little to do directly with coronal heating per se. Instead $\mathbf{B}_0$ sets the topology and Alfvén speed for the propagation of Poynting flux through the structure. In focusing on the macroscopic magnetic field $\mathbf{B}_0$, one can be distracted from properties more closely associated with coronal heating, such as unresolved plasma motions. All the scaling relations discussed earlier (Sect. 4.3, Fisher et al., 1998; Mandrini et al., 2000; Aschwanden, 2001) merely beg the central questions concerning the journey from large to dissipative scales (page 21). Scalings with $\mathbf{B}_0$ in these relations merely reflect the speed at which free energy can be generated by a given motion along flux bundles and transported into the corona.

Explanations for this "selectivity" based upon the best available observations abound, appealing naturally to the ingredients of coronal heating. Yet there is no agreement, owing perhaps to the ambiguities and challenges outlined in Chap. 3. Litwin and Rosner (1993) note that, with regard to active region loops,

> The relative rarity of coronal loops implies that, whatever the heating mechanism, it must have a relatively high threshold so that conditions for its onset are satisfied only in a small fraction of the available volume.

Litwin and Rosner (1993) then focus upon the implications of relatively "thick" observed loops (as reported in 0.7" resolution X-ray data), given far smaller scales of dissipation and cross-field kinetic transport coefficients.

In seeking signatures of the dynamics at the base of both bright and dark "loops" in active regions, several groups have correlated these coronal structures with various phenomena observed lower, in the photosphere. For example, Tiwari and colleagues argued that both the presence of convection, suppressed in sunspot umbrae, and flux "cancellation"[3] are needed to produce bright plasma loops rooted in apparently unipolar and multipolar photospheric foot-points, respectively. Otherwise bundles of magnetic flux are not loaded with plasma. Falconer and associates argued that surface velocity shear near polarity inversion lines are needed, and Chitta et al. (2017) suggested that bright coronal "loops" (bundles) are rooted at locations with minor small-scale but persistent opposite-polarity magnetic elements very close to a dominant polarity, based upon their very high angular resolution, balloon-based vector spectropolarimetry. But all these conclusions depend at least in part on the quality of photospheric polarimetry obtained, because of the difficulty of detectability of minority polarities with small fluxes (Wang, 2016; Barczynski et al., 2018). The gap between photospheric and coronal conditions as well as line-of-sight confusion may also affect such qualitative, associative statements.

Does the existence of such bright plasma bundles which are spatially separate from neighboring darker structures suggest that the small-scale magnetic components $\mathbf{b}$ are not solely responsible for heating, that $\mathbf{B}_0$ plays a major role? Not at all, heating requires not just a guide field but also a suitable velocity field to generate a Poynting flux (Eq. 2.26), and as yet we do not know the velocity

[3] This is not a physical process, but the disappearance of a magnetic beneath noise levels from a magnetogram image.

fields that can generate $\mathbf{b}_\perp$ at the top of the chromosphere. We do know that the photospheric fields advected and compressed by photospheric convection expand into the chromosphere, so that we should not be surprised if the base of coronal loops has macroscopic structure imparted by photospheric magnetic flux and development of velocity fields through the chromosphere (see Sects. 3.9 and 6.2). The onset of heating in active regions clearly has a macroscopic component which defines the "outer" (and perhaps smaller) scales represented in Fig. 3.1.

In fact, the mix of bright and dim bundles leads to even more interesting conclusions. In long-lived active regions, dim bundles have lower densities (observably), and higher magnetic fields (because of cross-field force balance), than the bright bundles. Dimmer bundles therefore have higher Alfvén speeds. Thus, here, all other factors being equal, is an *inverse* relationship between $\mathbf{B}_0$ and heating, in stark contrast to those many analyses based upon bright bundles alone (Fisher et al., 1998; Mandrini et al., 2000; Aschwanden, 2001, see page 24). On this basis it is tempting to conclude again that, in the absence of current sheets associated with $\mathbf{B}_0$,

> $\mathbf{B}_0$ plays a secondary role to $\mathbf{b}$, in heating the corona.

But without information on the Poynting flux at the coronal base, we cannot make firm conclusions, merely speculate. For example, the lower magnitude of $\mathbf{B}_0$ inside high pressure and bright bundles may make it more susceptible to the slow rotational, translational and compressive motions induced at the coronal base, leading to continuous current systems, internal braiding and viscous dissipation. The same problems arise in assessing if the basic difference between dark and bright bundles arises from current sheets living longer than dynamical times. Information-rich data are absent.

DKIST, offering an unprecedented angular resolution for spectropolarimety, as well as coronagraphic capabilities, will hopefully reveal constraints on Poynting fluxes even in the darkest interiors of sunspot umbrae. DKIST's high angular resolution will hopefully enable the above picture to be probed through measurements of the three components of $\mathbf{u}$ and $\mathbf{B}$ over active regions, giving us credible Poynting flux vectors $\mathbf{S}$ upwards into the corona, for the first time. Another path forward is outlined in Sect. 6.3.3. Steps along this line have already been taken Judge (2023), adding support to the idea that the problem of selectivity arises because coronal heating is part of a *self-regulating* driven system (Litwin and Rosner, 1993; Uzdensky, 2007; Einaudi et al., 2021).

Coupled with an EUV imager observing structures below say 50 km, the answer to the question central to this section should be forthcoming. However, it is going to be difficult using EUV resolutions similar to current capabilities, such as HiC (200 km, see Fig. 1.5).

5.7 Clues from Transient Changes

In a laboratory experiment, we might wish to set up a plasma heated by some mechanism, and then switch on or off the source of energy sustaining the plasma suddenly, in order to observe its response. The Sun offers such sudden changes macroscopically in the form of flares and other rapid events resulting from large-scale topological changes driven by rapid reconnection. The time scales for decay and recovery of solar plasmas can be estimated observationally, and then related to the underlying cooling and heating processes.

The causes of the sudden onset of flare emission lie beyond the scope of this book, even though such behavior should inform us eventually about coronal heating. As we have already noted, flares occur in spite of the continued loss of magnetic free energy leading to steady heating in active regions. In our naïve conceptual separation (Eq. 2.25), the free energy on scales of $\mathbf{B}_0$ builds independently of the smaller scale variations $\mathbf{b}$ responsible for "'steady" heating. Nevertheless, sudden changes in the thermal state of the active corona have been observed since the 1970s, as dimmings over significant fractions of active regions. These are known as "coronal dimmings" (Webb et al., 2000), or "transient coronal holes" (Rust, 1983), which are:

> short-lived ($\leq$ 2 days) regions of dimmed X-ray intensity sometimes observed in association with coronal mass ejections. These features, first discovered from SKYLAB observations, can occur in magnetic unipolar regions on either side of the X-ray post-eruptive arcades (Kahler and Hudson, 2001).

They arise as closed field regions expand rapidly or become open to interplanetary space, after a large-scale reconfiguration. The $\approx$40% typical drop in intensities occurs over tens of minutes (Dissauer et al., 2018), but can occur within a few minutes (a striking example is shown in Figures 3 of Zhang et al., 2017). Observations have confirmed that the drops in intensity are faster than conductive and radiative cooling times, pointing to a dynamical expansion and density drop, as the primary cause (Hudson et al., 1996), sometimes accompanied by measurable drops in relative temperature (Attrill et al., 2010). Times for recovery to a pre-dimming state are typically half a day, sometimes longer. A typical case is shown in Fig. 5.6. The case reported by Rust (1983) had a recovery time of about 10 hours. Attrill et al. (2010) discussed recovery times and reviewed mechanisms for the recovery in terms of phenomena at observable scales, such as the re-filling of large coronal loops with hot plasma along the line-of-sight, the contraction of the boundaries, re-linking of post-CME magnetic lines of force to produce new closed magnetic structures, and the re dispersal of magnetic fields via diffusion across the surface enhanced by "interchange reconnection". The latter process describes the jumping of magnetic lines of force embedded in the corona across the solar surface, to produce pre-CME-like conditions. These macroscopic pictures are necessarily qualitative in the absence of information-rich magnetic data. Again, we lack the data with which to assess suggestions for the evolution of magnetized coronal plasmas.

The important point is that the dense photosphere, which is believed to initiate the sequence of events leading to steady coronal heating, can hardly be affected

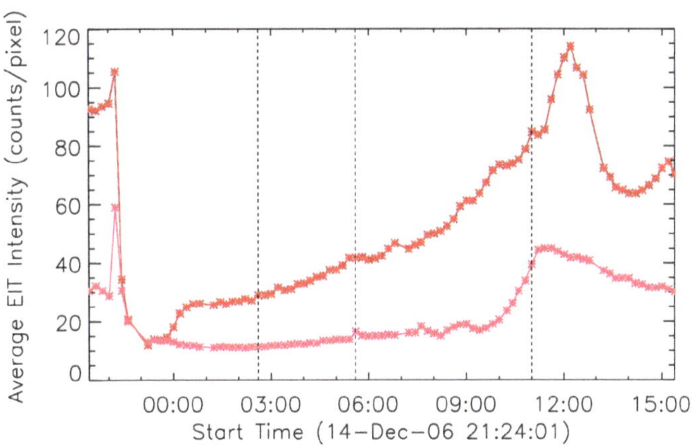

Fig. 5.6 The evolution of brightnesses of two areas undergoing a coronal dimming are shown (Attrill et al., 2010). While the drops in brightness are larger than typical cases (~40%), the evolution is representative of other dimmings. After a sudden drop in intensity during the dynamic release of a CME, the coronal emission slowly recovers only after 10 hours

by a CME launch (the thermal radiative relaxation time of the photosphere is mere seconds). Even massive CMEs ($m \approx 2\ 10^{13}$ kg), assumed to come from an area of a transient coronal hole ($\sim 30 \times 30\,\mathrm{Mm}^2$, perhaps an under-estimate) have a mass per unit area on average over >1000 times smaller than the photosphere. Magnetoconvection, delivering and radiating energy on time scales of seconds, continues throughout these large-scale events which have dynamical times of minutes to hours. Thus the recovery of *transient coronal holes must concern the physics of the middle chromosphere on up*. Locally, time scales of the chromosphere are minutes, less at the top of the chromosphere (see Table 1.2). This implies that the recovery must involve non-local effects, either from across the solar surface or from above (the photosphere being unaffected by the CME). Either way, novel observations of such processes are needed to determine the recovery of the Poynting flux and perhaps $\mathbf{B}_0$, and the observations suggested for the dark and bright loop bundles will surely also shed light on the chromospheric recovery of the Poynting flux over transient coronal holes. This is yet another demonstration that in order to understand coronal heating, we must first understand the magnetic fields emerging from the chromosphere. This issue will be the subject of Sect. 6.2.

A quite different coronal phenomenon is "coronal rain" (Schrijver, 2001; Müller et al., 2005; Antolin et al., 2015; Antolin, 2020), a subject receiving much recent attention with a long history particularly in observations from the Hα line (reviewed by, for example, by Bray et al., 1991). Two phenomena are observed, one occurring within non-flaring active regions and another after flaring. The latter is beyond our current discussion. Active region coronal rain has been increasingly visible as higher quality data (temporal, angular resolutions) have been obtained, seen above the solar limb as emission in cool plasma draining downwards, and on the disk as cooler plasma which absorps/scatters light at EUV and other wavelengths. Using data from Antolin et al. (2015), the raining plasma drops about 5 10^6 kg sec^{-1} across an active region area of 10^3 Mm^{-2} in amorphous blobs of plasma which appear to have densities between 10^{16} and 10^{17} m^{-3} at velocities of around 10^2 km s^{-1}. The geomteric widths are close to observational limits of a few hundred km, lengths somewhat larger. With a lifetime of about 3 minutes, we find that at any given time they will occupy about 3% of the volume of the active region corona. The volume occupied by steadily heated loops is difficult to estimate because of projection effects, but if data from Fig. 3.4 are representative then crudely perhaps 20–100% of the corona is filled with plasma. Coronal rain would therefore occur between 3 and 15% of the time.

Its meaning in terms of coronal heating is a currently a subject of research, because its origins can be diverse. Considerable work on numerical hydrodynamic and MHD experiments has revealed yet another sobering result. Enhanced heating can, counter-intuitively, translate into runaway cooling and rain. Heating that is concentrated closer to the solar surface naturally drives thermal instabilities as plasma densities increase higher up (Antiochos and Klimchuk, 1991, radiation losses varying as density squared, so positive excursions in density lead to instability, if average heating varies roughly at a lower power of density). Such instability was anticipated by Parker (1953) and Field (1965). Positive (and of course negative) changes in heating lead to local plasma cooling and downward flow in multi-dimensional MHD calculations (Fang et al., 2013, 2015).

Thus the precise meaning of this non-linear dynamical sequence of events in the context of coronal heating does not yet have a unique interpretation. Nevertheless if must be part of the toolkit needed to solve an important part of the problem. It is another example of a self-regulating system (e.g. Einaudi et al., 2021). We might speculate that the small fraction of the corona undergoing raining at any time is compatible with the idea of a mostly steady mechanism of coronal heating. High quality movies obtained during 2022 from the EUI instrument on the Solar Orbiter mission reveal remarkably stable coronal active region structure, with rare intermittent variations (Auchère et al., 2023). Analyzing these data in detail, Judge (2023) has argued that the bulk of the corona is probably heated by pico-flares, or steadily via ion viscosity associated with unobservably small variations in $\mathbf{b}$ and $\mathbf{u}$, with at least some compressive motions along $\mathbf{B} = \mathbf{B}_0 + \mathbf{b}$.

5.8 Stable or Unstable Current Sheets?

The dynamics of current sheets within the corona has been an important theme of this story. The hydromagnetic sheets of interest (à la Parker, 1988) are formed as a small misalignment in angle between large scale magnetic field vectors brought into contact. If the phenomenology of MHD turbulence is essentially correct (Appendix C), similar current sheets may form as the result of the intriguing dynamics of aligned and intermittent turbulence in a strong guide field (Schekochihin, 2022). In ideal MHD they can be considered as discontinuous changes in components of a small magnetic field **b** in a direction perpendicular to a stronger, uniform guide field $\mathbf{B}_0$ (see Fig. 4.2. The misalignment angles rendered in panel (b), and Fig. 3.6 are exaggerated for illustration). These configurations stand in contrast to Sweet-Parker configurations where there is no guide field, and instead oppositely-directed field lines make contact at a flux surface, as shown in the upper panel of Fig. 1.9. Any sheets with large enough aspect ratios are unstable (Pucci and Velli, 2010; Uzdensky and Loureiro, 2016):

> to understand the onset of magnetic reconnection, one needs to consider the growth of the tearing instability in a current layer as it is being formed. Uzdensky and Loureiro (2016)

To clarify, in the process of forming current sheets by whatever mechanism (topological constraints and/or MHD turbulence), the subsequent dynamics is then, if viscous dissipation is weak, accompanied by enhanced reconnection as smaller plasmoid structures form. Sheets, plasmoids and rapid reconnection are all part of the picture. Once a current sheet reaches a specific aspect ratio, plasmoids disrupt the sheet and reconnection proceeds rapidly, with an inflow speed of order $0.1c_A$ (Cassak et al., 2017). Such rates are inferred from measurements in situ in the magnetosphere, and remotely on the Sun. It must be emphasized that, here, c_A refers to the Alfvén speed of the reconnecting field not of the guide field. Thus speeds far less than the large-scale Alfvén speed will characterize the dynamics of these processes. This is encouraging, since coronal emission lines are surprisingly narrow, typically $\xi_0 \approx$ 10–40 km s^{-1} (Appendix A). Even if such motions are attributed wholly to turbulence, such speeds lie far below "the" Alfvén speeds of 10^3 km s^{-1}. In fact, the absence of such speeds in quiescent coronal plasma is yet another indication than neither $\mathbf{B}_0$ nor large-scale reconnection thereof plays an active part in steady coronal heating. Observations of *fast* reconnection were previously explained through anomalous resistivity. However, the paradigm has since changed to plasmoid-driven reconnection (Uzdensky and Loureiro, 2016; Mallet et al., 2017).

It is important to understand the conditions under which this instability arises in the solar corona, given evidence for current sheets surviving long enough to be observed (e.g. Lipartito et al., 2014), as well as for eventual turbulent disruption of the heliospheric current sheet (Cheng et al., 2018; French et al., 2019). The latter studies concern specifically the large-scale heliospheric sheet, presumably with just a weak or absent guide field. Consider the sheets postulated by Lipartito et al. (2014) in the chromosphere, evidently threaded by a strong guide field aligned along obvious chromospheric "fibrils" (although see Martínez-Sykora et al., 2016). They are stable enough to be observed over tens of seconds and longer. Is this observation consistent with the growth rate estimates of plasmoid disruption? The observed sheets have an aspect ratio of $\gtrsim$10:1, a length of at least 4 Mm. They form in the upper chromosphere where the mass density is near 10^{-10} kg m^{-3}, the Alfvén speed of order 100 km s^{-1}, a magnetic diffusivity of about 3 10^3 m^2 s^{-1}, a near-isotropic viscous diffusivity ten times smaller, and a Lundqvist number of 10^8. The growth rate of the tearing instability is $\gamma \sim a^{-4/3} S^{-1/3} (c_A/L) \gtrsim 0.001\,\mathrm{s}^{-1}$ (Pucci and Velli, 2010). Thus, such chromospheric sheets are stable for up to 20 minutes before disruption. Polarimetric measurements of these structures should therefore reveal strong linear polarization measurements parallel to the long axes of the dark structures.

A similar calculation for conditions in the heliospheric current sheet yields $\gamma \approx 0.03\ \mathrm{s}^{-1}$. Thus, disruption of the heliospheric sheet occurs there in less than a minute, and one would therefore expect abundant plasmoid structures within this sheet. Linear polarization measurements, sensitive to the directions of magnetic fields in the plane-of-the-sky independent of angular resolution, should reveal that sheet-like magnetic fields have been disrupted by multiple plasmoids. They should show negligible linear polarization signals in magnetic dipole coronal lines, such as those of Fe^{12+} observed by French et al. (2019).

In short, remarkably, both of these different predicted outcomes are consistent with the observations of Lipartito et al. (2014) and French et al. (2019) respectively. The polarization measurements used by the latter are information-rich, refuting the possibility of an undisrupted equatorial current sheet resolved at an adequate resolution of a few arc seconds. Such results suggest a promising path forward for assessing roles of current sheets, plasmoid instabilities and plasmoid-driven reconnection in driving coronal dynamics towards dissipation scales.

The hierarchy of scales listed in Table 4.1 suggests that, in the MHD turbulence problem, there is less than a decade of scales over which the plasmoid instability may dominate the dynamics of current sheets. Unless compressible motions and viscosity swamp the dynamics on larger scales, aligned turbulent cascades will survive down to scale of meters before destruction by plasmoids. Then collisionless effects will determine dissipation before kinetic resistivity or viscosity become dynamically important. Disruption by plasmoids, formed from the turbulent components $\mathbf{b}$, is not observable through the type of polarization measurements analyzed by French et al. (2019), which reflect instead plasmoids formed within the equatorial current sheet. The fact that both $\mathbf{b}$ and $\mathbf{u}$ lie essentially within the plane perpendicular to $\mathbf{B}_0$ may, however lead to measurable anisotropies in line widths as active regions transit the solar disk. The lack of reports of such variations suggests that anisotropies might be small, under-resolved or absent. If the coronal $\mathbf{B}_0$ vectors are close to orderly potential fields, this might imply that MHD turbulence is not a dominant dynamical component of the journey of energy from large to dissipation scales. Again, EUV spectral observations across the solar disk with $\mathcal{R} \gtrsim 20{,}000$ appear very worthwhile (Sect. 5.9). Anisotropies can naturally arise from other processes, the point here is that we have a specific prediction from a specific model which can be tested against data.

Yet again, the *absence* of an expected observational signature (in this case linear polarization) is of prime importance to make progress. This re-emphasizes the need for robust theoretical predictions, for more reasons than application of Bayesian methods to establish meaningful priors.

5.9 High Signal-to-Noise Coronal Line Profiles

A striking property of coronal lines measured to date is the narrowness of the emission lines. In plasmas with Alfvén speeds $c_A \gtrsim 1$ Mm s^{-1}, we observe non-thermal line widths of order 10–40 km s^{-1}, i.e. typically $\lesssim$2% of the Alfvén speed. These motions are also subsonic, like almost all spectral lines in the Sun except for lines formed in plasmas with little inertia in the transition region. The literature reports diverse values of coronal line widths (Del Zanna and Mason, 2018, see also Appendix A), even in the quiet Sun. By itself this suggests that new, more accurate measurements would be valuable, to understand if these variations suffer from line-of-sight broadening, from the use of spectrometers of only modest resolution, variable sky variations or perhaps other systematic problems. But there are also compelling theoretical reasons to measure with greater accuracy the profiles of coronal emission lines.

For example, if magnetic reconnection is a dominant aspect of coronal heating, we would expect to observe signatures of the accompanying dynamics, determined largely by the Alfvén speed c_A. Let us consider how line widths may inform us about a specific mechanism. In the nanoflare scenario, the reconnecting component of the magnetic field is **b**, not the far larger guide field $\mathbf{B}_0$. The latter does alter the dynamics of the reconnection (e.g. Stanier et al., 2017), but the energy density available for acceleration of plasma is $b^2/2\mu_0$. Thus we expect to see plasma exiting the reconnection region approaching the Alfvén speed associated with **b** only:

$$c_A \lesssim \frac{b}{\sqrt{\mu_0 \rho}}$$

where the upper limit applies when the conversion to kinetic energy is 100%. Let us adopt for simplicity the parameters from Parker's (1988) original paper applied to active regions, then $B_0 = 2000\,\mathrm{km\,s^{-1}}$, $b/B_0 \approx \arctan 14° = 0.24$, so that $c_A \lesssim 480\,\mathrm{km\,s^{-1}}$. Active region coronal line widths ξ_0 are typically 20 km s^{-1} (Del Zanna and Mason, 2018), with some X-ray measurements exceeding those made at EUV wavelengths. The factor 20 discrepancy would seem to refute outright the nanoflare picture using Parker's estimates! But small-scale reconnection-driven flows leading to lines with $\xi_0 \sim 480\,\mathrm{km\,s^{-1}}$ may have been missed, having peaks 20 times smaller than typical line cores carrying the same energy, and so lying beneath noise levels. Such high-speed flows encountering ambient plasma would tend to compress and form at least some slow shocks for which plasma at speeds of 100 km s^{-1} would be observable as the shocks convert K.E. to heat and perhaps accelerated particles by the Fermi process. Evidence for such behavior has not yet been found. Important also is the numerical result that the balance between magnetic and kinetic energy depends on the line-tied conditions at the footpoints in the chromosphere (Rappazzo et al., 2008, $b_\perp$ dominating over $u_\perp$ except at the smallest scales). Strong guide fields $\mathbf{B}_0$ also inhibit inverse cascades.

Nevertheless, these considerations reveal the need for high signal-to-noise measurements of coronal line profiles with spectral resolutions of at least 20,000 (i.e. with pixels of 7.5 km s^{-1} in Doppler units), extending perhaps as far as $\pm 10^3$ km s^{-1} around the center of each line.

5.10 Waves in Inhomogeneous Plasmas

The "turbulence" and "topological dissipation" pictures have dominated our discussion. In the first case non-linear interactions drive the small scales, in the second MHD waves deliver native stresses to form intense current sheets as the Maxwell stress tensor, subject to a fixed magnetic topology, demands (Parker, 1994). Both of these mechanisms intrinsically begin with long wavelength Alfvén waves, outer scale drivers of smaller scales driven by non-linear interactions, which can then dissipate. But the first mathematically "complete" model of coronal heating originally developed by Ionson (1978) was also the first to make explicit use of inhomogeneities in Alfvén speed, as the basis for developing a theory of coronal heating (Ionson, 1978, see Sect. 4.6). By including inhomogeneities from the outset, Ionson showed how a linear theory for wave motions couples to an Alfvén continuum to generate internal surface waves, and kinetic Alfvén waves which dissipate rapidly. Predictions from such calculations appear robust in that the important wave modes are linear, except at the crucial point where non-linearities finally introduce dissipation upon permitting non-ideal effects at the very end (nicely recapitulated by Davila, 1987). Davila discusses the observability of small dissipation scales, running in to the same questions concerning the coupling of incompressible transverse with compressible parallel modes discussed in Sect. 5.2. Depending on the nature of the small-scale dynamics, Davila concluded that the narrow resonance layers may be on scales between 1 km and up to 250 km.

Davila's calculations suggest that observations below 100 km angular resolution will reveal an answer to the details of the resonant absorption process, and whether ultimately compressive viscosity might be sufficient to explain coronal heating in steady, long-lived plasmas (Hollweg, 1985, 1986). The addition of observations of varying magnetic fields and fluid motions, with assessments of Poynting fluxes at the top of the chromosphere will also help determine the viability of these models, even if the volumes of resonance layers lie spatially well below current or foreseeable instrumental capabilities. Solar research along these lines will also help inform about the behavior of laboratory plasmas in large tokamaks since the first measurements of dissipation involving resonant absorption (McPherson and Pridmore-Brown, 1966).

While not a resonant phenomenon, phase mixing of Alfvén waves is a mechanism also relying on inhomogeneous wave speeds that might, in principle, readily lead to heating in magnetic bundles that are open to interplanetary space. However this has been contested by Parker (1991). Early work (summarized by Mok and Einaudi, 1990) pointed to the importance of inhomogeneities in Alfvén speeds in low-β

coronal plasmas to create much more efficient dissipation than plasmas with an unnaturally homogeneous Alfvén speed:

> ... the inhomogeneity of the medium makes possible the existence of modes whose damping length is of the order of the height of the corona.

Parker's objection is that when inhomogeneities are recognized along one dimension (say along one axis x when the homogeneous magnetic field is $(0, 0, B_0)$), the inevitable inhomogeneities along the orthogonal (y) direction (even in current sheets) alter the nature of the wave propagation problem. Those Alfvén waves propagating along z which in 2D (x, z) become out-of-phase because the Alfvén speed is a non-trivial function of x, tend to generate surface waves in 3D instead of generating huge gradients in velocity and magnetic field $\mathbf{b}$ along the x-direction. This question, among others, can be addressed yet again with high angular resolution observations and spectra, the latter required to obtain spatial power spectra of perpendicular dynamics. If phase mixing dominates heating of plasma in magnetic bundles open to the interplanetary medium, the power spectra will peak at high spatial frequencies (see the dot-dashed line sketched in Fig. 3.1), as was found qualitatively by Dere (1989) for transition region plasma, compromised as it is by lack of knowledge of the nature of transition region plasma (Sect. 5.1, 5.12).

Numerical experiments of wave motions in inhomogeneous magnetized MHD systems have led to further insight into both phase mixing and resonant absorption (Howson et al., 2019, 2020). A particularly clear presentation of the physical processes is found in the paper by Howson et al. (2019). The degrees of magnetic complexity as well as thermodynamic inhomogeneity have large influences on the outcome of such experiments. Non-linear effects, even in relatively ordered systems perturbed by transverse wave fields exhibit almost all conceivable outcomes-generation of slow- and fast-like compressive modes, cross-field transport, build up of large velocity gradients (vorticities) in resonant regions, including viscous and current dissipation, enhancement and reduction of dissipation, development of sheet-like structures, aligned velocity perturbations in the perpendicular direction.

Nevertheless, various results seem to arise that qualitatively make sense. For example, a complex background magnetic field encourages phase mixing throughout the plasma, driving larger local alternating currents and vorticities. The picture of Goldreich and Sridhar (1995) of counter-propagating interacting waves is observed in the experiments, but phase-mixing also induces an energy cascade. At sufficiently high frequencies, global resonances along field lines (like harmonics on piano wires) occur to produce efficient wave heating, otherwise the system tends not to inject sufficient Poynting flux to power coronal heating. The authors conclude that, using drivers inspired from photospheric motions, there is insufficient Poynting flux to heat the corona, and that significant heating only occurs when accompanied by large amplitude velocity oscillations greatly in excess of observed values of 20 km s^{-1}.

These are important results worthy of further numerical experimentation. Clearly, the dependences found on the driving motions again beg for measurements of

Poynting fluxes at the top of the chromosphere, which is expected to modify dynamically the fluxes entering the corona for multiple reasons (Sect. 3.2).

5.11 Remote Sensing of Kinetic Coronal Physics

We have seen that in the corona, strong viscous dissipation via can occur through motions with field-aligned compressions ($\partial u_\parallel / \partial x_\parallel \neq 0$). From Table 4.1 we see that non-compressive conditions, classical dissipation (electrical resistance and non-compressive viscosity) does not occur before ion inertial lengths are reached (1–10 m), whether the small scales are generated by turbulence or wave processes. The onset of "ideal" tearing (Pucci and Velli, 2010) will induce the rapid reconnection of plasmoids with motions on scales of about the inertial lengths for motions perpendicular to the guide field.

Here we consider the case where compressions are negligible, and that kinetic effects must be explored (Table 4.1). Then, kinetic effects must ultimately be implicated in dissipation of magnetic free energy. For example, Alfvén waves become dispersive, kinetic Alfvén waves when scales perpendicular to $\mathbf{B}_0$ are formed below ion gyroradii ($\lesssim 10^1$ m). Kinetic Alfvén waves formed the primary dissipation structure in Ionson's (1978) resonant absorption model of coronal heating. Magnetic reconnection below 10 m is driven not by resistivity but by separate ion and electron dynamics. Signatures of ion-cyclotron kinetic heating have been discovered in radiation emerging from plasma $\geq \frac{1}{2} R_\odot$ (>5 pressure scale heights) above the limb photosphere (Kohl et al., 2006). But evidence for kinetic effects within denser quiet coronal plasma or in active region loops is scant or absent, outside of flares.

Classical diffusive processes (calculated separately for ions and electrons by, e.g. Braginskii, 1965), formally apply over distances larger than mean free paths, a few hundred km in the corona. Braginskii discussed some kinetic effects involving collisions within and between the electrons and ions. Viscosity primarily heats ions, and electrical resistance heats electrons. Under conditions listed in Table 1.2, relaxation times for small departures from Maxwellian velocity distribution functions are fractions of a second between electrons, seconds between ions, and a minute or so between them. For plasmas higher in the corona, say $2R_\odot$, these relaxation times can be multiplied by 1000. If present, super-thermal particles in the tails of distributions can take far longer to isotropize/equilibrate because Coulomb collision cross sections vary as impact energy ε^{-2}. But unless there are efficient, rapid particle acceleration processes, coronal plasmas in quiet and active regions are likely to be dominated by thermal particle populations. If more extreme departures driven by strong electric fields, the electrons may "runaway" to higher energies (Dreicer, 1959), a process with considerable evidence supporting it, in flaring conditions. The "velocity filtration" picture of Scudder (1992a,b) applies to non-flaring conditions, demanding a strong higher energy tail in the upper chromosphere in order to produce sufficient particles in the overlying corona. At least one of the predictions appears inconsistent with the observed distribution of abundances because of varying charges and masses (Meyer-Vernet, 2007), as we saw in Sect. 4.2.2.

Another source of non-thermal particle distributions can arise when local thermal populations are transported, for example by conductive heat fluxes associated with steep temperature gradients, particularly when particle mean free paths exceed temperature scale lengths. These ideas, have nevertheless prompted searches for spectral signatures of "hot" coronal electrons transported to be mixed with thermalized transition region plasma (i.e. these coronal electrons form high energy tails). Lie-Svendsen et al. (1999) solved Boltzmann transport equations for electrons under a test particle approximation. They found that asymmetric departures from Boltzmann distributions under quiet and active solar conditions occur at the 10^{-4} level of the core population of electrons, making detections by remote sensing very unlikely. In a detailed review by Dudík et al. (2017), the evidence for such tails is weak outside of flaring plasma, and what there is remains contingent upon a variety of assumptions, several ad-hoc parameters, and reasonable alternative explanations.

In contrast, several pressure scale heights above the coronal base, researchers predicted the generation of *anisotropic ion distribution functions* via cyclotron heating by high frequency Alfvén waves, later to find supporting evidence with an ingenious use of spectroscopy (Noci et al., 1987). Emission lines can be excited within the higher corona by the scattering of disk radiation formed lower down in the Sun's atmosphere. The intensity of scattered emergent radiation depends on the local coronal ion distribution functions. In particular, the widths and Doppler shifts of ion distribution functions along the radial direction change the integrals of incident radiation over wavelength and direction which determine the scattered, emergent line intensities. Observations of such lines were acquired by the mid 1990s with the UVCS instrument on SoHO, the evidence supporting the statement that (Kohl et al., 1998):

> ...for $r \geq 2.2R_{\odot}$, $w_{\perp}$ must be larger than $w_{\parallel}$ and ... $w_{\perp}$ for O^{5+} ions is larger at most radii than $w_{\perp}$ for H^{0+} atoms, this implies that neither thermal motions ... nor common bulk-fluid motions dominate the unresolved perpendicular [line width] speeds $w_{\perp}$.

The essential picture of a larger linewidth $w_{\perp}$ perpendicular to the almost radial magnetic field relative to the parallel $w_{\parallel}$ component is not yet fully resolved. Cranmer (2009), Noci et al. (1987), Kohl et al. (1998), and Cranmer et al. (2008) contend that evidence for anisotropic distribution functions of O^{5+} is strong, contrasting with the work of Raouafi and Solanki (2004), Raouafi and Solanki (2006), and Raouafi et al. (2007). Both works involve significant modeling of light scattered from a radially stratified, expanding corona, incorporating isotropic and anisotropic ion velocity distribution functions. Cranmer et al. (2008) state that models must include anisotropy to account for an extended set of UVCS observations. Given the potential importance of consequences of anisotropic motions for coronal heating, at least in the solar wind, it deserves further discussion.

So here, let us give Kohl and colleagues the benefit of the doubt. Cranmer et al. (1999) constructed an elaborate model of the multi-fluid, multi- and anisotropic-temperature solar wind ions of H^+ (protons) and O^{5+} ions based on the dissipation of high-frequency (10–10,000 Hz) ion-cyclotron resonant Alfvén waves which differentially heat and accelerate ions. The H Lα spectrum is assumed to reflect the dynamics of coronal protons, because even though H Lα is a transition between bound states of neutral hydrogen, the upper bound state is populated by recombination by collisions with much lighter electrons. The emitting hydrogen atoms essentially retain the kinetic energy of the hot protons.

The origin of the required high frequency oscillations remains a subject for debate and future exploration. The fluctuations of magnetic waves and turbulence discussed so far have little power at high frequencies (power spectra varying as wavenumber k to a negative power between -1 and -2). Does the Poynting flux emerging from the upper chromosphere contain dynamically significant kHz frequency power? Or are other non-linear effects within the coronal plasma capable of generating such waves? In the solar wind, model predictions appear to refute the notion that MHD turbulence converts low-frequency Alfvén wave power into high frequency dissipation (see the quote on page 86 from Squire et al., 2022). Their proposed resolution involves ion cyclotron wave generation by the low-frequency Alfvén waves which are imbalanced (a greater upward than downward flux) and modulated by a "helicity barrier", which reverses the direction of energy propagation, in a clear modification of the ARLToR conjecture. Readers should refer to Squire et al. (2022) for details.

These are some of the kinetic effects which may successfully account for the remotely sensed and in-situ measurements of solar wind plasma, potentially down to $10R_\odot$, to be sampled by Parker Solar Probe. Conditions in the denser corona are more collisional, but it is clear that studies of the wind will inform on coronal heating, and vice-versa. Current wind probes, massive computers and turbulence theory seem well poised to make significant advances (see, e.g. Kumar et al., 2023). Observational tools for the deeper corona are more blunt, less information-rich, but perhaps offer opportunities for creative re-investment going well beyond current instrumental capabilities. The need for measurements of Poynting fluxes entering the corona from the chromosphere is, yet again, pressing.

5.12 Trouble with "the Transition Region"

We have encountered puzzling behavior in the emission from ions predominantly emitted from plasmas between 2×10^4 K and 5×10^5 K. This behavior presents the problem of coronal heating with yet more difficulties. We know that chromospheric/coronal interfaces exist. Two different interfaces were found in one theoretical model (Gabriel, 1976). We don't know how much emission the plasmas within these interface contribute to the observed emission, either on average or on spatial and/or temporal averages. We don't even know if long-lived transition region interfaces exists, or if what we observe is from plasmas continuously changing in response to chromospheric and coronal dynamics.

Until these questions are resolved, our attempts to connect the corona to its sources of mass, momentum and energy are forever contingent upon interpretations of what these emissions correspond to. Here is a brief, subjective discussion pointing to the nature of the challenges.

Unlike coronal emission, elementary models of emission from transition region plasmas essentially fail to account for the intensity spectra, observed for decades, under standard assumptions (as discussed in depth by Lang et al., 1990). No single emission measure distribution can account for the brightness of emission lines of diverse ions formed between 10^4 and 10^6 K. Emission lines of ions in the first column of abundant elements in the periodic table- lines from ions which are iso-electronic with Li, Na and K, tend to be brighter than expected by a factor of 2–3, particularly those such as C^{3+} whose ions are abundant near 10^5 K (Judge et al., 1995). Lines of He and He^+ also are also systematically too strong by even larger factors (Jordan, 1975; Laming and Feldman, 1992; Pietarila and Judge, 2004; Judge and Pietarila, 2004). Landi et al. (2002) identified systematic problems with various ions formed in the corona near 1.3 MK. It is clear that no combination of abundances variations or standard assumptions can account emission from spectral lines spanning the entire transition region. This kind of discrepancy applies to a significant fraction of all lines in the entire spectrum, begging the question: Are these assumptions essentially invalid? Do we understand the basic properties of these plasmas? How might one disprove that ionization equilibrium is valid, given the many diverse assumptions which must also be made (see, for example, section 11 of Lang et al., 1990)?

Some analyses seem spectacularly consistent using such assumptions. For example, the figures of Landi and Feldman (2008) were inferred by the authors to reveal that just three isothermal plasmas along a line of sight which sampled plasmas with very different connections and histories to the surface beneath. Are such results unreasonably good? Or perhaps equivalently, we might wonder if our tools are just too blunt? Judge (2010) studied the limits to which we can determine if a plasma might be isothermal by replacing Dirac δ-functions with Gaussian functions. The important points to emphasize are that multiple assumptions are required for spectroscopy to reveal plasma properties at all. These are not all defensible from first principles, and we may not be able to assess which are robust and which are violated merely from observations (ill-posedness). It seems that, without some deeper thinking and perhaps a slice of luck, we might face real difficulties in assessing what we can and cannot learn about plasmas from remote sensing. In this situation we must surely focus on identifying the keenest tools and the most tractable questions in order to advance beyond our current understanding.

Unlike coronal images, transition region plasmas exhibit a curious variety of morphologies in images and imaging spectroscopy (see Judge, 2021, for a recent overview of these problems, including earlier analyses). We simply have no agreement on the basic configurations of the solar plasmas emitting transition region lines. It is probable that some of these emissions amplify signatures of dynamics in coronal plasma, as well as of chromospheric plasma. But from which particular plasma configurations? Where does the conductive flux from the corona go if not into radiation (Anderson-Huang, 1998)? What observations of transition region plasma can help define the lower boundary conditions for coronal heating?

The growing litany of articles highlighting other particular difficulties includes the inability for 1D heat conduction to account for the "lower" transition region (e.g. Jordan, 1980), effects of optical depth on emission lines emitting above/below the Lyman continuum at 91.1 nm (Mariska and Withbroe, 1975; Mariska et al., 1978), a magnetostatic model of conductively heated transition region plasma compatible with the geometry of observed structures (Gabriel, 1976), a picture of "unresolved fine structure" denying the energetic field-aligned connection between coronal and transition region plasmas (Feldman, 1983, 1987, 1998), physical models based upon extended, extremely narrow current carrying filaments (Roumeliotis, 1991; Ashbourn and Woods, 2001), models in which cross-field heat conduction by ions heat sheaths around spicules (Athay, 1990; Ji et al., 1996, see the physical model of Ashbourn and Woods, 2001), a model in which turbulent energy transport delivers heat to the lower transition region (Cally, 1990, again see Ashbourn and Woods, 2001 for a physical model), a multi-faceted controversy over the role of cool loops in contributing to the emitted radiation (Antiochos and Noci, 1986; Dowdy et al., 1986; Cally and Robb, 1991; Hansteen et al., 2014; Judge, 2021).

This battlefield is strewn with the bodies of partially-surviving models because our data are *information-poor*. This is not to say that we have little data, we have vast quantities of UV data from the sequence of spacecraft experiments flown over five decades. Instead, the information contained is very often *redundant* with earlier data. For several decades we have analyzed the same types of data with newer instruments: intensity imaging, intensity spectroscopy, and intensity imaging spectroscopy. Our theoretical creativity has exploded in explaining curious behavior, in part because the data acquired are information-poor, permitting a greater latitude in the kinds of models which might be consistent with the data and laws of physics. It is remarkable, for example, how the magnetic structure of the transition region has been speculated upon so spectacularly given the total lack of measurements of magnetic fields of direct relevance to the chromosphere-corona interface (e.g. Gabriel, 1976; Dowdy et al., 1986; Antiochos and Noci, 1986; Dere et al., 1987; Innes et al., 1997; Hansteen et al., 2014; Judge, 2021). The *only* direct property on magnetic fields from intensity measurements is the morphology observed in images, because of the frozen-field condition. Even morphology is subject to line of sight interpretation problems.

Information poverty may not, however, be the only problem. *The entire concept of a permanent transition region may need re-consideration.* We know that the dynamically evolving plasmas between $10^{4.3} < T_e < 10^{5.7}$ K can readily be driven from both below and above. Consider the literature on "explosive events", which has matured with the advent of the IRIS mission and novel suborbital measurements (Rust and Kankelborg, 2019): the diversity of line profiles suggests that these are phenomena mostly occurring separately from any pre-existing, steady component of the corona-chromosphere interface. Hansteen (1993) demonstrated how modest changes in thermodynamic conditions in the corona can lead to dramatic signatures at transition region temperatures. Heggland et al. (2009) revealed how reconnection within the chromosphere leads to explosive signatures in transition-region lines, recently studied further by Wargnier et al. (2023). If we believed that the transition region is a pre-existing structure where events such as reconnection can happen, we would have to explain why such typically narrow regions shown in Fig. 1.4 would be a preferred location for such dynamic events. After all such a structure spans just 100 km of an atmosphere comprising a stratified layer 2000 km thick, and a far larger tenuous coronal plasma containing magnetic free energy above. Transition region plasmas are perhaps not an entity waiting to be perturbed, instead it may be dynamic in its very nature. Perhaps only the averaging over finite volumes gives us the impression of quasi-permanence (Sect. 3.3.1).

Remarkably though, we find puzzling statements which seem to insist that a permanent transition region exists and is a primary region for providing heat to the overlying corona

> this cycle of hot upflows and cool downflows clearly points to the transition region as primary heating source. Aschwanden et al. (2007).

How then are we to proceed? A high resolution EUV instrument operating with angular resolutions better than 100 km may be able to resolve some if these painful issues, or it may, if the region proves so intransigent, introduce more confusion. But this is of no consequence scientifically, either outcome would mark progress. Again, high spectral resolution ($\gtrsim$30,000) has great advantages for measuring turbulent and other motions. As already discussed, the spectral region near 62.5 nm has strong lines of O^{4+} and Mg^{8+}, which may be natural targets for new missions. Unfortunately, future planned missions will not have adequate spectral resolutions to study many of the signatures of coronal heating related to wave motions, current sheet dynamics and MHD turbulence (see Table 5.1).

We have provided multiple new arguments that measurements of vector magnetic fields $\mathbf{B}_0$ at the base of the corona (i.e. the top of the chromosphere) should be of a high priority. These measurements can be achieved with several spectral lines to be observed by the operational instruments at DKIST. While we have emphasized the role of $\mathbf{b}$ and associated dynamics in driving coronal heating, it is mediated by the large-scale field $\mathbf{B}_0$ which defines the Alfvén speed and, partly, the Poynting flux which is important for generating the conditions under which $\mathbf{b}$ and its dissipation occurs. These are challenging observations, fraught with difficulties, but recent intense work (Judge et al., 2021, 2022) has identified the potential for spectropolarimetry at near UV wavelengths (250–281 nm) to answer such questions as: how does the chromosphere change the Poynting flux entering the corona? What magnetic configurations correspond to observed transition region structures? How is plasma between 10^4 and 10^5 K heated, if the plasma above 10^5 K is satisfactorily explained by heat conduction (Cally, 1990; Jordan, 1992)?

Chapter Summary

We have identified novel possible lines of inquiry, prompted by various observations in the light of modern theories of MHD turbulence, topological dissipation, and MHD waves.

Evidence has accumulated that observable guide fields $\mathbf{B}_0$ play passive roles in coronal heating, fixing only Alfvén speeds c_A and the frozen topology.

What do facts related to oscillations, fuzzy loops, the absence of high energy electrons, observed current sheets and other highlighted phenomena tell us about coronal heating?

Do compressions aligned along the total magnetic field $\mathbf{B}$ dissipate free energy via ion viscosity before kinetic scales are reached? What then is the role of the plasmoid instability and kinetic physics?

Is MHD turbulence a dominant mechanism for generating dissipative scales in the corona? Is there evidence for flare-like heating?

Is there a permanent transition region? What can lines from transition region plasmas tell us about coronal heating?

These questions point to the need for EUV data with angular resolutions $\lesssim$ 100 km, and spectral resolutions $\gtrsim$ 20,000, requiring an unusually stable platform for new observations.

Chromospheric spectropolarimetry is needed to define evolving Poynting fluxes at the corona base. Measurements of full vectors $\mathbf{u}$ and $\mathbf{B}$ are the goal.

The next chapter will discuss how we might exploit these weaknesses in the Sun's armor, which hide signs of the heating processes.

References

Anderson-Huang, L.S.: The solar quiet chromosphere-corona transition region. Space Sci. Rev. **85**, 203–213 (1998)

Antiochos, S.K., Klimchuk, J.A.: A model for the formation of solar prominences. Astrophys. J. **378**, 372 (1991). https://doi.org/10.1086/170437

Antiochos, S.K., Noci, G. The structure of the static corona and transition region. Astrophys. J. **301**, 440 (1986)

Antolin, P.: Thermal instability and non-equilibrium in solar coronal loops: from coronal rain to long-period intensity pulsations. Plasma Phys. Control. Fusion **62**, 014016 (2020)

Antolin, P., et al.: The multithermal and multi-stranded nature of coronal rain. Astrophys. J. **806**, 81 (2015). https://doi.org/10.1088/0004-637X/806/1/81

Arnaud, J.: The analysis of Fe XIV 5303 coronal emission line polarization measurements. Astron. Astrophys. **116**, 248–254 (1982)

Arnaud, J., Newkirk, G.: Mean properties of the polarization of the Fe XIII 10747Å coronal emission line. Astron. Astrophys. **178**, 263–268 (1987)

Aschwanden, M.J.: An evaluation of coronal heating models for active regions based on Yohkoh, SOHO, and TRACE observations. Astrophys. J. **560**, 1035–1044 (2001). https://doi.org/10.1086/323064

Aschwanden, M.J., et al.: The coronal heating paradox. Astrophys. J. **659**, 1673–1681 (2007)

Ashbourn, J.M.A., Woods, L.C.: Energy transport in the solar transition layer. Proc. R. Soc. Lond. Ser. A **457**, 1873 (2001)

Athay, R.G.: Chromospheric Fine Structure. Reidel, Dordrecht (1974)

Athay, R.G.: Chromospheric fine structure. In: Physics of the Sun. Volume 2, vol. 2, pp. 51–69 (1986)

Athay, R.G.: A model of the solar chromosphere-corona transition region based on classical thermal conduction. Astrophys. J. **362**, 364 (1990). https://doi.org/10.1086/169272

Attrill, G.D.R., et al.: Revealing the fine structure of coronal dimmings and associated flows with hinode/EIS. Implications for understanding the source regions of sustained outflow following CMEs. Sol. Phys. **264**, 119–147 (2010). https://doi.org/10.1007/s11207-010-9558-8

Auchère, F., et al.: Image enhancement with wavelet-optimized whitening. Astron. Astrophys. **670**, A66 (2023). https://doi.org/10.1051/0004-6361/202245345

Barczynski, K., et al.: Emission of solar chromospheric and transition region features related to the underlying magnetic field. Astron. Astrophys. **619**, A5 (2018). https://doi.org/10.1051/0004-6361/201731650

Bartoe, J.D.F., Brueckner, G.E.: New stigmatic, coma-free, concave-grating spectrograph. J. Opt. Soc. Am. **65**, 13–21 (1975)

Berger, T.E., et al.: Hinode SOT observations of solar quiescent prominence dynamics. Astrophys. J. Lett. **676**, L89 (2008). https://doi.org/10.1086/587171

Billings, D.: The Solar Spectrum - the Coronal Spectrum. D. Reidel Publ. Co. (1965)

Braginskii, S.I.: Transport processes in a plasma. Rev. Plasma Phys. **1**, 205–311 (1965)

Bray, R.J., et al.: Plasma Loops in the Solar Corona. Cambridge Astrophysics Series. Cambridge University Press, Cambridge (1991)

Breu, C., et al.: A solar coronal loop in a box: energy generation and heating. Astron. Astrophys. **658**, A45 (2022). https://doi.org/10.1051/0004-6361/202141451

Cally, P.S.: Turbulent thermal conduction in the solar transition region. Astrophys. J. **355**, 693–699 (1990)

Cally, P.S., Robb, T.D.: Stability, structure and evolution of cool loops. Astrophys. J. **372**, 329–335 (1991)

Carlsson, M., et al.: A publically available simulation of an enhanced network region of the sun. Astron. Astrophys. **585**, A4 (2016)

Cassak, P.A., Liu, Y.H., Shay, M.A.: A review of the 0.1 reconnection rate problem. J. Plasma Phys. **83**, 715830501 (2017). https://doi.org/10.1017/S0022377817000666

Charbonneau, P.: Dynamo models of the solar cycle. Living Rev. Sol. Phys. **17**, 4 (2020)

Charvin, P.: Étude de la Polarisation Des Raies Interdites de la Couronne Solaire. Application uu Cas de la Raie Verte Λ5303. Ann. Astrophys. **28**, 877 (1965)

Cheng, X., et al.: Observations of turbulent magnetic reconnection within a solar current sheet. Astrophys. J. **866**, 64 (2018). https://doi.org/10.3847/1538-4357/aadd16

Chiba, S., et al.: Observation of the solar corona using radio scintillation with the Akatsuki spacecraft: difference between fast and slow wind. Sol. Phys. **297**, 34 (2022). https://doi.org/10.1007/s11207-022-01968-9

Chitta, L.P., et al.: Solar coronal loops associated with small-scale mixed polarity surface magnetic fields. Astrophys. J. Suppl. **229**, 4 (2017). https://doi.org/10.3847/1538-4365/229/1/4

Chitta, L.P., et al.: Solar coronal heating from small-scale magnetic braids. Astron. Astrophys. **667**, A166 (2022). https://doi.org/10.1051/0004-6361/202244170

Cranmer, S.R.: Coronal holes. Living Rev. Sol. Phys. **6**, 3 (2009). https://doi.org/10.12942/lrsp-2009-3

Cranmer, S.R., Field, G.B., Kohl, J.L.: Spectroscopic constraints on models of ion cyclotron resonance heating in the polar solar corona and high-speed solar wind. Astrophys. J. **518**, 937–947 (1999). https://doi.org/10.1086/307330

Cranmer, S.R., Panasyuk, A.V., Kohl, J.L.: Improved constraints on the preferential heating and acceleration of oxygen ions in the extended solar corona. Astrophys. J. **678**, 1480–1497 (2008). https://doi.org/10.1086/586890

Culhane, J.L., et al.: The EUV imaging spectrometer for hinode. Sol. Phys. **243**, 19–61 (2007). https://doi.org/10.1007/s01007-007-0293-1

Davila, J.M.: Heating of the solar corona by the resonant absorption of Alfvén waves. Astrophys. J. **317**, 514–521 (1987)

de Pontieu, B., et al.: The interface region imaging spectrograph (IRIS). Sol. Phys. **289**, 2733–2779 (2014)

de Pontieu, B., et al.: The multi-slit approach to coronal spectroscopy with the multi-slit solar explorer (MUSE). Astrophys. J. **888**, 3 (2020)

de Pontieu, B., et al.: A new view of the solar interface region from the interface region imaging spectrograph (IRIS). Sol. Phys. **296**, 84 (2021)

Del Zanna, G., Mason, H.E.: Solar UV and X-ray spectral diagnostics. Living Rev. Sol. Phys. **15**, 5 (2018). https://doi.org/10.1007/s41116-018-0015-3

Dere, K.P.: Turbulent power and dissipation in the solar transition zone. Astrophys. J. **340**, 599 (1989). https://doi.org/10.1086/167420

Dere, K.P., et al.: Discrete subresolution structures in the solar transition zone. Sol. Phys. **114**, 223–237 (1987)

Dissauer, K., et al.: Statistics of coronal dimmings associated with coronal mass ejections. I. Characteristic dimming properties and flare association. Astrophys. J. **863**, 169 (2018). https://doi.org/10.3847/1538-4357/aad3c6

Dowdy Jr., J.F., Rabin, D., Moore, R.L.: On the magnetic structure of the quiet transition region. Sol. Phys. **105**, 35–45 (1986)

Doyle, J.G., et al.: Nonthermal velocities in the solar transition zone and corona. Sol. Phys. **173**, 243–258 (1997). https://doi.org/10.1023/A:1004981825463

Dreicer, H.: Electron and ion runaway in a fully ionized gas. I. Phys. Rev. **115**, 238–249 (1959). https://doi.org/10.1103/PhysRev.115.238

Dudík, J., et al.: Nonequilibrium processes in the solar corona, transition region, flares, and solar wind (Invited Review). Sol. Phys. **292**, 100 (2017). https://doi.org/10.1007/s11207-017-1125-0

Einaudi, G., et al.: Energetics and 3D structure of elementary events in solar coronal heating. Astrophys. J. **910**, 84 (2021)

Falconer, D.A., et al.: Neutral-line magnetic shear and enhanced coronal heating in solar active regions. Astrophys. J. **482**, 519–534 (1997). https://doi.org/10.1086/304114

Fang, X., Xia, C., Keppens, R.: Multidimensional modeling of coronal rain dynamics. Astrophys. J. **771**, L29 (2013)

Fang, X., et al.: Coronal rain in magnetic arcades: rebound shocks, limit cycles, and shear flows. Astrophys. J. **807**, 142 (2015). https://doi.org/10.1088/0004-637X/807/2/142

Feldman, U.: On the unresolved fine structures of the solar atmosphere. Astrophys. J. **275**, 367 (1983)

Feldman, U.: On the unresolved fine structures of the solar atmosphere. II - the temperature region 200,000 - 500,000 K. Astrophys. J. **320**, 426 (1987)

Feldman, U.: On the unresolved fine structures of the solar atmosphere. III. Elemental abundances consideration. Astrophys. J. **507**, 974–977 (1998)

Field, G.B.: Thermal instability. Astrophys. J. **142**, 531 (1965). https://doi.org/10.1086/148317

Fisher, G.H., et al.: Coronal heating in active regions as a function of global magnetic variables. Astrophys. J. **508**, 885–898 (1998)

Fludra, A., et al.: First observations from the SPICE EUV spectrometer on Solar Orbiter. Astron. Astrophys. **656**, A38 (2021). https://doi.org/10.1051/0004-6361/202141221

Fort, B., Martres, M.J.: On the structure of the quiet corona near quiescent prominences. Astron. Astrophys. **33**, 249–255 (1974)

French, R.J., et al.: Spectropolarimetric insight into plasma sheet dynamics of a solar flare. Astrophys. J. Lett. **887**, L34 (2019). https://doi.org/10.3847/2041-8213/ab5d34

Gabriel, A.: A magnetic model of the solar transition region. Philos. Trans. R. Soc. Lond. **281**, 339 (1976)

Goddard, C.R., et al.: A statistical study of the inferred transverse density profile of coronal loop threads observed with SDO/Aia. Astron. Astrophys. **605**, A65 (2017). https://doi.org/10.1051/0004-6361/201731023

Goldreich, P., Sridhar, S.: Toward a theory of interstellar turbulence. II. Strong alfvenic turbulence. Astrophys. J. **438**, 763 (1995)

Grossmann-Doerth, U., Schmidt, W.: Chromospheric fine structure revisited. Astron. Astrophys. **264**, 236–242 (1992)

Guarrasi, M., Reale, F., Peres, G.: Coronal fuzziness modeled with pulse-heated multi-stranded loop systems. Astrophys. J. **719**, 576–582 (2010). https://doi.org/10.1088/0004-637X/719/1/576

Gudiksen, B.V., Nordlund, Å.: An *ab initio* approach to the solar coronal heating problem. Astrophys. J. **618**, 1020–1030 (2005)

Habbal, S.R., et al.: Thermodynamics of the solar corona and evolution of the solar magnetic field as inferred from the total solar eclipse observations of 2010 July 11. Astrophys. J. **734**, 120 (2011)

Hansteen, V.: A new interpretation of the redshift observed in optically thin transition region lines. Astrophys. J. **402**, 741–755 (1993)

Hansteen, V., et al.: The unresolved fine structure resolved: IRIS observations of the solar transition region. Science **346**, 1255757 (2014)

Harrison, R.A., et al.: The coronal diagnostic spectrometer for the solar and heliospheric observatory: experiment description and calibration. Metrologia **32**, 647–652 (1996)

Heggland, L., de Pontieu, B., Hansteen, V.H.: Observational signatures of simulated reconnection events in the solar chromosphere and transition region. Astrophys. J. **702**, 1–18 (2009). https://doi.org/10.1088/0004-637X/702/1/1

Henriques, V.M.J., et al.: Umbral chromospheric fine structure and umbral flashes modelled as one: the corrugated umbra. Astron. Astrophys. **642**, A215 (2020). https://doi.org/10.1051/0004-6361/202038538

Hollweg, J.V.: Viscosity in a magnetized plasma: physical interpretation. J. Geophys. Res. **90**, 7620–7622 (1985). https://doi.org/10.1029/JA090iA08p07620

Hollweg, J.V.: Viscosity and the Chew-Goldberger-Low equations in the solar corona. Astrophys. J. **306**, 730–739 (1986)

Horbury, T.S., Forman, M., Oughton, S.: Anisotropic scaling of magnetohydrodynamic turbulence. Phys. Rev. Lett. **101**, 175005 (2008). https://doi.org/10.1103/PhysRevLett.101.175005

House, L.L.: Coronal emission line polarization. Sol. Phys. **23**, 103–119 (1972)

House, L.L.: The theory of the polarization of coronal forbidden lines. Publ. Astron. Soc. Pacific **86**, 490 (1974)

Howson, T.A., et al.: Magnetohydrodynamic waves in braided magnetic fields. Astron. Astrophys. **629**, A60 (2019). https://doi.org/10.1051/0004-6361/201935876

Howson, T.A., de Moortel, I., Reid, J.: Phase mixing and wave heating in a complex coronal plasma. Astron. Astrophys. **636**, A40 (2020)

Hudson, H.S., Acton, L.W., Freeland, S.L.: A long-duration solar flare with mass ejection and global consequences. Astrophys. J. **470**, 629 (1996). https://doi.org/10.1086/177894

Innes, D.E., et al.: Bi-directional plasma jets produced by magnetic reconnection on the Sun. Nature **386**, 811–813 (1997). https://doi.org/10.1038/386811a0

Ionson, J.A.: Resonant absorption of Alfvénic surface waves and the heating of solar coronal loops. Astrophys. J. **226**, 650–673 (1978)

Ji, H.S., Song, M.T., Hu, F.M.: A unified model for the Sun's lower and upper transition regions. Astrophys. J. **464**, 1012 (1996)

Jordan, C.: The intensities of helium lines in the solar EUV spectrum. Mon. Not. R. Astron. Soc. **170**, 429–440 (1975)

Jordan, C.: The energy balance of the solar transition region. Astron. Astrophys. **86**, 355–363 (1980)

Jordan, C.: Modelling of solar coronal loops. Mem. Soc. Astron. Ital. **63**, 605–620 (1992)

Judge, P.G.: Coronal emission lines as thermometers. Astrophys. J. **708**, 1238–1240 (2010)

Judge, P.G.: Solar magnetic fields: history, tragedy or comedy? In: Almeida, J.S., González, M.J.M. (eds.) Cosmic Magnetic Fields, Canary Islands Winter School of Astrophysics, pp. 13–46. Cambridge University Press, Cambridge (2018). https://doi.org/10.1017/9781316160916

Judge, P.G.: Magnetic connections across the chromosphere-corona transition region. Astrophys. J. **914**, 70 (2021)

Judge, P.G.: Steadiness of coronal heating. Astrophys. J. **957**, 25 (2023)

Judge, P.G., Pietarila, A.: On the formation of extreme-ultraviolet helium lines in the Sun: analysis of SOHO data. Astrophys. J. **606**, 1258–1275 (2004)

Judge, P.G., et al.: On the failure of standard emission measure analysis for solar EUV and UV irradiance spectra. Astrophys. J. Lett. **455**, L85–l89 (1995)

Judge, P., et al.: Measuring the magnetic origins of solar flares, coronal mass ejections, and space weather. Astrophys. J. **917**, 27 (2021)

Judge, P., et al.: Optimal spectral lines for measuring chromospheric magnetic fields. Astrophys. J. **941**, 159 (2022). https://doi.org/10.3847/1538-4357/aca2a5

Kahler, S.W., Hudson, H.S.: Origin and development of transient coronal holes. J. Geophys. Res. **106**, 29239–29248 (2001). https://doi.org/10.1029/2001JA000127

Kiepenheuer, K.O.: Solar activity. In: Kuiper, G.P. (ed.) The Sun, pp. 322–465. Chicago University Press, Chicago (1953)

Kohl, J.L., et al.: UVCS/SOHO empirical determinations of anisotropic velocity distributions in the solar corona. Astrophys. J. Lett. **501**, L127 (1998)

Kohl, J.L., et al.: Ultraviolet spectroscopy of the extended solar corona. Astron. Astrophys. Rev. **13**, 31–157 (2006). https://doi.org/10.1007/s00159-005-0026-7

Koza, J.: Chromospheric fine structure didactically. Central Eur. Astrophys. Bull. **38**, 39–52 (2014)

Kumar, P., et al.: New evidence on the origin of solar wind microstreams/switchbacks. Astrophys. J. Lett. **951**, L15 (2023). https://doi.org/10.3847/2041-8213/acd54e

Laming, J., Feldman, U.: A burst model for line emission in the solar atmosphere. I. XUV lines of He I and He II in impulsive flares. Astrophys. J. **386**, 364–370 (1992)

Landi, E., Feldman, U.: The thermal structure of an active region observed outside the solar disk. Astrophys. J. **672**, 674–683 (2008)

Landi, E., Klimchuk, J.A.: On the isothermality of solar plasmas. Astrophys. J. **723**, 320–328 (2010). https://doi.org/10.1088/0004-637X/723/1/320

Landi, E., Feldman, U., Dere, K.P.: CHIANTI-an atomic database for emission lines. V. Comparison with an isothermal spectrum observed with SUMER. Astrophys. J. Suppl. **139**, 281–296 (2002). https://doi.org/10.1086/337949

Lang, J., Mason, H.E., McWhirter, R.W.P.: The interpretation of the spectral line intensities from the CHASE spectrometer on spacelab 2. Sol. Phys. **129**, 31–81 (1990)

Lazarian, A., Vishniac, E.T.: Reconnection in a weakly stochastic field. Astrophys. J. **517**, 700–718 (1999). https://doi.org/10.1086/307233

Lie-Svendsen, Ø., Holzer, T.E., Leer, E.: Electron heat conduction in the solar transition region: validity of the classical description. Astrophys. J. **525**, 1056–1065 (1999)

Lipartito, I., et al.: The solar chromosphere observed at 1 Hz and 0."2 resolution. Astrophys. J. **785**, 109 (2014)

Litwin, C., Rosner, R.: On the structure of solar and stellar coronae - loops and loop heat transport. Astrophys. J. **412**, 375–385 (1993)

Malanushenko, A., et al.: The coronal veil. Astrophys. J. **927**, 1 (2022)

Mallet, A., Schekochihin, A.A., Chandran, B.D.G.: Disruption of sheet-like structures in Alfvénic turbulence by magnetic reconnection. Mon. Not. R. Astron. Soc. **468**, 4862–4871 (2017). https://doi.org/10.1093/mnras/stx670

Mandrini, C.H., Démoulin, P., Klimchuk, J.A.: Magnetic field and plasma scaling laws: their implications for coronal heating models. Astrophys. J. **530**, 999–1015 (2000)

Mariska, J.T., Withbroe, G.L.: Analysis of EUV limb-brightening observations from Atm. I - model for the transition layer and the corona. Sol. Phys. **44**, 55–68 (1975)

Mariska, J.T., Feldman, U., Doschek, G.A.: Measurements of extreme-ultraviolet emission-line profiles near the solar limb. Astrophys. J. **226**, 698–705 (1978)

Marsch, E., Tu, C.Y.: Solar wind and chromospheric network. Sol. Phys. **176**, 87–106 (1997). https://doi.org/10.1023/A:1004975703854

Martínez-Sykora, J., et al.: On the misalignment between chromospheric features and the magnetic field on the Sun. Astrophys. J. Lett. **831**, L1 (2016). https://doi.org/10.3847/2041-8205/831/1/L1

McPherson, D.A., Pridmore-Brown, D.C.: Density gradient effects on Alfvén wave propagation in a cylindrical plasma. Phys. Fluids **9**, 2033–2042 (1966). https://doi.org/10.1063/1.1761562

Meyer-Vernet, N.: Basics of the Solar Wind. Cambridge University Press, Cambridge (2007)

Mok, Y., Einaudi, G.: Comments on compressible effects on Alfven normal modes in nonuniform plasmas. Astrophys. J. **351**, 296 (1990). https://doi.org/10.1086/168465

Morton, R.J., Tomczyk, S., Pinto, R.: Investigating Alfvénic wave propagation in coronal open-field regions. Nat. Commun. **6**, 7813 (2015). https://doi.org/10.1038/ncomms8813

Müller, D.A.N., et al.: High-speed coronal rain. Astron. Astrophys. **436**, 1067–1074 (2005). https://doi.org/10.1051/0004-6361:20042141

Nikolsky, G.M.: The observation of the chromospheric fine structure by the 53-cm Lyot coronagraph. Sol. Phys. **12**, 379–390 (1970). https://doi.org/10.1007/BF00148021

Noci, G., Kohl, J.L., Withbroe, G.L.: Solar wind diagnostics from Doppler-enhanced scattering. Astrophys. J. **315**, 706 (1987). https://doi.org/10.1086/165172

Noglik, J.B., et al.: Deconstructing active region AR10961 using STEREO, Hinode, TRACE, and SOHO. Astrophys. J. **703**, 1923–1938 (2009). https://doi.org/10.1088/0004-637X/703/2/1923

Parker, E.N.: Instability of thermal fields. Astrophys. J. **117**, 431 (1953). https://doi.org/10.1086/145707

Parker, E.N.: Nanoflares and the solar X-ray corona. Astrophys. J. **330**, 474–479 (1988)

Parker, E.N.: Heating solar coronal holes. Astrophys. J. **372**, 719 (1991)

Parker, E.N.: Spontaneous Current Sheets in Magnetic Fields with Application to Stellar X-rays. International Series on Astronomy and Astrophysics. Oxford University Press, Oxford (1994)

Pascoe, D.J., Wright, A.N., De Moortel, I.: Propagating coupled Alfvén and Kink oscillations in an arbitrary inhomogeneous corona. Astrophys. J. **731**, 73 (2011). https://doi.org/10.1088/0004-637X/731/1/73

Pietarila, A., Judge, P.G.: On the formation of the resonance lines of helium in the Sun. Astrophys. J. **606**, 1239–1257 (2004)

Przybylski, D., et al.: Chromospheric extension of the MURAM code. Astron. Astrophys. **664**, A91 (2022). https://doi.org/10.1051/0004-6361/202141230

Pucci, F., Velli, M.: Reconnection of quasi-singular current sheets: the "ideal" tearing mode. Astrophys. J. Lett. **780**, L19 (2010). https://doi.org/10.1088/2041-8205/780/2/L19

Querfeld, C.: A near-infrared coronal emission-line polarimeter. In: SPIE Optical Polarimetry–Instrumentation and Applications, vol. 122, pp. 200–208 (1977)

Querfeld, C.: The formation and interpretation of the Fe XIII 10747 A coronal emission line. Astrophys. J. **255**, 764–773 (1982)

Querfeld, C.W., Smartt, R.N.: Comparison of coronal emission line structure and polarization. Sol. Phys. **91**, 299–310 (1984)

Rachmeler, L.A., et al.: The high-resolution coronal imager, flight 2.1. Sol. Phys. **294**, 174 (2019)

Raouafi, N.E., Solanki, S.K.: Effect of the electron density stratification on off-limb O VI line profiles: how large is the velocity distribution anisotropy in the solar corona? Astron. Astrophys. **427**, 725–733 (2004). https://doi.org/10.1051/0004-6361:20041203

Raouafi, N.E., Solanki, S.K.: Sensitivity of solar off-limb line profiles to electron density stratification and the velocity distribution anisotropy. Astron. Astrophys. **445**, 735–745 (2006). https://doi.org/10.1051/0004-6361:20042568

Raouafi, N.E., Harvey, J.W., Solanki, S.K.: Properties of solar polar coronal plumes constrained by ultraviolet coronagraph spectrometer data. Astrophys. J. **658**, 643–656 (2007). https://doi.org/10.1086/510286

Rappazzo, A.F., et al.: Nonlinear dynamics of the Parker scenario for coronal heating. Astrophys. J. **677**, 1348–1366 (2008). https://doi.org/10.1086/528786

Rempel, M.: Numerical simulations of quiet Sun magnetism: on the contribution from a small-scale dynamo. Astrophys. J. **789**, 132 (2014). https://doi.org/10.1088/0004-637X/789/2/132

Rempel, M.: Extension of the MURAM radiative MHD code for coronal simulations. Astrophys. J. **834**, 10 (2017)

Rosner, R.: A brief introduction to coronal "loops". In: Russell, C.T., Priest, E.R., Lee, L.C. (eds.) Physics of Magnetic Flux Ropes. Washington DC American Geophysical Union Geophysical Monograph Series, vol. 58, p. 189 (1990). https://doi.org/10.1029/GM058

Rosner, R., Tucker, W.H., Vaiana, G.S.: Dynamics of the quiescent solar corona. Astrophys. J. **220**, 643 (1978)

Roumeliotis, G.: Joule heating as an explanation for the differential emission measure structure and systematic redshifts in the Sun's lower transition region. Astrophys. J. **379**, 392 (1991)

Rust, D.M.: Coronal disturbances and their terrestrial effects. Space Sci. Rev. **34**, 21–36 (1983). https://doi.org/10.1007/BF00221193

Rust, T., Kankelborg, C.C.: Quiet-Sun explosive events observed in He II λ304 with MOSES-06. Astrophys. J. **877**, 59 (2019). https://doi.org/10.3847/1538-4357/ab12e2

Sahal-Bréchot, S.: Calculation of the polarization degree of the infrared lines of Fe XIII in the solar corona. Astrophys. J. **213**, 887–899 (1977)

Schekochihin, A.A.: MHD turbulence: a biased review. J. Plasma Phys. **88**, 155880501 (2022). https://doi.org/10.1017/S0022377822000721

Schrijver, C.J.: Catastrophic cooling and high-speed downflow in quiescent solar coronal loops observed with TRACE. Sol. Phys. **198**, 325–345 (2001). https://doi.org/10.1023/A:1005211925515

Scudder, J.D.: On the causes of temperature change in inhogeneous low-density astrophysical plasmas. Astrophys. J. **398**, 299–318 (1992a)

Scudder, J.D.: Why all stars should possess circumstellar temperature inversions. Astrophys. J. **398**, 319–349 (1992b)

Shimizu, T., et al.: The solar-C (EUVST) mission: the latest status. In: Society of Photo-Optical Instrumentation Engineers (SPIE) Conference Series, 114440N (2020). https://doi.org/10.1117/12.2560887

Singh, J. et al.: Exploring the outer emission corona spectroscopically by using visible emission line coronagraph (VELC) on board ADITYA-L1 mission. Adv. Space Res. **64**, 1455–1464 (2019). https://doi.org/10.1016/j.asr.2019.07.007

SPICE Consortium et al.: The solar orbiter SPICE instrument. An extreme UV imaging spectrometer. Astron. Astrophys. **642**, A14 (2020). https://doi.org/10.1051/0004-6361/201935574

Squire, J., et al.: High-frequency heating of the solar wind triggered by low-frequency turbulence. Nat. Astron. **6**, 715–723 (2022). https://doi.org/10.1038/s41550-022-01624-z

Stanier, A., et al.: The role of guide field in magnetic reconnection driven by island coalescence. Phys. Plasmas **24**, 022124 (2017). https://doi.org/10.1063/1.4976712

Tam, K.V., et al.: Coronal heating in multiple magnetic threads. Astron. Astrophys. **580**, A122 (2015). https://doi.org/10.1051/0004-6361/201525995

Tiwari, S.K., et al.: New evidence that magnetoconvection drives solar-stellar coronal heating. Astrophys. J. Lett. **843**, L20 (2017). https://doi.org/10.3847/2041-8213/aa794c

Tomczyk, S., et al.: Alfvén waves in the solar corona. Science **317**, 1192 (2007)

Tomczyk, S., et al.: An instrument to measure coronal emission line polarization. Sol. Phys. **247**, 411–428 (2008)

Tripathi, D., et al.: Active region loops: hinode/extreme-ultraviolet imaging spectrometer observations. Astrophys. J. **694**, 1256–1265 (2009). https://doi.org/10.1088/0004-637X/694/2/1256

Uzdensky, D.A.: The fast collisionless reconnection condition and the self-organization of coronal heating. Astrophys. J. **671**, 2139–2153 (2007) https://dx.doi.org/10.1086/522915

Uzdensky, D.A., Loureiro, N.F.: Magnetic reconnection onset via disruption of a forming current sheet by the tearing instability. Phys. Rev. Lett. **116**, 105003 (2016). https://doi.org/10.1103/PhysRevLett.116.105003

Van de Hulst, H.C.: The chromosphere and the corona. In: Kuiper, G.P. (ed.) The Sun, pp. 207–321. Chicago University Press, Chicago (1953)

Wang, Y.M.: The ubiquitous presence of looplike fine structure inside solar active regions. Astrophys. J. Lett. **820**, L13 (2016). https://doi.org/10.3847/2041-8205/820/1/L13

Wargnier, Q.M., et al.: Multifluid simulations of upper-chromospheric magnetic reconnection with helium-hydrogen mixture. Astrophys. J. **946**, 2, 115 (2023). https://doi.org/10.3847/1538-4357/acbfb1

Webb, D.F., et al.: The origin and development of the May 1997 magnetic cloud. J. Geophys. Res. **105**, 27251–27260 (2000). https://doi.org/10.1029/2000JA000021

Wicks, R.T., et al.: Power and spectral index anisotropy of the entire inertial range of turbulence in the fast solar wind. Mon. Not. R. Astron. Soc. **407**, L31–L35 (2010). https://doi.org/10.1111/j.1745-3933.2010.00898.x

Wilhelm, K., et al.: SUMER - solar ultraviolet measurements of emitted radiation. Sol. Phys. **162**, 189–231 (1995). https://doi.org/10.1007/BF00733430

Zhang, Q.M., Su, Y.N., Ji, H.S.: Pre-flare coronal dimmings. Astron. Astrophys. **598**, A3 (2017). https://doi.org/10.1051/0004-6361/201629477

Chapter 6
Fresh Approaches

Having undertaken this wide-ranging review, we can now offer a perspective on the current state of research related to coronal heating. Necessarily subjective, and intentionally thought-provoking, some may find the words uncomfortable reading, but the sentiments are sincere and based upon years of experience, observations and community interactions. They are written in the hope that researchers might benefit from our experience, our mistakes, and be able to make up their own minds concerning future work.

In the mature yet unsolved problem of coronal heating, science as usual has thoroughly mixed with technology, especially computation. The "business" of science also entails service to society, the ability to offer "deliverables", indeed these are common requirements for a successful career. Introduction of metrics and the eternal pressure to publish has led to over-specialization and a morass of contingent publications apparently related to, but rarely not central to the coronal heating problem. A summary of such issues has bee compared creatively to financial investment strategy by Loeb (2010). Yet some voices have declared that fundamental problems in space-related science are now understood, so that emphasis should be been given to understanding systems, even systems of systems. All this in the face of the fundamental, philosophical and practical problems highlighted here, which face any and all researchers generally interested in really solving this problem.

Given these concerns, the goal of this chapter is to suggest solutions to the challenges identified in earlier chapters. We will return to some basic questions which many may well consider already answered. They may be, but the dependence of so many earlier works on contingencies has led us to conclude that we are not making much progress. Too much contingency leads to too much doubt, and as scientists we are obliged to adopt a skeptical position. Being skeptical from the outset of a new research endeavor is a healthy starting point, helping us to identify those approaches which might close the door on the mature problem at hand.

In an attempt at an *aide-memoir* to condense our thesis, we might be forgiven in indulging in yet another clumsy metaphor:

P. Judge, J. A. Ionson, *The Problem of Coronal Heating*, Astrophysics and Space Science Library 470, https://doi.org/10.1007/978-3-031-46273-3_6

> We have assembled large fractions of the puzzle, although we have not yet been able to see the full picture. Our unfinished puzzle resembles a crudely-made jigsaw with isolated islands of connected pieces sitting within the readily-assembled and fully connected straight-edged frame (the "low hanging fruit"). Scattered around are remaining pieces which appear almost exactly the same shape but which are different colors. It turns out though that our puzzle solvers are all color-blind, also suffering from tunnel vision and myopia. Their vision is not sharp, the field of view frequently awkwardly narrow. The color-blindness enables them to assemble the puzzle, appearing to solve it once and for all. But when compared with solutions of others, all are different but all agree that the puzzle is solved. Islands within the puzzle have been assembled correctly by many because, even though visually challenged, detailed focusing on small areas has led to clear, unique and arguably correct connections. However, the final puzzle remains quite different every time it is completed.
>
> Advice to impaired puzzle solvers might read: First, with the agreement of all, assemble carefully the outermost frame, a firm foundation within which the rest of the puzzle can be assembled. Second, obtain glasses to correct for the myopic and tunnel-vision, and lastly, acquire those new technology filters to correct color-blindness.

This metaphor is meant to be taken lightly. But the remedies have loose parallels in solar physics. The straight-edged frame encompasses the entire puzzle, and is relatively straightforward to piece together, giving confidence and forming a foundation for starting the puzzle (≡ fundamental physics). Second, while our ability to correct for myopia has been around for centuries, glasses can be restrictive (sometimes it is more convenient to narrow our view). Lastly it is only in recent years that technology has advanced sufficiently to correct color-blinded vision (≡ breakthrough advances in remote-sensing technology).

6.1 A Perspective on Methodologies

We might appropriately consider some methodological, but fictional words of wisdom:

> "... when you have eliminated the impossible, whatever remains, however improbable, must be the truth"—Arthur Conan Doyle, *The Sign of the Four*

Unfortunately, absolutes in fiction are rarely reflected in nature. For example, how would Holmes and Watson decide that they have eliminated *all* possible impossibilities? Holmes misses a tenet of modern philosophy of epistemology: the "truth" is never accessible to us. Instead our theories are always provisional, forever awaiting a potential refutation signaling a change of paradigm, reminding us of Wheeler's words of wisdom in Chap. 4, to get the mistakes over with as quickly as possible.

The coronal heating problem is further encumbered with the challenges of Chap. 3 and practical and methodological weaknesses highlighted in Chap. 4. Following developments in the philosophy of science late in the twentieth century, we are obliged to accept that Popper's refutation principle is equally as absolute, as "black and white", as Holmes' truth. Thus our call to accept refutation as the basis for advancement requires attention.

While a messy and loosely defined an endeavor as science is without a fully acknowledged "scientific method", it does offer freedom simply to explore. Coronal physics is an observationally-driven science, at least in part because non-linear and non-local effects occurring over a formidable range in physical scales lead to an overwhelming range of possible phenomena arising from first-principles approaches. Given this, we must identify methods incorporating fully the importance of observations, yet recognizing that clean refutations are likely naïve if not downright impossible. Implementation is not as straightforward as it might sound (see Pascoe et al. 2017, for an informative discussion related to coronal "loops"). For example the number of "priors"—assumptions or models to be tested—may be intractably large when little is known about the system. The prior information must also be carefully assembled and applied, otherwise it can impose unwanted biases into the probabilities (C. Kankelborg, private communication 2022, see also Ross 2022). But in principle Bayes' theorem informs us how much we should believe in a given hypothesis based on the evidence available. More evidence might be classified as information-rich by increasing some probabilities at the expense of others, or information-poor if probabilities do not change or an ensemble of them reflects increased redundancy among competing hypotheses.

In using Bayesian methods, as for frequentist approaches, it can be helpful to reduce the scope, or the realm, of possible models related to the problem of interest. Naturally, most research papers adopt this kind of narrowing in scope. At the risk of being guilty of myopia, the solar research reviewed so far suggests that we might usefully consider four different macroscopic situations, within which to seek microscopic processes leading to heating. First are those leading to a steady average state of the quiet corona, second those that lead to a steady fast solar wind from magnetically "open" coronal holes, third those that cause steady (on dynamical time scales) states within active regions. Lastly there are flares and CMEs, which, in spite of ongoing heating, build up vast energies within coronal plasma to release magnetic stresses and energy suddenly, witnessed as flares and CMEs. Ultimately, of course, such limiting of the scope of problems must be regarded as just steps towards the broader more general goal of understanding electrodynamic coupling.

For example, the causes of explosive behavior were excluded from this book in the desire for brevity. But nevertheless, these are cases where the phenomena observed actually do impact our understanding of steady coronal heating. There are very few indicators of an imminent flare or CME, in fact, many studies have fruitlessly sought indicators of such events for decades (reviewed by Judge et al. 2021, see the corona/flare prediction work of Leka et al. 2023). This "negative" result in prediction in fact gives us a critically important data point: coronal heating continues unabated in an active region, even when it is approaching a major change of state. The build up of magnetic energy on huge scales, for large X-class flares and CMEs, requires storage and sudden release of 10^{25} J. This energy would require several days to be lost to steady coronal heating at 10^4 W m^{-2} . Yet this energy is released in flares over minutes. The rate of flare energy release is over 1000 times that of that released to maintain the uneventful active corona. What should we conclude?

We can, for example, hypothesize that *steady heating mechanisms have little to do with large-scale changes in the measurable changes in magnetic fields, and consequent large-scale dynamics, and flares*, and then try to test it. Given the failure of modern research to deliver conclusive results, this hypothesis is not unreasonable. One possibility is that coronal heating is dominated by viscous dissipation of compressible motions along **B**, triggered by non-linearities in the dynamics of wave motions and/or turbulent eddies and current sheets. This scenario would not require any magnetic reconnection, seemingly required for flaring on all scales. Instead it may naturally occur through dissipation of free energy in velocity gradients generated through MHD turbulence before all the intricacies of the "plasmoid mediated aligned turbulence" picture come into play (see Appendix C).

An alternative more popular hypothesis is that all manifestations of coronal heating are nothing more than an ensemble of flares, spanning a huge range of scales (Lu and Hamilton 1991). This idea is appealing in that studies of flares at resolvable scales might, with a little faith in extrapolation, suffice to explain the entire problem of coronal heating (Bogachev et al. 2020). But what evidence is there to support or refute these two different ideas? This question deserves the more direct attention that it has received from the solar community.

We have often noted that the absence of a signal can be equally important as its presence (Sects. 3.3, 4.5, 4.10.3, 5.2, 5.8, and 5.12). A related example is the observation encapsulated in Fig. 5.4, showing the remarkably narrow distributions of brightnesses of the quiet and active Sun and coronal hole plasmas. Some people might say this is obvious, others might be surprised. But in a literature flooded with articles focused on changes with time and emphasis on spatial fine structure, this "boring" observation is wanting of an explanation, yet it speaks directly to a surprisingly structure-less state of the heated plasma over enormous distances. Strong and frequent variations from median brightnesses are absent.

More sobering is the myth, attributed to (fictional?) Nasreddin, purveyor of satirical anecdotes in the thirteenth century Muslim World. Nasreddin tells of a husband losing a ring inside his house. When his wife asks why he is looking in the garden when the ring was lost inside, he replies, "because there is much more light out here". Awareness of fallacies is naturally part of scientific approaches, helping avert confirmation bias. The point here is that sometimes one should ask

> "why did the Sun NOT do ...?",

as well as asking

> "why did the Sun do ...?"

The celebrated Michelson-Morley experiment produced a null result (Michelson and Morley 1887). Ostensibly it was a failure (unlike the celebrated experiment of Miller and Urey 1959), no aether was uncovered within which light waves could propagate through the universe! Yet it was central to the revolution in physics that came in 1905.

One might expect the first step in a scientific investigation should be to identify the questions to be asked. But this step is often out-of-step with modern pressures. Owing perhaps to culture or to protect careers, instead emphasis is put upon tried-and-true measures of success. Consequently, some publications fail to ask what needs to be done to solve a specific problem. Too often a somewhat formulaic and philosophically bankrupt approach involves data acquisition and processing, comparison with output from a computer code, followed by a highly specialized publication with results contingent upon untested or untestable ideas. This approach certainly leads to advancement of careers. The reader should judge for her- or his-self if our understanding of coronal heating has advanced quiet as much as it might.

What then *can* be done to advance understanding? Bayesian methods possess one advantage that they *require* one to assess the probability of a hypothesis (which could be an explicit model, part model, or even qualitative idea, if carefully formulated). Thus, one avoids results which are "not even wrong". They may well be inconclusive (no particular hypothesis being favored), but even then the researcher has learned something! The conclusion from this outcome would be that different hypotheses are required, better data, or both.

These general thoughts prompt us to condense ideas into the following suggestions for future research:

1. Refutation of specific models must be a guiding principle.
2. In "simple" systems, refutation can require just one critical observation (such as the precession of Mercury's orbit). But in non-linear dynamical systems, multiple solutions (bifurcations and attractors) must be expected, and perhaps emergent properties. Therefore we should not rely on the strategy of finding the *single* simplest model compatible with the data from any given observation (Occam's Razor).
3. Instead, each hypothesis, prediction, attractor or convergent state might be used as a prior for the probabilistic assessment of its likelihood of compatibility. Bayesian methods should be applied to multiple models of multiple observations of various types of coronal structure.
4. Novel instrumentation and observational programs should be designed to refute multiple models. Numerical experiments are an excellent tool available to the experimenter to identify the least ambiguous signals emerging from a 3D, time-dependent experiment which survive integrations to produce synthetic data. In this manner observations rich in *relevant* information content, and as tightly connected to the models as possible can be identified[1]

[1] As a counter-example, predictions of global white-light and EUV emission prior to total eclipses are often made using MHD models (see Boe et al. 2022 for a recent example). Being replete with multiple assumptions, ad-hoc prescriptions for extrapolation of magnetic fields and plasma densities, and brushing all small scales under numerical transport coefficients, little can be concluded by comparing prediction with observation without a great deal of faith. Worse still, the synthetic EUV data do resemble the corona on global scales. The authors of this book, at least, do not know how to interpret the meaning of this resemblance.

5. In turn, *robust* quantitative predictions reflecting essential properties of multiple models must be sought. One example might be spectral line profile asymmetries associated with the MHD range of scales comprising a turbulent cascade, phase mixing or resonant absorption. Another may be the assessment of ensembles of observations of flares against attractor and convergent solutions found in numerical experiments or analytically.
6. Machine learning tools (ML) are attractive for applications to many aspects of the problem. The agnostic nature of ML algorithms means that applications are limited firstly by our collective imagination. From schemes for speeding up intensive numerical calculations, to the expansion of Bayesian assessment of multiple, coupled hypotheses, it is easy to imagine developments revolutionizing solar science, among many others. It can be hoped that risks associated with subjectivity, confirmation bias, under-sampling, and the challenges associated with analyzing overwhelmingly large, diverse datasets can be addressed. Such tools might also be applied before embarking on a given study, to assess the richness of information made available through new and existing instruments delivering new or un-analyzed data, thereby minimizing potentially redundant and expensive machine operations. Wheeler might have appreciated the availability of tools allowing us to get the mistakes over with as quickly as possible.

Confirmation bias is the tendency of people to favor information that confirms or strengthens their beliefs or values, difficult to dislodge once affirmed. It can be particularly difficult to eliminate in remote sensing when there exist null spaces in inverse problems, because the data used can never inform about certain aspects of a system. Confirmation bias reflects an unfortunate, but natural tendency for society to get positive outcomes, to "validate" models, and it encourages authors, editors and referees to feel like progress is ongoing.

The downsides of confirmation bias are evident in the published literature, they are real and present. It is an "open secret" that it remains more difficult to get negative results published. Yet zeros in data, and negative or null results can be every bit as important as positive results. Another aspect of this stealthy bias can be put in check by asking the question, "are our inferences a case of *unreasonable agreement*, given the complexity of the system under study?" The corona observed above the solar limb appears to be consistent with plasmas at just a few discrete electron temperatures (see Page 5.4, Landi and Feldman 2008). Again, Bayesian methods can help avoid bias by delivering posterior probabilities for hypothetical prior emission measure distributions. Indeed, Kashyap and Drake (1998) successfully applied this method to active region coronal data from the SERTS rocket experiment (Thomas and Neupert 1994), finding weaker evidence for discrete temperatures, which anyway were different from those of Landi and Feldman (2008).

We should remember that entirely un-directed research ("fishing expeditions") can be critically important, particularly perhaps for a mature but challenging subject like solar physics. Darwin could not easily have anticipated the richness of phenomena laid out for him to observe as he traversed the globe in the Beagle. The pressing need to observe as much of the journey of magnetic free energy from large to dissipative scales in the heating problem suggests that fishing with bigger telescopes in space, seeing more of the "inertial range" (using the turbulence language), just to "see what is there" over the scale horizon, is highly desirable.

The rapid developments in Machine Learning and Artificial Intelligence are already yielding fascinating new results in solar physics. They are so rapid and with such a general range of applications that we can do no justice to the emerging field, and instead can point to specific examples in prediction (e.g. Panos and Kleint 2020; Panos et al. 2021; Kleint and Panos 2022), and to broader reviews (Armstrong and Fletcher 2019; Asensio Ramos et al. 2023). Such general methods can of course be applied to enhance data processing, to models varying from numerical experiments of the whole solar atmosphere to MHD turbulent scenarios, and surely we can expect progress in understanding coupling across scales, from MHD to kinetic scales, development of fast algorithms enabling large advances over standard techniques. The possibilities are summarized nicely in the closing words of Asensio Ramos et al. (2023):

> "...we are also starting to witness a huge revolution in solar physics in which deep learning models are informed with physical models. A good synergy can be obtained if the physical laws that we currently use to interpret our observations are used together with neural networks to approximate the most complex parts of the process."

This optimism should perhaps be tempered by the insight from Tsiklauri (Page 3.6).

6.2 How Does the Chromosphere Control Coronal Physics?

Research on coronal heating is rightly driven in the framework of electrodynamic coupling across the solar atmosphere (Chap. 4). A large literature has emerged based upon available measurements of components of magnetic field vectors, which are overwhelmingly from the photosphere (e.g. Fisher et al. 1998; Mandrini et al. 2000; Lites 2000; Wiegelmann and Sakurai 2021). These measurements are then related to the thermodynamic response of the overlying corona, the plasma properties which are frozen into the coronal extensions of measured photospheric fields (e.g. Kazachenko et al. 2022). The chromosphere has frequently been considered somewhat as an afterthought, perhaps because it is geometrically thin, even though physically it is a complicated region of the Sun. But some have valiantly sought to explore the effects of the chromosphere on the corona and its dynamic behavior, because extrapolations from the photospheric "surface" have been found to depart from "fine structure" in chromospheric observations, assuming that this fine structures trace bundles of magnetic lines of force (Wiegelmann et al. 2008; de la Cruz Rodríguez and Socas-Navarro 2011; Wiegelmann et al. 2014). Here we encounter, unfortunately, not just weakness in methodology and analysis of information-poor data, but more serious issues. Quoting from an abstract from an article from 2012:

> "...since the photospheric magnetic field does not in general satisfy the force-free condition, some kind of processing is required to assimilate data into the model"

we read that some modern ideas (in this case, data assimilation) actually violate a central rule of scientific methodology: editing the data to fit a model. At its core, this work has changed the photospheric data to fit a force free model, when photospheric fields are self-evidently not force-free, and when even the weaker low chromospheric fields are also not force free (Metcalf et al. 1995). Data editing must have no place in natural science. This departure from good practice is perhaps driven by the lack of information-rich data while being under pressure from societal demands to provide "deliverables", i.e. something of perceived value to society, not necessarily natural science.

Improved methods in this particular problem have included magnetostatic modeling, but because of the gulf between photosphere and corona, even the most recent work fails in serious ways to provide useful lower boundary conditions for the corona (Vissers et al. 2022). The authors found systematic differences in vector fields approaching an order of magnitude in the near-horizontal chromospheric fields, tens of percent in the vertical fields. Inversions (ill-posed, of course) produced fine structure in the magnetic fields not seen in magnetostatic models. Ironically, this result seems related to a well established observation in that the magnetic structure above sunspots depends not only on lower and horizontal boundary conditions, but also on *upper* magnetic boundary conditions (Rempel 2012), i.e. within the corona itself! Here perhaps is another reminder that the entire solar atmosphere may be self-regulating, the stresses in the coronal magnetic field having measurable influences on the atmosphere below. In this picture, the very idea that the chromosphere can set lower boundary conditions to set the stage for coronal heating may need re-visiting. There is no contradiction with cause-and-effect as measured by looking to energy losses (Table 1.3), because magnetic stresses dominate both in upper chromosphere and corona and global constraints from the fixed topology influence the plasma dynamics equally in both types of plasma.

In order to explore this possibility, we can assume the hypothesis that such self-regulation is, to first order, absent, and test it. Even then, other potentially tricky methodological problems persist. A recent article asks if the force-free assumption matters for the extrapolation of photospheric magnetic fields into the corona. The bigger question is what might be the benefit of knowing the answer, especially when the conclusions are highly contingent and conditional, qualitative ("rough agreement"), derived from an elaborate analysis which also conflates observations with the numerical experiment. The methodology seems akin to swimming in warm but dangerous shark-infested waters, the swimmers allured to them by the comfort of insight, societal interest, and apparent (but dubious) connection to the "real Sun" in comparing chromospheric fine structures with extrapolations.

Many other works assume (perhaps implicitly) that the chromosphere is sufficiently well understood, on the basis of measured photospheric magnetism and Maxwell's equations alone, when constructing large-scale and global models of the emerging magnetic field. The chromosphere is frequently ignored completely (e.g. Gary 1989; de Rosa et al. 2009; Boe et al. 2022). The impression gained from reading these and related articles is that they are informing something about coronal heating through a comparison of observed and computed magnetic morphology. But this must be treated with skepticism, considering the issues addressed throughout this book, and the naïve prescriptions used in the MHD models for the lower boundary conditions and ad-hoc prescriptions used for heating rates. If our $B^\flat$ conjecture proves useful, the morphology of guide fields $\mathbf{B}_0$ is essentially irrelevant to the steady heating problem, playing only passive roles. Given the (sobering?) reality of Fig. 1.5, namely that our observations of photosphere and corona are desperately mis-matched, and the focus on instruments more targeted towards large scale magnetically-driven plasma dynamics, the situation is perhaps not surprising. The purpose of the TRACE mission two decades ago (Handy et al. 1999) was indeed to trace structures and record the connections through the chromosphere (TRACE measured narrow-band intensities of UV and EUV emission formed from upper photosphere to the corona). This proved difficult if not impossible for many TRACE observations (Berger et al. 1999), for reasons which are now clear (Fig. 1.5).

More recently, the IRIS mission ("Interface Region Imaging Spectrograph" Lemen et al. 2011) combined UV spectroscopy, slit-jaw imaging and "advanced" MHD models (Gudiksen et al. 2011, updates from Nóbrega-Siverio et al. 2020) in pursuit of understanding the chromosphere and its upward extension into the corona (see the review by de Pontieu et al. 2021). But, like all previous missions, no direct measurements of magnetic fields which reach across this "interface region" were possible, except occasionally through supporting ground-based observatories targeting spectral lines sensitive to magnetic fields in the chromosphere (e.g. Judge et al. 2010; Aschwanden et al. 2016; Vissers et al. 2022). The first direct measurement of systematic changes in chromospheric magnetic fields came only in 2017 (Kleint 2017, see Sect. 4.10.2), recorded by the IBIS imaging spectropolarimeter (Cavallini 2006), over the course of an X-class flare. A commonly held scenario based upon the step-wise contraction of overlying coronal loops (Hudson 2000) produced changes in line-of-sight magnetic field of the opposite sign to those observed (Kleint 2017). Clearly much ground will need covering before we can claim an understanding even of the flaring chromosphere and corona, where changes in chromospheric magnetic fields are large and accessible to future observations with DKIST and other facilities.

Seen in terms of energy fluxes, coronal heating results from a small (1–10%, in the quiet Sun) spill-over of the energy used to sustain the chromosphere (see Tables 1.3, B.1, also Athay 1976). This is not the same idea that there is only a chromospheric heating problem (Aschwanden et al. 2007), instead it is a statement that the overwhelming fraction of energy needed to sustain the atmosphere against energy losses is deposited in the chromosphere. Seen in this light, *the coronal heating problem can surely be considered as solved only after the chromosphere itself is quantitatively understood*. If, additionally, self-regulation is important, the chromosphere and corona need to be treated as part of the same system from the outset. Such coupling provided a solution to an earlier conundrum in one dimensional models of the solar wind. Hansteen et al. (1997) showed that a theoretical problem of extreme mass-flux dependence of the wind plasma temperature was an artifact of ignoring coupling of multi-fluids between chromosphere and corona. We suggest that coronal heating processes themselves may determine their own fate (Einaudi et al. 2021), perhaps explaining why only some flux bundles are heated at any given time (Sect. 5.6).

While the chromosphere and its connections to the photosphere has been the subject of many studies (see the review by Carlsson et al. 2019), few have addressed how the chromosphere sets the electromagnetic lower boundary for the corona. Notable exceptions are articles by Beliën et al. (1999) and Prokopyszyn et al. (2021). The effect of the chromosphere on the upward Poynting flux of EM energy measured in the photosphere is basically not understood (e.g. Welsch 2015, found no relationship between the upward directed Poynting flux and chromospheric heating in plages). Multiple processes can actively modify the hydromagnetic state of the plasma as it emerges through the atmosphere, electric currents can be dissipated via rapid collisions including those between ions and neutrals ("Cowling conductivity", e.g. Arber et al. 2007, 2009; Nóbrega-Siverio et al. 2020). Multi-fluid physics of chromospheric reconnection, an explicit and zeroth-order effect on the upward Poynting flux to the corona, has recently been discussed by Wargnier et al. (2023). Additionally, the chromosphere hosts the magnetic $\beta = 1$ surface, because the average solar magnetic flux density in interplanetary space is a few times 10^{-4} T (Linker et al. 2017; Riley et al. 2019). Extrapolating down to the surface, it exerts a pressure of 0.07 Pa, which equals the plasma pressure somewhere within the chromosphere (Table 1.2). The chromosphere is therefore expected to host MHD wave mode conversion (Osterbrock 1961; Cally 2011; Cally and Khomenko 2018, 2019; Cally 2022) near the surface where sound and Alfvén speeds are equal (close to the surface where $\beta \approx 1$), and where the changing pressure balance causes the development of significant field-line curvature and current sheets (e.g. Gabriel 1976). Slow MHD waves efficiently shock in the highly stratified layers (Osterbrock 1961), which can nevertheless modify the upward Alfvén wave spectrum (Beliën et al. 1999).

In summarizing this section, *a concerted effort to measure the change in hydromagnetic state directly across the chromosphere with polarimetry*, no matter the challenges (e.g. Uitenbroek 2011), is surely warranted, both for the sake of understanding the chromosphere, as well as setting the lower boundary for the corona. Judge et al. (2021) have presented a possible way forward, through high signal-to-noise spectropolarimetry of strong lines of Fe, Fe^+ and Mg^+ between 259 and 281 nm. Many may consider this region of the solar spectrum to be too crowded with spectral lines. But effects of crowding are avoidable by using only the strongest lines, just as the very strong lines of Mg^+ have been successfully observed and analyzed in terms of chromospheric magnetic fields with the CLASP spectropolarimeter (Ishikawa et al. 2021; Rachmeler et al. 2022). Some initial data from the suborbital flights of CLASP appear to be compatible with calculations for the Mg^+ h and k lines. In concert with numerical MHD, radiative transfer models and novel algorithms (Rempel 2017; Manso Sainz et al. 2019; Štěpán et al. 2022), such work from space supported by ground-based measurements appears desirable and even necessary to address both the coronal heating problem, and serious deficiencies in predictions of flares, CMEs with space weather applications (e.g. Leka et al. 2019).

The history described above would suggest that the continued pursuit of thermodynamic properties of coronal plasmas through just imaging and spectroscopy will be an endeavor with diminishing returns. Over 5 decades of space-based imaging and spectroscopy of the chromosphere and corona, from the OSO series of satellites, through SKYLAB, Solar Maximum Mission, HRTS and other suborbital instruments, SoHO, SDO, IRIS and others have failed to deliver conclusive results on coronal heating. A next natural step will be to acquire polarimetric measurements across the chromosphere, extending as high as possible into the corona. In this way, the electrodynamic state in low β conditions will be probed. The interface between chromosphere and corona is thermodynamically inhomogeneous in space and time. But with accompanying traditional coronal observations, informed by numerical experiments, the nature of chromospheric-coronal coupling will become clearer, including the extent to which coronal heating is part of a self-regulating system.

This book will have been a success if readers appreciate that new approaches are needed, whether or not they are found in the pages here, to obtain information-rich data of painful and direct relevance to corona heating. Two kinds of new observations requiring instrument development are outlined in the next section.

6.3 Novel Instrumentation

6.3.1 A Theory-Driven, Observationally Tested Approach

Following the above considerations, we ask what the best new instruments might look like, by recognizing first the *theoretical* underpinning of coronal heating. This approach stands in contrast to an approach based upon *observable phenomena*, such as studies prompted by the properties of spicules or "flux cancellation", whatever these phenomena might be identified as, physically. Nevertheless our proposed methodology remains an observation-driven enterprise, observations—from an instrument or perhaps a numerical experiment—are the only way to assess the credibility of any hypothesis. The approach is to be firmly guided by theory, in order to identify the new observational tools needed to confront ideas emerging from fundamental physics. Unlike phenomenological approaches, theory can predict the presence of critical scales at which we expect changes in dynamics to become observable, for example. Let us illustrate our approach with an example.

Inherent non-linearities in the dynamical evolution across an enormous range of scales have denied to us unambiguous observable signatures of the journey of ordered energy into heat. Interestingly, section 7.3 of Schekochihin (2022) answers its own question about the elaborate phenomenological theory of anisotropic and aligned MHD turbulence:

> "Is this a falsifiable theory?"
> "...With considerable difficulty".

The answer is a refreshingly honest assessment of a theory which has recently undergone significant breakthroughs in understanding, it is not meant to be discouraging. Indeed, it offers a gauntlet to experimenters to pick up, and challenge. One question stands out that may well be soon within the reach of experimenters, it has been encountered several times already in this book. A major *potentially observable* difference in dissipation scales arises according to whether compressible or incompressible modes of fluid motion are produced during the evolution towards dissipative scales (Hollweg 1986). Specifically, compressive motions may be dissipated on scales of order 100 km by large viscosities associated with the ion fluid (Sect. 5.2). The conditional "may" used is required, but the prediction is worth exploring for several different reasons.

Evidence for compressive waves propagating downwards in lines formed near 10^5 K was presented by Hansteen (1993), Wikstøl et al. (1997), and Judge et al. (1998), but like all earlier work, the results, based upon information-poor data, can be debated. Compressible motions can be produced easily from transverse motions in situations where the Alfvén speeds vary across the dominant magnetic field $\mathbf{B}_0$ (Cargill et al. 2016; Howson et al. 2019, 2020; Goossens et al. 2021). In many cases the dissipation via viscosity is enhanced, but the results depend on details of the numerical experiments, since it is only the component of velocity gradients along $\mathbf{B}$, not $\mathbf{B}_0$ or $\mathbf{b}$, which is important. Compressions have been found in models of braided magnetic fields and those undergoing resonant absorption. Viscous dissipation is given by the sum of Eq. (D.2) for Q_0 and (D.4) for Q_1 which is $\omega^2\tau^2$ smaller. Given that the non-linear interaction terms in the equations of motion are of order u^2/c_A^2 and $u/c_A \approx \xi_0/c_A \sim 0.02$ with ξ_0 measured line widths (Appendix A and Table 1.2), then it seems possible that the compressional dissipation will dominate over dissipation through shear motions, because the latter have coefficients $\omega^2\tau^2 \sim 10^{10}$ times smaller. However, a stronger statement will require knowledge of the stress tensor components $\partial u_i/\partial x_j$, generally requiring numerical modeling such as that of Howson et al. (2020). But fortunately phenomenology allows us to estimate the magnitude of these terms (Appendix C) for the special case when MHD turbulence is the dominant non-linear mechanism generating small scales. Let us use the anisotropic but un-aligned results from Goldreich and Sridhar (1995), and set $\partial/\partial x = \partial/\partial y = k_\perp$, $\partial/\partial z = k_\parallel$. Then the total dissipation due to viscous ion heating, via MHD turbulence, is (Appendix D)

$$Q = Q_0 + Q_1 \sim \eta_0 \left[k_\parallel^2 u_\parallel^2 + \frac{1}{\omega^2\tau^2}\left(k_\parallel^2 u_\parallel^2 + 2k_\perp^2 u_\perp^2\right)\right] \quad \mathrm{W\,m^{-3}}. \tag{6.1}$$

Assuming all heating is supplied by Q_0, we have (Eq. (4.15)) which yields values for $k_\parallel$ and $k_\perp$. Thus perpendicular scales of 35 m or smaller will be controlled by ion viscosity. Any collisionless ion dynamics, occurring on scales of 10 m, would have to be in the sub-viscous range. The corresponding parallel length scale is 400 m, and perpendicular and parallel velocity amplitudes are just 90 and 25 m s^{-1} respectively, far below thermal speeds. *These scales and velocities mark a reliable prediction* since the relation between parallel and perpendicular scales is particularly robust in MHD turbulence (refer to section 5.2 of Schekochihin 2022).

We can also consider dynamical processes other than MHD turbulence which can generate shear and compressional motions in the coronal plasma. Let us assume more generally that ion viscous dissipation erases structure on all scales, whether induced by waves or turbulence, below a critical length ℓ_η which is somewhere well above the kinetic scales of $\sim$ 10 m. The idea has already passed one interesting test, there should be no accelerated particles because classical viscosity suffices to thermalize particle distribution functions at scales orders of magnitude larger than inertial lengths (Sect. 4.5). Then new observations with angular resolutions below ℓ_η will show no structure below that scale.

Now if MHD turbulence is not the dominant mechanism driving small, dissipative scales, then detailed computations of terms in Eq. (6.1) will be required. It is probable that other mechanisms will lead to different scales of viscous dissipation. An observation revealing the absence of structure below, say, 100 km would suggest that very different motions accompany the primary energy transport mechanisms than are developed by MHD turbulence. Therefore we should be open to observing the smallest scales that are practicable, even (or perhaps especially) if we believe that we will not reach dissipation scales.

6.3.2 *Observing Scales of Dissipation*

Length scales of $\sim 10^2$ km have been encountered in several different situations in this text. Davila (1987) estimated scales of $\sim 10^2$ km from arguments involving resonant absorption. The HiC EUV imaging instrument achieved an angular resolution approaching 200 km during a 2018 flight (Rachmeler et al. 2019). Williams et al. (2020a) suggest, through multiple Gaussian profile analysis, that spatial, cross-field profiles of observed "loops" are

> the result of multiple strands along the integrated line of sight that can be resolved, rather than being the result of even finer sub-resolution elements.

But of course, this amounts to extrapolation to sub-pixel scales, so it must be viewed accordingly. While Williams et al. (2020b) advocate for "a permanent solar observatory with the resolving power of Hi-C", here we argue, on the basis of known theory (viscous damping), as well as intriguing open theoretical questions in MHD, for the need to obtain data with a resolution of *at most* 100 km, preferably closer to or below 50 km.

Can an EUV instrument capable of resolving 100 km on the Sun's surface provide data to refute a dominant role for compressive viscous dissipation? Can such an instrument be built (cf. Kobayashi et al. 2014; Rachmeler et al. 2019)? Can it be stabilized against jitter to permit long enough integrations (Douglas et al. 2021)? In the 1990s a proposal to do ultra-high resolution solar work in space was considered but not selected for launch (Damé et al. 1991; Damé et al. 1994). Can such an instrument be flown? These are questions for detailed research. He we argue that such a mission *should* be flown.

But we can see that such an instrument, subject to overcoming these technical challenges, *will* refute this question, because we are assessing a theoretical prediction that the dissipation scale is *resolvable* by the new instrument. For simplicity, consider that the objects we observe are slender loops of equal brightness on a black background, which are sometimes blended along the line-of-sight (Figs. 2.2 and 5.5). Consider also that line-of-sight additions of brightness (Fig. 3.3) are not an issue, the elementary structures being sufficiently narrow that overlapping structures are minimal in number. The image shown in Fig. 6.1 illustrates a purely hypothetical structure, intended to represent several bright magnetic flux bundles in a force-free structure. Three outcomes of the hypothetical observation are possible:

1. Neither instrument resolves the elementary structures, each pixel has counts proportional to the area of a pixel (flux = intensity times area), and the images are diffuse and uniformly bright (not shown in the figure).
2. Both instruments resolve the elementary structures, again each pixel has counts proportional to the area of a pixel, images have similar contrasts.

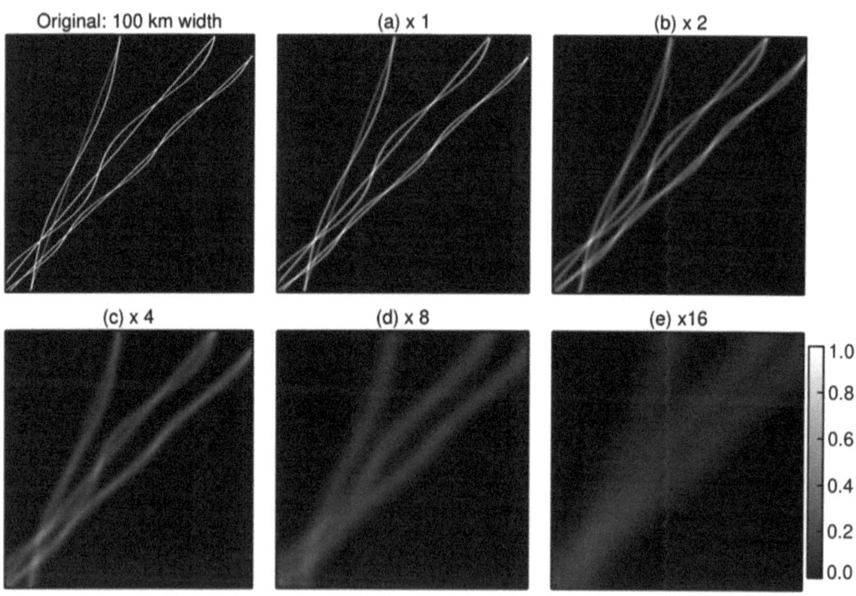

Fig. 6.1 An artificial image of heated flux bundles representing viscously heated loops with widths of 100 km. The top left panel shows an ideal, hypothetical solar image covering about 13×13 Mm of the surface. The other images (**a**)–(**e**) have been convolved with Gaussians of width 100 km times the factor listed at the top, and then binned in widths 1/2 of this (critically sampled). The counts per pixel are smaller when under-resolved. In the resolved image (panel a) the counts per pixel are highest, being optimally suited to measure structure at the actual solar scales ("Original" panel). *The zeros in the images are every bit as important as the non-zeros*

3. The new instrument resolves the structure, the other does not. The counts per pixel in each structure are higher with the new instrument, because fewer pixels are occupied by the emitting structures.

There is a "sweet spot" where the accumulated counts per pixel reach a maximum, just where the bright structures are resolved, i.e. where the pixel size matches the projected size of the structure on the Sun. The zeros in the resolved images contain information of as much interest as the non-zero pixels. In our quest to find the mechanisms for heat dissipation, the distribution of observable widths of heated features constitute a strong constraint. Again, compressive viscous heating is predicted to suppress structure on scales below 100–200 km.

For reference, the HiC instrument has a diffraction-limited resolution $1.22\lambda/D$ of 0.019" at $\lambda = 17$ nm, about 14 km on the Sun's surface. So far it has operated with pixels 0.129", covering an area of fifty times the diffraction limit. Hi-C operates as a "light bucket", i.e. under-sampling the resolution in order to acquire higher signals. Even at this resolution only some images were minimally affected by jitter during flight, revealing an important technical challenge. Thus, if the HiC image structures were under-resolved, with many pixels close to zero, the result is that a blurry image with few zeros is measured (panel e), whereas in reality there is a sharp image of a few bright pixels with many zeros (panel "Original").

6.3.3 The Lower Boundary Conditions of the Corona

How are we to improve knowledge of the nature of the drivers of coronal heating? Do we need to study the transition region, assuming that what we observe there is a *bone fide* permanent transition between chromosphere and corona, allowing energy and momentum and mass to flow through it, revealed to us through the radiation that we observe? Or is it something else entirely, that we cannot trust to inform us of what is entering the corona (e.g. Feldman 1983)? Does this even matter because it contains so little mass that it must do whatever the neighboring plasmas and their magnetic stresses demand? To treat the transition region as a fixed structure in which heating can occur, and through which waves pass or reflect assumes that the work done by the waves does not overwhelm and disrupt the structure. No matter what, the chromosphere is a bigger bully and will define, along with the downward heat flux from the corona, how the chromosphere-corona transition behaves (see the enormous vertical corrugation of the interface between the chromosphere and corona presented in the "realistic" models of Rempel 2017).

Chromospheric work in recent decades has focused almost exclusively on analyzing and synthesizing chromospheric data of the highest angular and time resolution possible. DKIST is on the verge of providing spectropolarimetric data for chromospheric lines with resolutions beneath the scales of the mean free paths of the emitted photons (roughly a pressure scale height of 10^2 km within the stratified chromosphere). But in recent years, a deluge of papers based on thermal (i.e. non-magnetic data) has been published, reviewed in exhausting detail by Carlsson et al. (2019) and de Pontieu et al. (2021). Results from the IRIS mission, which observes intensities through small spectral regions at UV and near-UV wavelengths, have flooded the literature with thermal data, generously augmented by speculative, contingent claims as to hydro-magnetic effects. Yet again, information-poor data of thermal properties are nevertheless used to claim advances in MHD waves, instabilities, jets, reconnection, non-thermal distributions of particles, electron beams, and a host of magnetically-driven phenomena sufficient to fill a zoo with new animals. The explosive growth of publications so strongly contingent on untested hypotheses and numerical calculations so far from solar conditions, reflects a worrying trend in solar physics. A failed search for occurrences of the word "refutation" and derivative/related words in the reviews reveals again the danger of confirmation bias. It is difficult to see where credible advances have been made in these complicated analyses, given some elementary contradictions already pointed out (e.g. Sect. 4.5).

In this book we deliberately chose to start not from the morass of chromospheric and transition region data from instruments or computer programs, but from simpler physical principles. To those deep in the modern activities of chromospheric and coronal work, this will certainly seem naïve and perhaps fruitless, given the sheer complexity of the staggering array of phenomena reported in recent years. Yet we have concluded that we must understand the chromosphere in order to understand the corona. Does this mean that we must explain all these detailed studies whose number grows sharply every year? We believe not. Thus we will restrict this short section about novel chromospheric instrumentation not only to simple physics, but also only to those physical processes which we consider likely to have a major impact on setting the scene for coronal heating, with all its apparent complexity, yet also simplicity (see Fig. 5.4).

So, with the continued patience of the reader, in trying to understand the lower boundary conditions for the corona, we return again to some basic ideas. The coronal plasmas are connected through some kind of thermal interface to the chromosphere. This interface may or may not be clearly visible as what we call "the transition region", it can easily be parallel to and perpendicular to the magnetic field vectors (Gabriel 1976). There certainly are field-aligned connections to the corona, the corona does not float like bubbles atop a deck of clouds of "unresolved fine structures" around 10^5 K, it is anchored magnetically through the dense chromosphere and into the massive photosphere beneath. Electrical conductivities are high enough for Alfvén's theorem and frozen topology condition to apply (although see an interesting discussion of the deepest, coolest regions of sunspots by Feldman 1993). Thus the photosphere-chromosphere-corona system can be studied as a fixed topological magnetic system subject to the vagaries of forcing from beneath and magnetic reconnection. This picture led naturally to Parker's Fundamental Theorem (Sect. 4.7). The transport of mass and energy occurs along lines of magnetic force, outside of reconnection events. Magnetically-dominated momentum transport is fully 3D, mediated by fast and intermediate wave modes, and is not restricted in the same manner. A major change in configuration arises somewhere in the chromosphere (outside of sunspot umbrae and penumbrae where it can occur in the photosphere), as the plasma pressure drops exponentially, according to hydrostatic stratification, but the magnetic pressure drops roughly algebraically, according to the well-known multipolar expansion of magnetic fields from sources under the photosphere. So we arrive at a first target for setting the lower coronal boundary condition:

> We must measure at least the plasma pressure and large-scale fields $\mathbf{B}_0$ across the many scale heights of the chromosphere, including those dynamical excursions naturally driven in the tenuous uppermost layers.

In identifying the $\beta = 1$ surface within the chromosphere (remembering to include neutrals in the plasma pressure owing to strong collisional coupling, see the footnote in Sect. 3.8), we set the stage for understanding the magnetic field expansion, field strengths and current sheets between strong- and weak-field regions even in the quiet Sun (see Gabriel 1976 for an uncomplicated example). This region is close to surfaces where conversion of MHD wave modes, is efficient. At heights near the $\beta = 1$ surface, the chromosphere therefore sets not only the steady state of the plasmas, but also the nature of the MHD wave field emerging into the upper chromosphere and corona. Above this surface, the β reduces further, usually to well below 1, and the Alfvén speed increases according to the reduction in pressure and the equation of state (relating pressure p to density ρ). The increase in Alfvén speed thereby permits a rapid increase in the opportunity for *reconnection* to occur while the atmosphere tries to reach a lowest energy state. If it occurs (with signatures similar to explosive events studied numerically by Heggland et al. (2009) and Wargnier et al. (2023), perhaps even including spicules), both the steady and wave fields of the system at the coronal base are modified. If this first target can be successfully hit, then we will better understand the steady state of the MHD system near the coronal base, if not also the wave field, and its "slow" evolution (compared to dynamical times in Table 1.2). *This would mark a significant improvement over using the photospheric magnetic fields as a lower boundary condition.*

Prompted by these and considerations based upon the evolution of $\mathbf{B}_0$ in the chromosphere and its effects on space weather, Judge et al. (2021, 2022) undertook a large study, culminating in a proposal for a new SMEX mission with the preliminary name Chromospheric Magnetism Explorer "CMEx". Figure 6.2 shows the plethora of atomic transitions all lying in a narrow near-UV wavelength region, which sample hydromagnetic conditions across the entire chromosphere. The CMEx configurations proposed make no attempt to explore thermal fine structure of the chromosphere beyond what an angular resolution of around 1" might deliver. After all, the plasma is in a low-β state where plasma contributes little to force balance, instead it reveals the plasma frozen within small bundles of flux within a larger, more uniform structure. The observed, assumed field-aligned thermal fine structure which naturally draws much attention yet says very little about the magnetic field. But, it may be important to assess the effects of changes in Alfvén speed as densities change across the dominant magnetic field vectors. Fortunately, DKIST

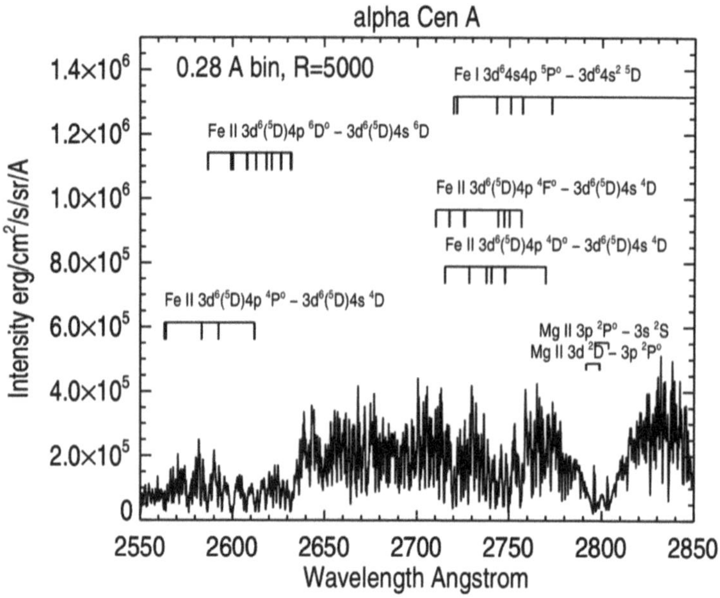

Fig. 6.2 Smoothed UV spectrum of α Cen A, obtained by the Hubble Space Telescope, displayed as a proxy for the mean solar intensity spectrum. The large number of lines of Fe ("Fe I") and Fe^{+} ("Fe II") are indicated and annotated with their multiplet terms. The Mg II spectrum is also annotated including the 3s–3p h and k lines, and the 3p–3d lines. Notice that the permitted Fe II lines ("Fe II", shown with solid bars with downward ticks) consist of the ^{6}D resonance lines (4s–4p) near 2600 Å, and quartet transitions at other wavelengths that occur between metastable (even-parity) quartet lower levels and odd-parity quartets of the same transition array (4s–4p). These atomic transitions are well understood and will potentially explore the middle and upper chromospheric hydromagnetic state reliably for the first time. From Judge et al. (2021)

offers opportunities to study the fine structure in relation to MHD processes and heating.

Various other arguments come into play in designing a mission to deliver such measurements and, rather unexpectedly, the near UV region is optimal. Trade-offs are inevitable, and minimum integration times limit the frequencies of waves which might be probed, but some wave motions could be measured with sampling intervals of a few seconds (along a single slit). The rationale behind this choice is that the scales relevant for driving coronal heating, for example by MHD turbulence, occur on the scales of Poynting fluxes generated by photospheric granules (Mm) and larger. In the chromosphere the magnetic structure expands to fill the available areas spanned by several Mm before meeting other magnetic fields collected into stable configurations (the calcium network) by supergranulation on scales of 20–30 Mm.

But this rationale does not hold up against all the evidence from the highest resolution observations, *if* one believes that these finer scale structures (many below 0.3", see for example Fig. 3.8) are abundant enough to affect the vastly larger volumes and masses of plasma in the corona. The following strawman arguments (*italicized*) reveal typical difficulties concerning the interpretation of such data. Not only do they illustrate the general problems and complicated, difficult arguments often associated with information-poor data, but they also illustrate the particular problem we face concerning the lower boundary conditions in the corona. Refer to Fig. 3.8.

> *The physical nature of these objects (spicules) is still under debate. But we can estimate their potential influence on the corona? We recall a calculation from (Athay and Holzer 1982), but use these new, Hinode data. Roughly* $\approx 10^5$ *such structures exist on the Sun at any time. Assuming that these are tube-like jets of plasma flowing into the corona, their cross-sections suggest an area of* $\approx 0.05\,Mm^2$ *each, amounting to about* 6×10^{-4} *of the area at the base of the corona. With a mass density of* $2\ 10^{-10}\ kg\ m^{-3}$*, upward velocity of* $50\,km\ s^{-1}$*, the total mass entering the corona is* $6\ 10^{-9}\ kg\ m^{-2}\,s^{-1}$*. The solar wind mass loss rate is* $\approx 2.5\ 10^{-10}\ kg\ m^{-2}\,s^{-1}$*, 20 times smaller. But how do such structures expand to fill the corona? Their cross sections appear in images to remain narrow well above where magnetic bundles have expanded. Measurements of magnetic properties within these structures have been published (e.g. Centeno et al. 2010), with average magnetic field strengths typical of those extrapolated downwards from in-situ measurements in interplanetary space (Linker et al. 2017). Additional expansion of plasma within these structures to fill the corona seems to be implausible. How then do these features, known to arise only from supergranular boundaries, then fill the corona?*

These arguments inform about the now-familiar reality, that information-poor data entail too many assumptions, unknowns, and premises in order to reach credible conclusions. In this particular case, the broadly accepted picture is that such spicules indeed supply the corona with mass, perhaps even play a role in heating the corona (de Pontieu et al. 2017), and the bulk of the injected mass returns to be seen as redshifted emission lines in the transition region (Athay and Holzer 1982; de Pontieu et al. 2007). But based upon what evidence? These questions will not be resolved without information-rich data aimed towards refuting hypotheses. Ultimately, it is to be hoped that by identifying data tailored to the problem, one of two conclusions will be refuted: (1) these are interesting phenomena by themselves with no significant influence on the entire overlying corona, (2) they provide the entire corona with mass momentum and energy and so cannot be ignored when studying coronal heating. We believe that this case remains open.

In moving forwards, Fig. 6.3 illustrates the idea of using many spectral lines to measure hydromagnetic conditions across the entire chromosphere, including spicules. The near UV region of the Sun's spectrum has long been regarded as a challenging region owing to the severe crowding of spectral lines (Kurucz 1991a,b, 1992). However, Judge et al. (2021) turned this perspective around in application to polarimetric measurements of the chromosphere. They showed that there are many lines of the iron spectrum, especially Fe^+, strong enough that they form within the chromosphere, dominate the opacity in the line cores, and which finely sample a large range in optical depth. Being in the UV below the 310 nm atmospheric cutoff, an observatory in space is highly desirable if not essential to fill large gaps and extend polarimetric measurements far higher in the Sun's atmosphere than can be reached using ground-based instruments. Judge et al. (2021) argued for a modest

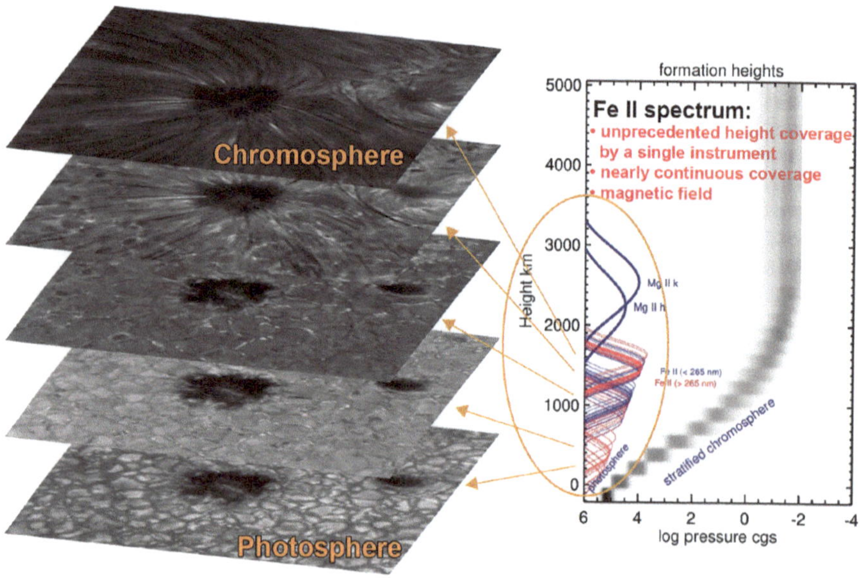

Fig. 6.3 A novel idea to span magnetic measurements from photosphere to the base of the corona using multiple spectral lines at UV wavelengths (259–281 nm). The figure is based upon the work of Judge et al. (2021). Left: an active region observed from the photosphere (bottom) to the mid chromosphere (top) using the IBIS instrument. Each image roughly corresponds to the formation heights of one of the spectral lines as indicated qualitatively with the yellow arrows. Right: the vertical axis shows the formation heights for the UV spectral lines. Only UV lines allow us to probe both the higher chromosphere and low transition region. Both the Zeeman and Hanle effects can be used to span a broad range of magnetic field strengths and configurations across the chromosphere using lines of Fe^+ (shown as Fe II)

telescope polarimeter, operating in low Earth orbit, to plug such gaps in essential capabilities.

These considerations lead us to fantasize further about a similar spacecraft instrument, placed far from Earth, designed to measure just the strong, first-order component of the polarized light in the Zeeman effect (Landi degl'Innocenti and Landolfi 2004). In measuring the line-of-sight component of the vector magnetic field, *from an additional vantage point far from the Earth-Sun line*, one can remove the degeneracy resulting from symmetries from measurements from Earth. Not only this, but a spectral line measured between 259 and 281 nm with a 0.4 m telescope has the same spatial resolution as that of DKIST with a 4 m aperture. Judge (2019); Judge et al. (2021) have shown that the photospheric and chromospheric spectra of neutral Fe and Fe^+ have UV and infrared lines which can be paired to measure true vector magnetic fields, without ambiguities, across the entire solar atmosphere. The "L4 and L5" Lagrange points would also benefit a society interested in space weather at Earth.

Chapter Summary

A glut of coronal data has revealed plasma properties after heating has occurred. Experiments geared towards space weather are generally ill-suited to the heating problem.

Recent decades have witnessed a huge growth in publication rates. Few have reached firm conclusions about coronal heating. Firm predictions should be sought and tested with novel data.

Hypothesis refutation and the acquisition of information-rich data must form the basis for future research. Bayesian methods offer remedies for confirmation bias, and for being "not even wrong".

Modern theories of coronal heating make important, specific predictions to be tested with novel instruments close to the limits of existing technology. Critical scales $\lesssim 100\,\text{km}$ are of particular interest.

The physics of the chromosphere must be understood through measurements of vector velocities and magnetic fields, before coronal heating mechanisms can be quantitatively assessed. Boundary conditions from photospheric measurements are not sufficient.

MHD and kinetic theories make different observable predictions depending on dissipation. These should be challenged through spectroscopy ($\mathcal{R} \geq 20{,}000$) and imaging down to critical viscous scales below 100 km, which will also more closely match scales accessible to DKIST.

Numerical experiments and observations suggest that the corona may be a self regulating system. With causes and effects mixed, boundary conditions might be just part of the story.

No instruments current or planned have capabilities which can address the most pressing theoretical questions raised in this book.

The questions raised here can be addressed using new instruments targeted towards coronal heating. They are sufficiently general to have relevance to studies of both laboratory and space plasmas. In particular, the theories of MHD turbulence are in need of observational scrutiny. Some fundamental questions involving both MHD and kinetic scales may be resolved with modest advancements in current instrumentation.

References

Arber, T.D., Botha, G.J.J., Brady, C.S.: Effect of solar chromospheric neutrals on equilibrium field structures. Astrophys. J. **705**, 1183–1188 (2009)

Arber, T.D., Haynes, M., Leake, J.E.: Emergence of a flux tube through a partially ionized solar atmosphere. Astrophys. J. **666**, 541–546 (2007)

Armstrong, J.A., Fletcher, L.: Fast solar image classification using deep learning and its importance for automation in solar physics. Sol. Phys. **294**, 80 (2019). https://doi.org/10.1007/s11207-019-1473-z

Aschwanden, M.J., et al.: The coronal heating paradox. Astrophys. J. **659**, 1673–1681 (2007)

Aschwanden, M.J., Reardon, K., Jess, D.B.: Tracing the chromospheric and coronal magnetic field with AIA, IRIS, IBIS, and ROSA data. Astrophys. J. **826**, 61 (2016). https://doi.org/10.3847/0004-637X/826/1/61

Asensio Ramos, A., et al.: Machine learning in solar physics. Living Rreviews in Solar Physics **20**, 4 (2023)

Athay, R.G.: The Solar Chromosphere and Corona: Quiet Sun. Reidel, Dordrecht (1976)

Athay, R.G., Holzer, T.E.: The role of spicules in heating the solar atmosphere. Astrophys. J. **255**, 743–752 (1982). https://doi.org/10.1086/159873

Beliën, A.J.C., Martens, P.C.H., Keppens, R.: Coronal heating by resonant absorption: the effects of chromospheric coupling. Astrophys. J. **526**, 478–493 (1999). https://doi.org/10.1086/307980

Berger, T.E., et al.: High-resolution imaging of the solar chromosphere/corona transition region. Astrophys. J. Lett. **519**, L97–l100 (1999)

Boe, B., et al.: The solar minimum eclipse of 2019 July 2. II. The first absolute brightness measurements and MHD model predictions of Fe X, XI, and XIV out to 3.4 $R_\odot$. Astrophys. J. **935**, 173 (2022). https://doi.org/10.3847/1538-4357/ac8101

Bogachev, S.A., et al.: Microflares and nanoflares in the solar corona. Phys. Uspekhi **63**, 783–800 (2020). https://doi.org/10.3367/UFNe.2019.06.038769

Cally, P.S.: Alfvén waves are easy: mode conversion in magnetic regions. In: Astronomical Society of India Conference Series, vol. 2, pp. 221–227 (2011)

Cally, P.S.: On the fragility of Alfvén waves in a stratified atmosphere. Mon. Not. R. Astron. Soc. **510**, 1093–1105 (2022)

Cally, P.S., Khomenko, E.: Fast-to-Alfvén mode conversion in the presence of ambipolar diffusion. Astrophys. J. **856**, 20 (2018)

Cally, P.S., Khomenko, E.: Fast-to-Alfvén mode conversion and ambipolar heating in structured media. I. Simplified cold plasma model. Astrophys. J. **885**, 58 (2019)

Cargill, P.J., de Moortel, I., Kiddie, G.: Coronal density structure and its role in wave damping in loops. Astrophys. J. **823**, 31 (2016). https://doi.org/10.3847/0004-637X/823/1/31

Carlsson, M., de Pontieu, B., Hansteen, V.H.: New view of the solar chromosphere. Ann. Rev. Astr. Astrophys. **57**, 189–226 (2019)

Cavallini, F.: Ibis: a new post-focus instrument for solar imaging spectroscopy. Sol. Phys. **236**, 415–439 (2006)

Centeno, R., Trujillo Bueno, J., Asensio Ramos, A.: On the magnetic field of off-limb spicules. Astrophys. J. **708**, 1579–1584 (2010)

Damé, L., et al.: A solar interferometric mission for ultrahigh resolution imaging and spectroscopy: SIMURIS. Adv. Space Res. **11**, 383–386 (1991). https://doi.org/10.1016/0273-1177(91)90135-7

Damé, L., et al.: The SIMURIS interferometric mission: Solar physics objectives and model payload. Adv. Space Research, **14**, 167–170 (1994). https://doi.org/10.1016/0273-1177(94)90177-5

Davila, J.M.: Heating of the solar corona by the resonant absorption of Alfvén waves. Astrophys. J. **317**, 514–521 (1987)

de la Cruz Rodríguez, J., Socas-Navarro, H.: Are solar chromospheric fibrils tracing the magnetic field? Astron. Astrophys. **527**, L8 (2011). https://doi.org/10.1051/0004-6361/201016018

de Pontieu, B., et al.: A tale of two spicules: the impact of spicules on the magnetic chromosphere. Publ. Astron. Soc. Jpn. **59**, 655 (2007)

de Pontieu, B., et al.: Observations and numerical models of solar coronal heating associated with spicules. Astrophys. J. Lett. **845**, L18 (2017). https://doi.org/10.3847/2041-8213/aa7fb4

de Pontieu, B., et al.: A new view of the solar interface region from the interface region imaging spectrograph (IRIS). Sol. Phys. **296**, 84 (2021)

de Rosa, M.L., et al.: A critical assessment of nonlinear force-free field modeling of the solar corona for active region 10953. Astrophys. J. **696**, 1780–1791 (2009)

Douglas, E.S., Tracy, K., Manchester, Z.: Fundamental limits on nanosatellite and cubesat telescope pointing: the impact of disturbances and photon noise. Front. Astron. Space Sci. **8**, 125 (2021). https://doi.org/10.3389/fspas.2021.676252

Einaudi, G., et al.: Energetics and 3D structure of elementary events in solar coronal heating. Astrophys. J. **910**, 84 (2021)

Feldman, U.: On the unresolved fine structures of the solar atmosphere. Astrophys. J. **275**, 367 (1983)

Feldman, U.: Is it justified to assume that 'everywhere in the sun's photosphere-corona domain the electric conductivity is HIGH?'; Or, what drives the solar upper atmosphere? Astrophys. J. **411**, 896–908 (1993)

Fisher, G.H., et al.: Coronal heating in active regions as a function of global magnetic variables. Astrophys. J. **508**, 885–898, 1998

Gabriel, A.: A magnetic model of the solar transition region. Philos. Trans. R. Soc. Lond. **281**, 339 (1976)

Gary, G.A.: Linear force-free magnetic fields for solar extrapolation and interpretation. Astrophys. J. Suppl. Ser. **69**, 323–348 (1989)

Goldreich, P., Sridhar, S.: Toward a theory of interstellar turbulence. II: Strong Alfvenic turbulence. Astrophys. J. **438**, 763 (1995)

Goossens, M., et al.: Mixed properties of magnetohydrodynamic waves undergoing resonant absorption in the cusp continuum. Astron. Astrophys. **646**, A86 (2021). https://doi.org/10.1051/0004-6361/202039780

Gudiksen, B.V., et al.: The stellar atmosphere simulation code BIFROST. Code description and validation. Astron. Astrophys. **531**, A154 (2011)

Handy, B.N., et al.: The transition region and coronal explorer. Sol. Phys. **187**, 229–260 (1999)

Hansteen, V.: A new interpretation of the Redshift observed in optically thin transition region lines. Astrophys. J. **402**, 741–755 (1993)

Hansteen, V., Leer, E., Holzer, T.: The role of helium in the outer solar atmosphere. Astrophys. J. **482**, 498–509 (1997)

Heggland, L., de Pontieu, B., Hansteen, V.H.: Observational signatures of simulated reconnection events in the solar chromosphere and transition region. Astrophys. J. **702**, 1–18 (2009). https://doi.org/10.1088/0004-637X/702/1/1

Hollweg, J.V.: Viscosity and the Chew-Goldberger-Low equations in the solar corona. Astrophys. J. **306**, 730–739 (1986)

Howson, T.A., et al.: Magnetohydrodynamic waves in braided magnetic fields. Astron. Astrophys. **629**, A60 (2019). https://doi.org/10.1051/0004-6361/201935876

Howson, T.A., de Moortel, I., Reid, J.: Phase mixing and wave heating in a complex coronal plasma. Astron. Astrophys. **636**, A40 (2020)

Hudson, H.S.: Implosions in coronal transients. Astron. Astrophys. Lett. **531**, L75–L77, 2000. https://doi.org/10.1086/312516

Ishikawa, R., et al.: Mapping solar magnetic fields from the photosphere to the base of the corona. Sci. Adv. **7** (2021)

Judge, P.G.: Chapter 5 - spectroscopy and atomic physics. In: Engvold, O., Vial, J.C., Skumanich, A. (eds.) The Sun as a Guide to Stellar Physics, pp. 127–155. Elsevier, Amsterdam (2019)

Judge, P.G., et al.: Evidence in support of the "nanoflare" picture of coronal heating from SUMER data. Astrophys. J. **502**, 981–996 (1998). https://doi.org/10.1086/305915

Judge, P.G., et al.: Fabry-Pérot versus slit spectropolarimetry of pores and active network: analysis of IBIS and Hinode data. Astrophys. J. **710**, 1486–1497 (2010)

Judge, P., et al.: Measuring the magnetic origins of solar flares, coronal mass ejections, and space weather. Astrophys. J. **917**, 27 (2021)

Judge, P., et al.: Optimal spectral lines for measuring chromospheric magnetic fields. Astrophys. J. **941**, 159 (2022). https://doi.org/10.3847/1538-4357/aca2a5

Kashyap, V., Drake, J.J.: Markov-Chain Monte Carlo reconstruction of emission measure distributions: application to solar extreme-ultraviolet spectra. Astrophys. J. **503**, 450–466 (1998). https://doi.org/10.1086/305964

Kazachenko, M.D., et al.: Invited review: short-term variability with the observations from the helioseismic and magnetic imager (HMI) onboard the solar dynamics observatory (SDO): insights into flare magnetism. Sol. Phys. **297**, 59 (2022). https://doi.org/10.1007/s11207-022-01987-6

Kleint, L.: First detection of chromospheric magnetic field changes during an X1-flare. Astrophys. J. **834**, 26 (2017)

Kleint, L., Panos, B.: Occurrence and statistics of IRIS bursts. Astron. Astrophys. **657**, A132 (2022). https://doi.org/10.1051/0004-6361/202142235

Kobayashi, K., et al.: The high-resolution coronal imager (Hi-C). Sol. Phys. **289**, 4393–4412 (2014). https://doi.org/10.1007/s11207-014-0544-4

Kurucz, R.L.: New opacity calculations. In: Crivellari, L., Hubeny, I., Hummer, D.G. (eds.) Stellar atmospheres: beyond classical models, pp. 441–448. NATO ASI Series C 341, KLuwer, Dordrecht (1991a)

Kurucz, R.L.: The solar spectrum. In: Solar Interior and Atmosphere, pp. 663–669. The University of Arizona Press, Arizona (1991b)

Kurucz, R.L.: Remaining line opacity problems for the solar spectrum. Rev. Mex. Astr, Astrophys. **23**, 1 (1992)

Landi, E., Feldman, U.: The thermal structure of an active region observed outside the solar disk. Astrophys. J. **672**, 674–683 (2008)

Landi degl'Innocenti, E.L., Landolfi, M.: Polarization in Spectral Lines. Astrophysics and Space Library, vol. 307. Kluwer Academic Publishers, Dordrecht (2004)

Leka, K.D., et al.: A comparison of flare forecasting methods. II. Benchmarks, metrics, and performance results for operational solar flare forecasting systems. Astrophys. J. Suppl. Ser., **243**, 36 (2019)

Leka, K.D., et al.: Properties of flare-imminent versus flare-quiet active regions from the chromosphere through the corona. II. Nonparametric discriminant analysis results from the NWRA classification infrastructure (Nci). Astrophys. J. **942**, 84 (2023). https://doi.org/10.3847/1538-4357/ac9c04

Lemen, J., et al.: The interface region imaging spectrograph (IRIS) NASA SMEX. In: AAS/Solar Physics Division Abstracts #42, vol. 289, p. 1512 (2011)

Linker, J.A., et al.: The open flux problem. Astrophys. J. **848**, 70 (2017)

Lites, B.W.: Remote sensing of solar magnetic fields. Rev. Geophys. **38**, 1–36 (2000)

Loeb, A.: Taking "the road not taken": on the benefits of diversifying your academic portfolio. arXiv:1008.1586 (2010)

Lu, E.T., Hamilton, R.J.: Avalanches and the distribution of solar flares. Astrophys. J. Lett. **380**, L89 (1991). https://doi.org/10.1086/186180

Mandrini, C.H., Démoulin, P., Klimchuk, J.A.: Magnetic field and plasma scaling laws: their implications for coronal heating models. Astrophys. J. **530**, 999–1015 (2000)

Manso Sainz, R., del Pino Alemán, T., Casini, R.: Magnetic field diagnostics with strong chromospheric lines. In: Belluzzi, L., Casini, R., Romoli, M., Trujillo Bueno, J. (eds.) Solar Polariation Workshop 8. Astronomical Society of the Pacific Conference Series, vol. 526, p. 145 (2019)

Metcalf, T.R., et al.: Is the solar chromospheric magnetic field force-free? Astrophys. J. **439**, 474–481 (1995)

Michelson, A.A., Morley, E.W.: On the relative motion of the earth and of the luminiferous ether. Sidereal Messenger **6**, 306–310 (1887)

Miller, S.L., Urey, H.C.: Organic compound synthesis on the primitive earth. Science **130**, 245–251 (1959). https://doi.org/10.1126/science.130.3370.245

Nóbrega-Siverio, D., et al.: Ambipolar diffusion in the BIFROST code. Astron. Astrophys. **638**, A79 (2020)

Osterbrock, D.E.: The heating of the solar chromosphere, plages, and corona by magnetohydrodynamic waves. Astrophys. J. **134**, 347 (1961)

Panos, B., Kleint, L.: Real-time flare prediction based on distinctions between flaring and non-flaring active region spectra. Astrophys. J. **891**, 17 (2020)

Panos, B., Kleint, L., Voloshynovskiy, S.: Exploring mutual information between IRIS spectral lines. I. Correlations between spectral lines during solar flares and within the quiet sun. Astrophys. J. **912**, 121 (2021). https://doi.org/10.3847/1538-4357/abf11b

Pascoe, D.J., et al.: Coronal loop seismology using damping of standing kink oscillations by mode coupling. II. Additional physical effects and bayesian analysis. Astron. Astrophys. **600**, A78 (2017). https://doi.org/10.1051/0004-6361/201629702

Prokopyszyn, A.P.K., Wright, A.N., Hood, A.W.: Line-tied boundary conditions can cause resonant absorption models to generate unphysically large boundary layers. Astrophys. J. **914**, 15 (2021). https://doi.org/10.3847/1538-4357/abf65b

Rachmeler, L.A., et al.: The high-resolution coronal imager, flight 2.1. Sol. Phys. **294**, 174 (2019)

Rachmeler, L.A., et al.: Quiet sun center to limb variation of the linear polarization observed by CLASP2 across the Mg II H and K lines. Astrophys. J. **936**, 67 (2022). https://doi.org/10.3847/1538-4357/ac83b8

Rempel, M.: Numerical sunspot models: robustness of photospheric velocity and magnetic field structure. Astrophys. J. **750**, 62 (2012). https://doi.org/10.1088/0004-637X/750/1/62

Rempel, M.: Extension of the MURAM radiative MHD code for coronal simulations. Astrophys. J. **834**, 10 (2017)

Riley, P., et al.: Can an unobserved concentration of magnetic flux above the poles of the sun resolve the open flux problem? Astrophys. J. **884**, 18 (2019)

Ross, K.: An Introduction to Bayesian Reasoning and Methods (2022). bookdown.org

Schekochihin, A.A.: MHD turbulence: a biased review. J. Plasma Phys. **88**, 155880501 (2022). https://doi.org/10.1017/S0022377822000721

Thomas, R.J., Neupert, W.: Extreme ultraviolet spectrum of a solar active region from SERTS. Astrophys. J. Suppl. Ser. **91**, 461–482 (1994)

Uitenbroek, H.: Detection of chromospheric magnetic fields: a forward modeling approach. In Kuhn, J.R., Harrington, D.M., Lin, H., Berdyugina, S.V., Trujillo-Bueno, J., Keil, S.L., Rimmele, T. (eds.) Solar Polarization 6. Astronomical Society of the Pacific Conference Series, vol. 437, 439 (2011)

Vissers, G.J.M., et al.: Active region chromospheric magnetic fields. observational inference versus magnetohydrostatic modelling. Astron. Astrophys. **662**, A88 (2022). https://doi.org/10.1051/0004-6361/202142087

Štěpán, J., del Pino Alemán, T., Trujillo Bueno, J.: Novel framework for the three-dimensional NLTE inverse problem. Astron. Astrophys. **659**, A137 (2022)

Wargnier, Q.M., et al.: Multifluid simulations of upper-chromospheric magnetic reconnection with helium-hydrogen mixture. Astrophys. J. **946**, 2, 115 (2023). https://doi.org/10.3847/1538-4357/acbfb1

Welsch, B.T.: The photospheric Poynting flux and coronal heating. Publ. Astron. Soc. Jpn. **67**, 18 (2015). https://doi.org/10.1093/pasj/psu151

Wiegelmann, T., Sakurai, T.: Solar force-free magnetic fields. Living Rev. Sol. Phys. **18**, 1 (2021). https://doi.org/10.1007/s41116-020-00027-4

Wikstøl, Ø., Judge, P.G., Hansteen, V.H.: Observational signatures of nanoflare heated solar stellar coronae. Adv. Space Res. **20**, 2289–2292 (1997). https://doi.org/10.1016/S0273-1177(97)00899-5

Wiegelmann, T., et al.: Can we improve the preprocessing of photospheric vector magnetograms by the inclusion of chromospheric observations? Sol. Phys. **247**, 249–267 (2008). https://doi.org/10.1007/s11207-008-9130-y

Wiegelmann, T., Thalmann, J.K., Solanki, S.K.: The magnetic field in the solar atmosphere. Astr. Astrophys. Rev. **22**, 78 (2014). https://doi.org/10.1007/s00159-014-0078-7

Williams, T., et al.: Evidence for and analysis of multiple hidden coronal strands in cross-sectional emission profiles: further results from NASA's high-resolution solar coronal imager. Astrophys. J. **902**, 90 (2020a). https://doi.org/10.3847/1538-4357/abb60a

Williams, T., et al.: Is the high-resolution coronal imager resolving coronal strands? Results from AR 12712. Astrophys. J. **892**, 134 (2020b)

Chapter 7
Closing Thoughts

So, why exactly *is* the Sun obliged to form a corona? Two ingredients, rotation and convection in the ionized plasmas in stellar interiors, appear to be *sufficient* to regenerate global-scale magnetic fields (Spruit 2011). These fields, emerging because of magnetic buoyancy in a gravitationally stratified object, control the dynamics and thermodynamics of the tenuous, outer regions of stars. This control is exerted through *electrodynamic coupling*. The coupling also has the remarkable effect of dramatically slowing down the rotation of the entire star over the main sequence lifetimes of the Sun and similar stars (those with masses $\approx 1 \pm 0.2\,M\odot$). This effect was first quantified by Skumanich (1972). This stellar "spin-down" occurs through two aspects of electrodynamic coupling (see, for example recent models of Chebly et al. 2023, and references therein). The magnetic heating leads to pressure gradients which help drive the solar wind, and the magnetic stresses deliver angular momentum to out-flowing wind plasma which is lost to the interstellar medium. The effect—slowing down stellar rotation on evolutionary time scales—is called "magnetic braking". Magnetically-controlled transport of angular momentum is consequential across astrophysics, for example in star formation (McKee & Ostriker 2007) and more exotic objects, such as active galactic nuclei (Begelman et al. 1984), accretion disks (Schramkowski & Torkelsson 1996; Blackman & Pessah 2009), and binary systems (Lai & Munõz 2023).

Returning to the problem of coronal heating, observations to date show that all convecting main sequence stars emit X-rays with a brightness always at least comparable to solar coronal holes (Schmitt 1997). This is true also for slowly rotating Sun-like stars which may have lost large-scale magnetic fields owing to spin-down, but which maintain a local dynamo (Cattaneo 1997; Judge et al. 2004). Indeed, Schmitt (2013) states

> "The conclusion from these studies must then be that coronal formation is indeed a universal phenomenon on solar-like stars."

P. Judge, J. A. Ionson, *The Problem of Coronal Heating*, Astrophysics and Space Science Library 470, https://doi.org/10.1007/978-3-031-46273-3_7

Young, rapidly-rotating Sun-like stars exhibit larger fractions of the stellar luminosity from coronal radiation than the middle-aged Sun (Güdel 2007). X-rays approach 0.1% of the luminosity, larger than chromospheric emission by a factor of three or greater. This level of emission is an order of magnitude above that for a typical active region when spread over the whole Sun (Table 1.3). It also marks a "saturated" regime where for the youngest stars (100 million years or younger) the corona's emission does not depend on age. Luminosities of coronae of older stars decay with time t as $t^{-1.5\pm0.3}$, chromospheric luminosities scale roughly as $t^{-1.5}$ (Skumanich 1972). Sun-like stars older than the Sun emit X-rays close to the Sun without active regions. These power-law scalings extend to stars roughly of solar age, after which evidence exists for weaker magnetic braking and through a broader scatter in rotation rates in older stars with masses mostly within $1 \pm 0.2 M_{\odot}$ (van Saders et al. 2016; Metcalfe et al. 2016). Güdel (2007) concludes that

> "The X-ray output is an excellent indicator of dynamo activity. The coronal output is strongly determined by parameters that control the magnetic dynamo."

In the context of heating mechanisms, a näive corollary of these words and Güdel's accompanying analyses is that the ARLToR conjecture applies, at least statistically, as judged by data typically plotted in a logarithmic plane. In the saturated regime, surface magnetic fields with energy density in equipartition with the Reynolds stresses due to rotation and convection cover almost the entire star, limiting the magnetic energy and its Poynting flux into the corona. As stars spin more slowly with age, the above power law implies that a deterministic fraction of this energy flux is deposited as heat in the corona. Naturally, the dissipation mechanisms in action are not determined from such data, without auxiliary strong and untested assumptions.

We conclude that rotation and convection, leading to buoyancy and magnetic flux emergence, are sufficient *ingredients* required for coronal heating. It is only the *recipe* that is not yet clear. Remarkably, only *two* aspects of MHD appear to guarantee the formation and maintenance of a million K corona.

The first aspect is that, once some process has generated ionized plasma in a stellar atmosphere, the free electrons are so mobile that the plasma evolves as an ideal MHD fluid. Meteorites, cosmic rays and a minimal physical level of electrodynamic coupling caused by a local dynamo (Judge 2020) can all seed hot plasma. But, and this is the second aspect, ideal MHD causes its own, non-ideal demise where ultimately collisions between the ordered motions of plasma charges lead to irreversible heating. Under ideal MHD conditions, the spontaneous formation of current sheets driven by untidy fluid motions (Parker 1994), like Shakespeare's Juliet, leads to its demise precisely because of its own perfection. Resonant absorption is also an ideal MHD process, leading also to dissipation. Recently, turbulence theorists have reached the same conclusion, for example Schekochihin (2022) notes that:

> "Thus, just like WT [weak turbulence], strong aligned turbulence too carries in it the seeds of its own destruction, making an eventual transition to some new state inevitable at sufficiently small scales."

The mathematics of ideal MHD, which applies to spontaneous current sheets, resonant absorption, and the inertial range of turbulent energy cascades, ultimately guarantees dissipation of ordered energy generated by the natural atmosphere beneath (Low 2023). So far as we are able to assess with current instruments, theoretical predictions within this framework, quantitatively uncertain as they are, have survived scrutiny. To the extent that they do not survive further more stringent tests, this constitutes, imprecisely but physically, the "solution to the heating problem". While quiet Sun and steady active region coronal plasmas can be regarded as "solved" within the framework, this does not constitute a full understanding of the problem. Only the ARLToR conjecture "saves the phenomena" at this juncture.

The hot solar wind plasma does not immediately fall under the above umbrella. The wind opens the magnetic field into interplanetary space, but its very existence requires heating (Parker 1958), and so far we have not presented a convincing argument for the development of dissipative scales in the wind plasma. Magnetic fields in the solar wind thread just once into the lower atmosphere. Thus the magnetic stresses in Parker's (1994) picture and in MHD turbulence (Rappazzo et al. 2008) do not readily build up through the tying of magnetic lines of force to two footpoints penetrating the photosphere. Instead, the theory of MHD turbulence appeals to *imbalanced* upward and downward propagating fluxes of Alfvén waves, but non-linear interactions still occur. Under inhomogeneous plasma conditions, phase-mixing and uniturbulence can occur (see page 141 Magyar et al. 2019; Ismayilli et al. 2022). A complete solution to the wind problem, heating and expansion, may or may not be found in these ideas. Certainly missions orbiting close to the Sun will add useful upper boundary conditions for the heating of solar wind plasmas.

The happy circumstance that three contemporaneous and current missions SDO, Solar Orbiter and Parker Solar Probe, and the recently launched Aditya-L1 mission (Seetha and Megala 2017) all observing different aspects of the solar corona, is certain to deliver opportunities to avoid some of the difficulties highlighted in Chap. 3. Combined with novel experiments to observe line profiles during eclipses (e.g. Samra et al. 2022), we may be poised to attack essential issues in coronal heating far more directly than has been achieved to date. The literature on coupling the low corona to in-situ measurements from Parker Solar Probe is only just emerging. In sampling fields and plasma within $\approx 30R_{\odot}$, imprints of the sources of the solar wind are more clearly revealed. Studies have uncovered connections between convective and jet-like structures within the deeper Sun's atmosphere, and in-situ measurements down to $10R_{\odot}$ (e.g. Bale et al. 2019, 2023; Kumar et al. 2023). Unanticipated magnetic "switchbacks", large-amplitude switches in the radial direction of the magnetic field (Mozer et al. 2020; Dudok de Wit et al. 2020; Telloni et al. 2022; Huang et al. 2023a,b) have been attributed to the evolution of Alfvén waves propagating upwards non-linearly, driven by various processes (Zank et al. 2020; Shoda et al. 2021; Magyar et al. 2021). This is an exciting state from which to move forward, potentially with interesting implications for coronal heating mechanisms. Such measurements augment chromospheric data from beneath with data from above ($R \geq 10R_\odot$), yet still within the sphere of influence of downward propagating Alfvén waves (Liu et al. 2021).

Flares and mass ejections have attracted the attention of the solar physics community as it responds to the needs to our society, which has become highly dependent on a stable space environment (Eddy 2009). Relatively few studies have set out to follow the energy down to dissipation scales. We have tried to convince the reader that, in any case, the coronal heating problem is largely independent of the modern discussions of large-scale reconfigurations driven by reconnection of large-scale fields. In doing so we have adopted the "ansatz" of Eq. (2.25) but without mathematics, to draw attention to the fact that studying the large-scales through observations is currently insufficient to test heating scenarios except in the crudest, energy-conserving sense. Again the ARLToR conjecture enforces the outcome of any given study, even though there exist physical processes contradicting it.

This point of view is heretical. Pontin and Priest (2022) have reviewed the role of magnetic reconnection in the Sun, giving a more conventional narrative. Drawing from their summary and outlook section, we read

> ... Reconnection now seems certain to contribute substantially to coronal heating. Several mechanisms have been identified, including Parker braiding... flux tube tectonics... and flux cancellation[1].

By putting reconnection first, this point-of-view implies that reconnection is the primary process of importance to converting magnetic free energy into heat. Yet in all the models we have examined energy is converted into heat via viscosity, electrically-driven Joule heating, via the rapid growth of steep gradients in phase-mixed regions, surface waves, and aligned anisotropic structures, most of which have little directly to do with reconnection. The formation of plasmoids drives rapid reconnection between them, and they may well dissipate kinetic energy of flows after reconnection. But in what direct sense might "reconnection heat the steady corona"? Nanoflares appear to be the commonly encountered phenomenological answer, and plasmoid formation an answer forthcoming from MHD and kinetic turbulence theories, which are exciting but new, largely untested and elaborate models. Recent MHD simulations including more realistic solutions to the energy equation are approaching the point at which observable quantities can be predicted (Sen et al. 2023).

The absence of tell-tail signatures of high energy particles associated with flaring, and perhaps very small coronal line widths seem to refute the overall idea of reconnection-driven "flare heating" on any scale. Reconnection of modest perpendicular components $\mathbf{b}_\perp$ across large guide fields $\mathbf{B}_0$ may be consistent with small sub-Alfénic Doppler line widths under line-tied conditions, when perpendicular kinetic energy is suppressed at the expense of magnetic energy (e.g. figure 3 of Rappazzo et al. 2008). In numerical reconnection calculations on MHD scales, heat is generated via the artifice of numerical diffusion of current sheets and shear layers,

[1] These "mechanisms" are in fact merely phenomena, requiring clearer physical attributions.

perhaps using hyper-diffusivities but always with a Prandtl number of order one. None of the the rather delicate dynamics of turbulence reviewed by Schekochihin (2022), is present in such numerical work, and the sub-grid-scale dynamics is only beginning to be explored. All things considered, it seems premature to declare that reconnection is intrinsic to steady coronal heating. Instead it seems best to test robust signatures of different modes of reconnection against critical observations. In the Sun, we might expect current sheets to form as a result of turbulence well below MHD scales, hand-in-hand with the plasmoid instability. Very little plasma can be heated across these tiny diffusion regions, and any accelerated bulk plasma or particles have just not been reported outside of flares. All of these open questions can be resolved, one way or another, by high resolution, and high signal-to-noise spectroscopic measurements of the EUV corona.

Throughout this book we have questioned if the coronal heating problem can be treated purely as a driven system, or if the solution to the problem also depends on the internal dynamics of the magnetized coronal plasmas themselves. We currently have no good explanation as to why only certain bundles of magnetic flux are selected at any given time for heating. If the explanation lies in self-regulation, this will require some re-thinking of the problem at a fundamental level.

But perhaps the most important messages to take from this book are not scientific. Begin with an open mind, a pencil-and-paper understanding of elementary plasma physics including MHD, identify clearly a problem to be solved, research it in the literature and work out how to solve it. Use refutation as a tool to advance resistant problems, and rejoice in publishing negative results. Not many working scientists will be able to do this even part-time. Given the current landscape for researchers, we acknowledge that this is not an approach likely lead to job security. But young researchers will benefit from this approach, perhaps just one morning, even hour, a week, away from the keyboards. Following the sentiments expressed by Avi Loeb to young scientists after a conference dinner (Loeb 2010), rewards surely await those who can succeed in this endeavor.

Let us give penultimate words to Pontin and Priest (2022), reflecting some of the questions we have re-iterated so far:

> In 3D reconnection in the various different regimes, what determines the rate of energy conversion, into which forms, and what is the mechanism by which the energy is ultimately dissipated?

In parting, following senitiments expressed by Pierrard et al. (2004), words of Zurbuchen (2007) express nicely the implications and applications which encompass the breadth and depth that solutions to the heating problem entail, as originally emphasized by Ionson (1978):

> "One of the most important problems in astrophysics is the coupling between [gravitationally bound] astrophysical objects and their magnetic field. For example, the interpretation of observed energetic phenomena near astrophysical objects, such as neutron stars, accretion discs, black holes, and stars with flares and activity cycles, relies on a solid understanding of the processes that control energy transmission from these objects into their surrounding magnetic field."

We shall not cease from exploration
and the end of all our exploring
will be to arrive where we started
And know the place for the first time.

—T. S. Eliot

References

Bale, S.D. et al.: Highly structured slow solar wind emerging from an equatorial coronal hole. Nature **576**, 237–242 (2019). https://doi.org/10.1038/s41586-019-1818-7

Bale, S.D. et al.: Interchange reconnection as the source of the fast solar wind within coronal holes. Nature **618**, 252–256 (2023). https://doi.org/10.1038/s41586-023-05955-3

Begelman, M.C., Blandford, R.D., Rees, M.J.: Theory of extragalactic radio sources. Rev. Mod. Phys. **56**(2), 255–351 (1984). https://doi.org/10.1103/RevModPhys.56.255

Blackman, E.G., Pessah, M.E.: Coronae as a consequence of large-scale magnetic fields in turbulent accretion disks. Astrophys. J. Lett. **704**(2), L113–L117 (2009). https://doi.org/10.1088/0004-637X/704/2/L113

Cattaneo, F.: The solar dynamo problem. In: Pijpers, F.P., et al. (eds.) Solar Convection and Oscillations and Their Relationship, pp. 201–222. Kluwer, Alphen aan den Rijn (1997)

Chebly, J.J. et al.: Numerical quantification of the wind properties of cool main sequence stars. Mon. Not. R. Astron. Soc. **524**(4), 5060–5079 (2023). https://doi.org/10.1093/mnras/stad2100

Dudok de Wit, T. et al.: Switchbacks in the near-sun magnetic field: long memory and impact on the turbulence cascade. Astrophys. J. Suppl. Ser. **246**, 39 (2020). https://doi.org/10.3847/1538-4365/ab5853

Huang, J., et al.: The structure and origin of switchbacks: parker solar probe observations. Astrophys. J. **952**(1), 33 (2023a). https://doi.org/10.3847/1538-4357/acd17e

Huang, J., et al.: The temperature, electron, and pressure characteristics of switchbacks: parker solar probe observations. arXiv e-prints, arXiv:2306.04773 (2023b). https://doi.org/10.48550/arXiv.2306.04773

Eddy, J.A.: The Sun, the Earth and Near-Earth Space: A Guide to The Sun-Earth System. NASA, Washington (2009)

Güdel, M.: The sun in time: activity and environment. Living Rev. Solar Phys. **4**, 3 (2007)

Ionson, J.A.: Resonant absorption of Alfvénic surface waves and the heating of solar coronal loops. Astrophys. J. **226**, 650–673 (1978)

Ismayilli, R. et al.: Non-linear damping of surface Alfvén waves due to uniturbulence. Front. Astron. Space Sci. **8**, 241 (2022). https://doi.org/10.3389/fspas.2021.769173

Judge, P.G.: Inevitable consequences of ion-neutral damping of intermediate MHD waves in sun-like stars. Mon. Notices Royal Astron. Soc. **498**, 2018–2029 (2020)

Judge,P.G., et al.: A comparison of the outer atmosphere of the "Flat Activity" Star τ Ceti (G8 V) with the Sun (G2 V) and α Centauri A (G2 V), Astrophys. J. **609**, 392 (2004)

Kumar, P. et al.: New evidence on the origin of solar wind microstreams/switchbacks. Astrophys. J. Lett. **951**, L15 (2023). https://doi.org/10.3847/2041-8213/acd54e

Lai, D., Munõz, D.J.: Circumbinary accretion: from binary stars to massive Bi- nary Black Holes. Ann. Rev. Astr. Astrophys. **61**, 517–560 (2023). https://doi.org/10.1146/annurev-astro-052622-022933

Liu, Y.D. et al.: Determination of solar wind angular momentum and Alfvén radius from Parker solar probe observations. Astrophys. J. Lett. **908**, L41 (2021). https://doi.org/10.3847/2041-8213/abe38e

Loeb, A.: Taking "The Road Not Taken": On the Benefits of Diversifying Your Academic Portfolio (2010)

Low, B.C.: Topological nature of the Parker magnetostatic theorem. Phys. Plasmas **30**, 012903 (2023). https://doi.org/10.1063/5.0124164

Magyar, N., Van Doorsselaere, T., Goossens, M.: Understanding uniturbulence: self-cascade of MHD waves in the presence of inhomogeneities. Astrophys. J. **882**, 50 (2019). https://doi.org/10.3847/1538-4357/ab357c

Magyar, N., et al.: Could switchbacks originate in the lower solar atmosphere? I. Formation mechanisms of switchbacks. Astrophys. J. **911**, 75 (2021). https://doi.org/10.3847/1538-4357/abec49

McKee, C.F., Ostriker, E.C.: Theory of star formation. Ann. Rev. Astr. Astrophys. **45**(1), 565–687 (2007). https://doi.org/10.1146/annurev.astro.45.051806.110602

Metcalfe, T.S., Egeland, R. van Saders, J.: Stellar Evidence That the Solar Dynamo May Be in Transition, Astrophys. J. **826**, L2 (2016)

Mozer, F.S. et al.: Switchbacks in the solar magnetic field: their evolution, their content, and their effects on the plasma. Astrophys. J. Suppl. Ser. **246**, 68 (2020). https://doi.org/10.3847/1538-4365/ab7196

Parker, E.N.: Dynamics of the interplanetary gas and magnetic fields. Astrophys. J. **128**, 664–676 (1958)

Parker, E.N.: Spontaneous Current Sheets in Magnetic Fields with Application to Stellar X-Rays. International Series on Astronomy and Astrophyics. Oxford University Press, Oxford (1994)

Pierrard, V., Lamy, H., Lemaire, J.: Exospheric distributions of minor ions in the solar wind. J. Geophys. Res. Space Phys. **109**, 118 (2004). https://doi.org/https://doi.org/10.1029/2003JA010069

Pontin, D.I., Priest, E.R.: Magnetic reconnection: MHD theory and modelling. Living Rev. Sol. Phys. **19**, 1 (2022)

Rappazzo, A.F. et al.: Nonlinear dynamics of the Parker scenario for coronal heating. Astrophys. J. **677**, 1348–1366 (2008). https://doi.org/10.1086/528786

van Saders, J.L. et al.: Weakened magnetic braking as the origin of anomalously rapid rotation in old field stars. Nature, **529**, 181 (2016)

Samra, J., Bryans, P., Cheimets, P., DeLuca, E., Franklin, J., French, R., Gall, A., Hannigan, J., Hertz, E., Judge, P., Lacatus, D., Madsen, C., Marquez, V., Paraschiv, A., Rivera, Y.: An airborne coronal emission surveyor (ACES) for total solar eclipse observations. Bull. AAS **54**(7) (2022). Retrieved from https://baas.aas.org/pub/2022n7i303p06

Schekochihin, A.A.: MHD turbulence: a biased review. J. Plasma Phys. **88**, 155880501 (2022). https://doi.org/10.1017/S0022377822000721

Schmitt, J.H.M.M.: Coronae on solar-like stars. Astron. Astrophys. **318**, 215–230 (1997)

Schmitt, J.H.M.M.: 50 (38) years of stellar X-ray astronomy . Mem. Soc. Astron. Ital. **84**, 532 (2013)

Schramkowski, G.P., Torkelsson, U.: Magnetohydrodynamic instabilities and turbulence in accretion disks. Astr. Astrophys. Rev. **7**(1), 55–96 (1996). https://doi.org/10.1007/PL00013286?

Seetha, S., Megala, S.: Aditya-L1 mission. Current Sci. **113**(4), 610 (2017). https://doi.org/10.18520/cs/v113/i04/610-612

Sen, S., Jenkins, J., Keppens, R.: 3D coupled tearing-thermal evolution in solar current sheets. arXiv e-prints, arXiv:2308.10210 (2023). https://doi.org/10.48550/arXiv.2308.10210

Shoda, M., Chandran, B.D.G., Cranmer, S.R.: Turbulent generation of magnetic switchbacks in the Alfvénic solar wind. Astrophys. J. **915**, 52 (2021). https://doi.org/10.3847/1538-4357/abfdbc

Skumanich, A.: Time scales for Ca II emission decay, rotational braking, and lithium depletion. Astrophys. J. **171**, 565–567 (1972)

Spruit, H.C.: Theories of the solar cycle: a critical view. In: Miralles, M.P., Sánchez Almeida, J. (eds.) The Sun, the Solar Wind, and the Heliosphere, 39. IAGA Special Sopron Book Series, vol. 4. Springer, Berlin (2011)

Telloni, D. et al.: Observation of a magnetic switchback in the solar corona. Astrophys. J. Lett. **936**(2), L25 (2022). https://doi.org/doi:10.3847/2041-8213/ac8104

Zank, G.P. et al.: The origin of switchbacks in the solar corona: linear theory. Astrophys. J. **903**, 1 (2020). https://doi.org/10.3847/1538-4357/abb828

Zurbuchen, T.H.: A new view of the coupling of the sun and the heliosphere. Ann. Rev. Astr. Astrophys. **45**, 297–338 (2007). https://doi.org/10.1146/annurev.astro.45.010807.154030

Appendix A
Coronal Line Widths

After decades of observations, the literature on the measurement and interpretation of coronal spectral line widths is not terribly conclusive. Reported values of random but non-thermal Doppler widths (ξ_0 defined below), deconvolved from observed data, vary from 0 to perhaps 40 km s^{-1} in quiet and active regions, with median values around 20 km s^{-1}. In chapter 13 of a comprehensive review of UV and EUV spectroscopy, Del Zanna and Mason (2018) discuss solar line width measurements made over the space era and earlier. In reviewing also the literature of coronal forbidden line spectra reported from the ground during eclipses and with coronagraphs, we conclude that the following text from Del Zanna and Mason (2018) applies in general:

> ... contradictory results have often been reported in the literature. Variations in the excess widths have often been attributed to a variety of reasons such as the choice of ion temperatures, non-equilibrium effects, curvature of the magnetic field, instrumental effects and solar cycle variations. On the observational side, we point out that for most instruments, measurements of line widths are subject to large uncertainties, some being systematic. For several instruments, the instrumental line width is a significant fraction of the observed width

Figure A.1 shows profiles of the strongest forbidden coronal lines observable from the ground, which are fully spectrally resolved. They are reported as being typical. The widths of the lines are listed as full-width at half-maximum intensity from Gaussian fits. The equivalent values of the non-thermal widths ξ_0 defined below are $25 - 35$ km s^{-1}.

The reader should be aware of differences in conventions used to specify "line width", from full width at half-maximum (FWHM), to half width at half maximum, to "1/e" width, even "$1/\sqrt{2}$e" width. There are also differences in "instrumental width" and "non-thermal line width", the latter being a parameter to be compared with predictions from MHD waves and/or turbulence. The most commonly used parameters are encapsulated in the following equations, which apply to profiles

P. Judge, J. A. Ionson, *The Problem of Coronal Heating*, Astrophysics and Space Science Library 470, https://doi.org/10.1007/978-3-031-46273-3

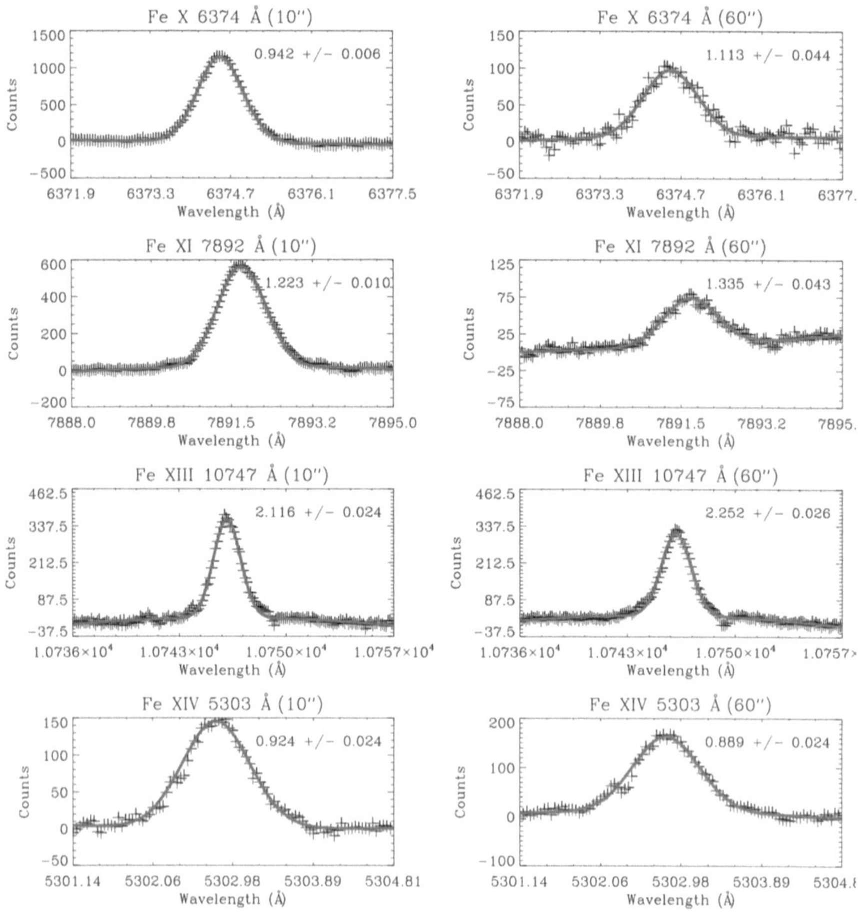

Fig. A.1 Typical, well-resolved forbidden emission line profiles from the corona are shown, from Krishna Prasad et al. (2013). These were obtained using the Norikura 25 cm coronagraph. The profiles shown include convolution with the instrument profile, but the derived values of FWHM have been corrected for this instrument contribution. The corresponding non-thermal parameters ξ_0 vary from 25 to 35 km s^{-1} for all lines. The instrument FWHMs were derived by the Krishna Prasad et al. using disk spectra of narrow photospheric features. For the 637.4 nm line, the FWHM appears to be at most 0.2 Å ($\equiv$ 9 km s^{-1}), judging from absorption features visible in Figure 1 of Singh et al. (2002)

of instrument, thermal plasma, and distribution of non-thermal speeds, which are assumed to be Gaussian functions:

$$v_\lambda = \frac{\lambda-\lambda_0}{\lambda_0}\, c \qquad \text{Doppler shift definition } v_\lambda \tag{A.1}$$

$$\phi(v) = \exp\left(-v^2/\sigma^2\right) \qquad \text{Gaussian line width definition } \sigma \tag{A.2}$$

$$W = 2\sqrt{\ln 2}\sigma \qquad \text{Full width half-maximum definition } W \tag{A.3}$$

$$W_{OBS}^2 = W_{INS}^2 + 4\ln 2\left\{\frac{2kT_i}{m_i} + \xi_0^2\right\} \qquad \text{Gaussian deconvolution,} \tag{A.4}$$

where all terms in the final equation are the square of Doppler shifts measured in m s^{-1}, derived from functions of wavelength through the first equation.

With these definitions, the measured values of ξ_0 of interest to heating problems can be summarized as follows. The bulk of measured values of ξ_0 reported lie in the range 10–35 km s^{-1}. Most measurements are made over significant volumes of plasma, integrating over regions large enough that any anisotropies due to MHD (e.g. turbulence, Appendix C) would be washed out. Averaging over wavelength is common especially in the relatively low spectral resolution instruments applied in the EUV and X-ray regions (for example, the "high resolution" SERTS EUV instrument, Thomas and Neupert, 1994 has $W_{INS} \approx 70$ km s^{-1}, EIS (Culhane et al., 2007) on the Hinode spacecraft (Kosugi et al., 2007) has $W_{INS} \approx 75$ km s^{-1}. These correspond to $\mathcal{R} \sim 4200$). Then the observed profiles W_{OBS} are dominated by instrumental line widths. Nevertheless, Coyner and Davila (2011) retrieved

> ... a strongly peaked distribution corresponding to a non-thermal velocity of 19-22 km s^{-1} in all three of the quiet-Sun, active region, and off-limb distributions.

At UV wavelengths, the SO82B spectrometer on SKYLAB had $W_{INS} \sim 12$ km s^{-1} ($\mathcal{R} \sim 25000$. At visible/infrared wavelengths, for example measurements from Norikura observatory, $W_{INS} \sim 10$ km s^{-1} (see Fig. A.1, or $\mathcal{R} \sim 30000$. There exists evidence that brighter regions have narrower spectral lines (e.g. Raju et al., 2001), but the statistical significance of this result is difficult to assess from the literature. Hara and Ichimoto (1999) studied data from Norikura in lines of Fe^{9+}, Fe^{13+} and Ca^{14+} from active regions, finding values from 10-26 km s^{-1}, the smallest widths measured were for Fe^{13+}.

In Sect. 3.5 of the main text, it is argued that to find signature asymmetries of MHD turbulent or wave motions, measurements of resolved coronal loops are needed to explore the variation of line profiles with the changing line-of-sight orientations of mean field $\mathbf{B}_0$, with spectral resolutions $\mathcal{R} \gtrsim 10000$. No planned space mission is capable of such measurements (MUSE, de Pontieu et al., 2022; Visible Emission Line Coronagraph VELC on the ADITYA–L1 spacecraft, Raghavendra et al., 2017; Singh et al., 2019). The DKIST can observe the corona only above the solar limb. It can be hoped that coronal line profiles can be acquired which might serve to probe such signatures, although the limb view makes tricky the ability to measure components of linewidths because the parallel and perpendicular directions of magnetic field are difficult to determine. It might be that a novel use of spectropolarimetry can help define the geometry (Judge et al., 2021; Paraschiv and Judge, 2022).

The major conclusion here concerning coronal line widths is that they all are observed to be sub-Alfvénic and subsonic, in quiet and active regions, outside of flares and CMEs. In Sect. 5.9, we argue that new high signal-to-noise line profiles with a resolution of at least 20,000 would potentially reveal vital information on coronal heating mechanisms. The power of such measurements to limit coronal heating processes has been elegantly elucidated in the MHD modeling work reported by van Ballegooijen et al. (2017).

Appendix B
Power Requirements of the Chromosphere and Corona

The first step in confronting data with magnetic model predictions is usually to assess if the mechanism has sufficient power, even though the corona requires little of the total mechanical energy available beneath the surface (see Sect. 1.7). Estimates have evolved significantly over time, so important updates are listed in Table B.1. We adopt the latest estimates.

Neither chromosphere nor coronal power needs can be directly measured owing to, respectively, the redistribution of energy by radiation and heat conduction. A combination of observations and models is required. Radiation lost from the chromosphere depends critically on effects of photon scattering in strong lines of Mg^+, Ca^+ and Fe^+, the latter spread over many lines across the UV and visible spectrum (e.g. Athay, 1976; Chapman, 1981). In the corona, it is unclear if the dominant H Lα transition should be included as a loss from the corona, or the chromosphere (Mariska, 1992, see Fig. 3.7).

The bulk of chromospheric plasma is hydrostatically stratified (Fig. 1.6). Radiation losses from the chromosphere occur mostly in the outward direction, where the optical distance to space is smallest. Inward-directed radiation in such strong lines is ultimately scattered outwards, or destroyed by collisions or photo-absorption (e.g. Ayres, 1979). Therefore the power requirements have historically been derived from 1D, hydrostatic atmospheric models which empirically match critical observations. Under steady conditions, the radiative losses integrated over chromospheric and coronal plasma yield the power requirements. Examples in Table B.1 include chromospheric models of (Vernazza et al., 1973, 1976, 1981; Anderson and Athay, 1989b). Each case included improved physical prescriptions and more complete treatments of radiation losses.

Coronal and transition region models were constructed by Burton et al. (1967) and Jordan (1976, 1980, 1988), again using the differential emission measure $\xi(T_e)$ (Sect. 3.7). Aside from H Lα (Fig. 3.7) these models are optically thin and so the integral of the product of emission measure and radiation loss functions (Fig. 3.7)

P. Judge, J. A. Ionson, *The Problem of Coronal Heating*, Astrophysics and Space Science Library 470, https://doi.org/10.1007/978-3-031-46273-3

Table B.1 Historical estimates of energy losses from the solar atmosphere

Source	Flux density W m^{-2}	Luminosity W	$F/F_\odot$	note
Photosphere				
Allen (1973)	6.27 10^7	3.83 10^{26}	1	$F_\odot$, $L_\odot$
Corona				
Biermann and Bruggencate (1947)	1 10^0	6 10^{18}	1.6 10^{-8}	H-only plasma
Woolley and Allen (1948)	1 10^1	6 10^{19}	1.6 10^{-7}	Includes metals
Withbroe and Noyes (1977)	1 10^2	6 10^{20}	1.6 10^{-6}	Conductive losses
	2 10^2	1 10^{21}	3 10^{-6}	
Withbroe (1988)	8 10^2	5 10^{21}	1.3 10^{-5}	quiet Sun
	5 10^2	3 10^{21}	8 10^{-6}	coronal hole
Chromosphere				
Osterbrock (1961)	3 10^4	2 10^{23}	4.8 10^{-4}	
Vernazza et al. (1981) Table 29	4.6 10^3	2.8 10^{22}	7 10^{-5}	no Fe
Anderson and Athay (1989b)	1.4 10^4	1 10^{23}	2.2 10^{-4}	Fe^+ dominant
				$\gtrsim$ 1000 lines

Withbroe and Noyes (1977) used direct measurements of UV, EUV and X-ray fluxes to estimate the radiation losses of the corona- by including the radiation losses from upper chromospheric and transition region material (including Ly-alpha) leads to a total radiative energy loss rate of 10^3 W m^{-2}. These data were used to construct some of Table 1.3 in Chap. 1

over depth yields the power requirements of transition region and corona. (e.g. Withbroe and Noyes, 1977; Withbroe, 1988) and compiled in Table 1.3.

These models were used to explore heating rates as a function of height in a hydrostatically stratified atmosphere, by comparing with magnetic wave modes (Jordan et al., 1984). But with crude estimates of Alfvén speeds, and so many assumptions needed to build such models, results were contingent upon untested assumptions. The model shown in Fig. 1.4 is a mix of the two approaches, attaching an emission measure model to the top of a chromospheric model (Pietarila and Judge, 2004).

Using just these techniques, one cannot follow the flow of energy which determines the atmospheric structure. Instead one may only seek compatibility or incompatibility of highly averaged power requirements with wave modes or other scenarios, with no attempt to discover the nature of the conversion of ordered energy to heat. As we already noted (Sect. 1.7), many mechanisms survive this first hurdle.

Appendix C
Phenomenology of MHD Turbulence

In the dynamic evolution towards dissipation, *MHD turbulence* has a large literature within which we may draw insight into what is an essentially non-linear system. The theory is usually conceived as "the interaction between two, oppositely-directed Alfvén waves" along the large-scale field $\mathbf{B}_0$. While this sounds like a highly special form of dynamics, it nevertheless gives up a general theory of small-scale dynamics driven on large scales, for the problem we have defined: $\mathbf{B}_0 \gg |\mathbf{b}|$, for which, outside of large-scale disruptions by reconnection, transverse Alfvén waves are the modes that can propagate the large-scale Poynting flux from the atmosphere below. Therefore we will summarize current understanding of what the theory can, and cannot, predict.

Beginning with Kolmogorov (1941), Schekochihin (2022) follows a timeline up to 2022 in the development of the understanding of MHD turbulence. For reference, Kolmogorov's local cascade of isotropic turbulence gives the familiar steady state energy, in the inertial range, with wavenumber k

$$E(k) \sim \varepsilon^{2/3} k^{-5/3}$$

Insight from Iroshnikov (1964) and Kraichnan (1965) marked the departure of MHD from Kolmogorov's turbulence in acknowledging the role of a guide field $\mathbf{B}_0$ with Alfvén speed $B_0/\sqrt{\mu_0 \rho}$. Recognizing that non-linear interaction times between magnetic "eddies" must vary as c_A^{-1}, the rate of injection of energy ε must enter dimensional analysis in the combination εc_A. Thus the kinetic energy spectrum in the inertial range of a cascade must be, by dimensional analysis,

$$E(k) \sim \; (\varepsilon c_A)^{1/2} \; k^{-3/2}, \tag{C.1}$$

called the IK scaling. Schekochihin's narrative then jumps to the next significant advance in 1995, with the work of Goldreich and Sridhar (1995). They explicitly acknowledge the anisotropic dynamics associated with the symmetries broken by

P. Judge, J. A. Ionson, *The Problem of Coronal Heating*, Astrophysics and Space Science Library 470, https://doi.org/10.1007/978-3-031-46273-3

the presence of a dominant $\mathbf{B}_0$. Variations parallel to $\mathbf{B}_0$ (scales $\sim \ell_\parallel$) are associated with Alfvén waves, with a characteristic time

$$\tau_{\rm A} \sim \frac{\ell_\parallel}{c_A}. \tag{C.2}$$

Perpendicular variations (scales $\ell_\perp \sim k_\perp^{-1}$) are set by non-linear interactions, whose characteristic time (and "eddy turnover time" τ_c) is

$$\tau_{\rm nl} \sim \frac{\ell_\perp}{\delta u_{\ell_\perp}} \tag{C.3}$$

and whose interactions are between Alfvén waves in the two directions along $\mathbf{B}_0$. (Here we have assumed that Alfvén speeds are, initially at least, homogeneous across the perpendicular direction, thus singling out non-linear terms instead of phase mixing or resonant absorption processes, for clarity). The perpendicular fluctuations are Alfvénic in character with equipartition between magnetic and kinetic energies: $|\delta\mathbf{b}_\perp|^2 = |\delta\mathbf{u}_\perp|^2$, with $\mathbf{b}$ in velocity units $\mathbf{b} \rightarrow \mathbf{b}/\sqrt{\mu_0\rho}$.

On the basis that the parallel dynamics only has a maximum time τ_c to interact with the Alfvén wave, Goldreich and Sridhar (1995) proposed that $\tau_c = \tau_{nl} = \ell_\parallel/c_A$, a state called dynamical "critical balance" (CB). Then, the perpendicular length for any $\ell_\parallel$ at which CB occurs is

$$\lambda_{\rm CB} \sim \varepsilon^{1/2}\left(\frac{\ell_\parallel}{c_A}\right)^{3/2}, \tag{C.4}$$

which is the length scale below which the turbulence along a cascade becomes *strong*. The characteristic time for energy transport through the cascade in $k_\perp$ should also be $\approx \ell_\perp/\delta u_\perp$, so that equating this with the CB conjecture yields

$$E(k_\perp) \sim \varepsilon^{2/3} k_\perp^{-5/3}. \tag{C.5}$$

This is Kolmogorov's spectrum recovered but only for $k_\perp$ not k. Parallel to $\mathbf{B}_0$, CB requires that $\mathbf{B}_0$ enter the phenomenon only through setting the scale $\ell_\parallel = \tau_{nl} c_A$, simply the distance traveled by an Alfvén wave in time τ_{nl}, which is the time needed for an eddy to break up non-linearly. This is arguably the maximum distance over which any perturbation can remain correlated. Then

$$k_\parallel \lesssim \left(\frac{\varepsilon^{1/3} k_\perp^{-5/3}}{c_A}\right) k_\perp \tag{C.6}$$

so that the *anisotropy is also scale dependent* with the maximum value of $k_\parallel \ll k_\perp$. The smallest scales develop only perpendicular to $\mathbf{B}_0$. An example of turbulent structures obtained from RMHD calculations is shown in Fig. C.2. The form of Eq. (C.6) is not, however, universally accepted.

Further developments relied on analytical and numerical work on *reduced MHD* (Strauss, 1976), an incompressible prescription solving for non-linear packets of Alfvén waves, parallel and anti-parallel to a strong field $\mathbf{B}_0$ with perturbations described by $k_\parallel$ and $k_\perp$ ($k_\parallel \ll k_\perp$). The solutions are decoupled from compressive modes, thus providing a minimal mathematical description of non-linearities likely to be present close to equilibrium states in the corona. Information in RMHD propagates along $\mathbf{B}_0$ at speed c_A, thus $\ell_\parallel \sim c_A \tau_{nl}$ represents the largest parallel length of an eddy with lifetime τ_{nl}.

In HD turbulence, the system must conserve energy, but in RMHD the system has two invariants, the *total energy* and the *cross-helicity*:

$$\frac{\langle |\mathbf{u}_\perp|^2 + |\mathbf{b}_\perp|^2 \rangle}{2} \quad \text{and} \quad \langle \mathbf{u}_\perp \cdot \mathbf{b}_\perp \rangle \tag{C.7}$$

respectively. The latter is a measure of the energy imbalance for Alfvén waves propagating parallel and anti-parallel to $\mathbf{B}_0$. Naively, we might expect cross-helicity to be statistically near zero for a loop excited by motions at two footpoints- and a significant fraction of the energy when fields are driven by just one footpoint motion (perhaps in coronal holes). These limits are of *balanced* and *unbalanced* MHD turbulence respectively.

The CB argument can be justified considering an ensemble of high-frequency Alfvén waves which is weakly excited ($\tau_{nl} \gg \ell_\parallel / c_A$, i.e. far from critical balance). The non-linear cascade will develop smaller scales until the critical balance occurs at $\tau_{nl} \sim \ell_\parallel / c_A$, whence it will be in the strong-turbulence (non-linear) dynamical regime. The energy spectra in the parallel directions "+" and "-" along $\mathbf{B}_0$, are (cf. Eq. C.5)

$$E^\pm(k_\parallel) \sim \frac{\varepsilon^\pm}{c_A} k_\parallel^{-2}, \tag{C.8}$$

with $\varepsilon^\pm$ are separate energy spectra for the two directions along $\mathbf{B}_0$. This result is deemed by Schekochihin (2022) to be a simple, robust result in MHD turbulence, compared with the equivalent wavenumber spectra perpendicular to the field (Eq. (C.5)). In extending the latter to include imbalanced turbulence (i.e. dependence on the $\pm$ superscript), Schekochihin (2022) shows that

$$k_\parallel^\pm \sim \left(\frac{\tilde{\varepsilon}^{\mp\,1/3} k_\perp^{-5/3}}{c_A} \right) k_\perp, \qquad \tilde{\varepsilon}^\pm \equiv \frac{(\varepsilon^\pm)^2}{\varepsilon^\mp} \tag{C.9}$$

which is the scale- and $\pm$ direction- dependent version of (C.6). Note the weak (1/3 power) dependence of $k_\parallel^\mp$ on $\tilde{\varepsilon}$, thereby reducing differences in the parallel wavenumbers of imbalanced MHD turbulcnce ($\varepsilon^+ \neq \varepsilon^-$). These relations have been shown to be in qualitative agreement both in RMHD calculations and for measurements in the solar wind.

Such results are important in exploring the role of incompressible turbulence in heating the corona (even imbalanced turbulence where motions are generated from one footpoint): potential observables include $\mathbf{B}_0$ and the kinetic power in plasma motions $\mathbf{u}(k_\parallel)$ and $\mathbf{u}(k_\perp)$.[1] The condition of applicability $|\mathbf{B}_0(\mathbf{r})| \gg |\mathbf{b}|$ immediately suggests that we will never sense remotely components of $\mathbf{b}$, because crude line-of-sight-averaged measurements of $\mathbf{B}_0$ are only now becoming feasible on any regular basis (Paraschiv and Judge, 2022). Instead, we might use statistical averages of line widths ξ_0 which are observed to be $\ll$ estimates of c_A, to probe properties of the kinetic energy $E^\pm(k_\parallel)$ and $E^\pm(k_\perp)$, with accompanying observations of the direction of $\mathbf{B}_0$.

The next advancements in understanding arose from numerical experiments in RHMD beginning in 2001, whose power spectra suggested a $k_\perp^{-3/2}$ rather than the $k_\perp^{-5/3}$ scaling. Interpretation of the calculations led to further departures in isotropy in the "azimuthal" plane, i.e. perpendicular to $\mathbf{B}_0$. Nonlinear effects were postulated as leading to dynamic (i.e. short-lived) *alignment* between $\mathbf{u}_\perp$ and $\mathbf{b}_\perp$, breaking previously assumed cylindrical symmetry around $\mathbf{B}_0$. The net effect is to increase the interaction time over the isotropic value $\tau_{nl} = \lambda/\delta u_\perp$ from Goldreich and Sridhar (1995) to the anisotropic times $\tau_{nl}^\pm$ by a factor of $(\sin\theta_\lambda)^{-1}$, where θ_λ is the ratio of the isotropic size of the "eddies" λ (without alignment) to the length of the eddy along a given aligned direction, determined by the dynamical interaction between two, oppositely directed Alfvén fluctuations. Strong alignments correspond to longer interaction times, because aligned features by definition have smaller gradients, and weaker interactions. These tend to shear the interacting waves in order to reduce the cross-$\mathbf{B}_0$ gradients in turbulent speed and magnetic field (Chandran et al., 2015). Boldyrev (2006) estimated these parameters

$$E(k_\perp) \sim (\varepsilon c_A)^{1/2} k_\perp^{-3/2}, \quad \text{with} \tag{C.10}$$

$$k_\parallel \sim \left(\frac{\varepsilon k_\perp^{-3}}{c_A^3}\right)^{1/2} k_\perp. \tag{C.11}$$

for incompressible motions and balanced turbulence. This turbulent spectrum for $k_\perp$ almost returns to the isotropic IK form of Eq. (C.1), and the parallel $k_\parallel$ can be compared with Eq. (C.6). Beresnyak (2011) threw a wrench into the Boldyrev work, pointing to an inconsistency in the ordering of terms in the RHMD formulation itself, and also in the embedded symmetries in the rescaling of variables that must be respected for RMHD to be applied. Beresnyak predicted that alignment disappears at asymptotically small scales. The jury remains out on these delicate but important

[1] However, to observe scale-dependent anisotropy, one must measure all parallel correlations along $\mathbf{B}_0 + \mathbf{b}$ not just $\mathbf{B}_0$, which will be problematic if $|\mathbf{b}|$ is not small compared to $\mathbf{B}_0$.

issues (compare the scalings of Eqs. (C.10) and (C.11) with Eqs. (C.5), (C.6), and (C.8)). Schekochihin points to the need for any heuristic picture to respect re-scaling symmetries in the RMHD equations, i.e. invariance of the form of the equations when scaled by a minimal set of scaling factors (one for energy, one for length).

Between 2015 and 2017, significant advances arose through recognition of the role of *intermittent* structures (i.e. those rarities which have a large effect on transport) within the above phenomenology (Chandran et al., 2015; Mallet et al., 2015; Mallet and Schekochihin, 2017). These authors emphasized the importance of distributions of the occurrence rate of turbulent structures in 3D, both parallel to ($\ell_{\|}$) and transverse to ($\lambda,\ \xi$) the guide field $\mathbf{B}_0$ (see Fig. C.1). The arguments and numerical work exhibit qualitative differences from earlier scalings, for instance:

$$\tau_{\rm nl} \sim \left(\frac{\ell_{\|}}{\varepsilon c_A}\right)^{1/4} \lambda^{1/2}, \tag{C.12}$$

greatly in excess of the value of $\tau_{\rm nl} = \ell_{\|}/c_A$ for the critically balanced G95 case. The preferential geometric alignment associated with the shearing interaction of oppositely directed Alfvén waves is evident in the extended dimensions of ovals projected onto the planes along the $\ell_{\|}$ and ξ dimensions in Fig. C.1. The

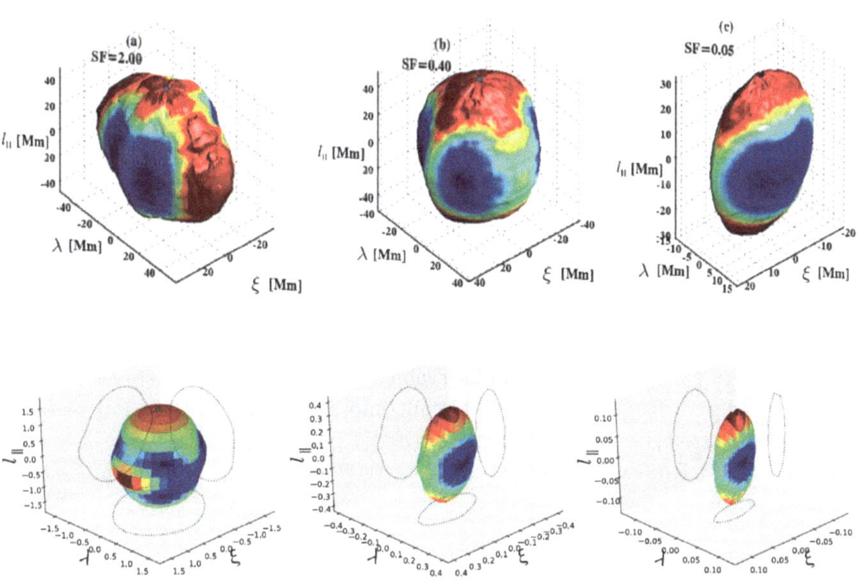

Fig. C.1 Locally 3D-anisotropic structures in the solar wind (upper row) and numerical simulations (lower row). The colored iso-surfaces show the local shapes of turbulent "eddies" in a representation where the lengths of $l_{\|}$ ("z-axes") have been shrunk dramatically to highlight the perpendicular scales (ξ, the eddy length along the aligned axis, and λ, the non-aligned component). From Schekochihin (2022)

alignment increases from large scales (left column) to smaller scales (middle and right column).

The rapidly forming intermittent, aligned structures (Fig. C.1) are current sheets (mentioned in Sect. 4.9), and thus if they approach the conditions for instability will be subject to rapid tearing (Loureiro et al., 2007). In 2017, two independent groups addressed the question of at what scale the aligned cascade must be disrupted by these processes (Uzdensky and Loureiro, 2016; Mallet et al., 2017; Loureiro and Boldyrev, 2017). But awkward questions remain concerning tearing: could velocity field gradients (local shear) suppress plasmoid instabilities? Is the theory of plasmoid growth developed really appropriate for aligned cascade structures (Fig. C.1)? Given that $\tau_{\rm nl}$ is so small, Schekochihin (2022) argues that shear effects should be small, then when the tearing growth rate $\gamma \sim 1/\tau_{\rm nl}$ (compare this with Eq. (4.17)), the cascade is disrupted. The disruption scale, found by equating the length scale λ_D for which $\tau_{\rm nl} = \lambda^2/\eta$, is

$$\lambda_{\rm D} \sim \left(\frac{\rm Rm^*}{(1+{\rm Pm})^{10}}\right)^{2/21} \lambda_\eta, \qquad \lambda_\eta \equiv \eta^{2/3}\left(\frac{L_\parallel}{\varepsilon c_A}\right)^{1/6}, \tag{C.13}$$

where Pm is the Prandtl number (ν_0/η) and $\rm Rm^*$ is the magnetic Reynolds number evaluated at the CB scale, $\lambda_{CB} = \varepsilon^{1/2}\left(L_\parallel/c_A\right)^{3/2}$. Here $L_\parallel$ is the outer scale (length of the structure). Then, with $\varepsilon = \delta u^2\left(c_A/L_\parallel\right)$:

$${\rm Rm^*} = \frac{\lambda_{CB}\delta u}{\eta} = \frac{\varepsilon}{\eta}\left(\frac{L_\parallel}{c_A}\right)^2, \tag{C.14}$$

Note the very weak dependence of $\lambda_{\rm D}$ on the Reynolds number, this means that many plasmas will have $\lambda_{\rm D} \lesssim 10\lambda_\eta$. In the quiet solar corona we can estimate the order of events, noting that

$$\epsilon \sim \frac{\delta u^2 c_A}{L_\parallel} \sim \frac{\xi_0^2 c_A}{L_\parallel}, \qquad \Longrightarrow \qquad {\rm Rm^*} \sim \left(\frac{\xi_0}{c_A}\right)^2 S \tag{C.15}$$

using values of ξ_0, c_A and $L_\parallel$ from observables (Table 1.2). Then[2] $\lambda_{CB} \sim 800$ km, $\varepsilon \sim 10^7$ W kg^{-1}, $\rm Rm^* \sim 2.8\ 10^{10}$. For incompressible fluctuations $\rm Pm \sim 0.14$, $\lambda_\eta \sim 0.3$ meters, $\lambda_D \sim 3$ m. Similarly for active conditions, we have $\lambda_{CB} \sim 700$ km, $\varepsilon \sim 7\ 10^7$, $\rm Rm^* \sim 2\ 10^{10}$. Then $\lambda_\eta \sim 0.04$ meters, $\lambda_D \sim 0.25$ m. In both

[2] The heating rate of $\varepsilon \sim 10^7$ W kg^{-1} corresponds to 10^{11} erg g^{-1} s^{-1}, exactly the number arrived at by Anderson and Athay (1989a) from observations of coronal energy losses and geometry. However, this means only that the ratio c_A/L adopted for the quiet Sun is what is needed to maintain the energy losses for turbulence with an observed rms amplitude of ξ_0 in the quiet corona, without reference to dissipation. In other words, this is merely another statement that "there is sufficient energy in Alfvén waves with $\xi_0 \sim 20$ km s^{-1} to power the corona" (Sect. 1.5), a statement offering no advancement of existing knowledge.

cases there is only about a decade in scale over which plasmoid formation might dominate, on scales of meters and less.

The important conclusion here is that in the quiet and active corona, the aligned, intermittent cascade extends down to scales well below mean free paths of 10^5 meters, the incompressible cascades would approach disruption scales of meters which, while well above resistive and viscous dissipation scales, are less than to ion inertial lengths. Therefore MHD instabilities might not occur before the aligned MHD cascade is terminated by collisionless effects, including fast (but not plasmoid-driven) reconnection. But as stressed by Schekochihin (2022), this theory is hotly debated and at present computer resources are too limited to perform required calculations. But if true, this amounts to a considerable simplification in the quest for mechanisms leading to dissipation as heat to for the solar corona. Schekochihin (2022) devotes almost the whole of his chapter 7 to tearing-mediated turbulence ($\lambda_D \gg \lambda_\eta$ and the ion inertial length), even asking if it might be a falsifiable theory (as usual, the answer is "maybe"). He concludes that

> Pending all this validation and verification, the tearing-mediated cascade remains a beautiful fantasy

Further, the emergent power spectra that might reveal the underlying dynamics (ideal/aligned or tearing-dominated) may also be complicated by the influence of the reconnection that follows plasmoid growth.

In solar and heliospheric physics, "reconnection" has tended to become synonymous with a fairly restricted set of ideas. Large plasmas, in a regime of extremely high Lundquist numbers $S \gtrsim 10^{12}$, initially evolve ideally into states that live long enough to form "X-points", "separator surfaces", where ideal current sheets form. Such sheets are ultimately unstable to reconnection following classical tearing mode, plasmoid instabilities (Sect. 5.11) given a small but finite resistive, viscous or kinetic route to dissipate the magnetic free energy. In solar physics these phenomena are regularly observed only on large scales ($\gtrsim$ 1 Mm) by witnessing the changing morphology of hot plasma frozen into bundles of magnetic flux, for which, nevertheless, $S \gg 10^4$. The free energy is released via "fast reconnection" (with a rate independent of S) that is sometimes triggered only above large thresholds sufficient to re-configure the global magnetic topology, and explain such phenomena as flares and CMEs, with energies between $10^{17} - 10^{25}$ J. Associated with these events are high energy particles as evidenced by hard X-ray and radio emissions. One envisages changes in large-scale topology (such as in the illustrations of Kulsrud, 2011).

However, above kinetic scales ($\sim$ 10 m) the systems are scale-invariant, following almost ideal MHD. Thus, in contrast to much of the solar literature, much of the discussion of reconnection in the MHD turbulence arena also includes scales down to resistive or viscous dissipation lengths, which themselves lie below particle mean free paths, below even scales of ion inertial lengths where kinetic physics kicks in. The MHD turbulence community seeks to understand, as one example, the reasons *why* such large systems must reconnect at a fast rate, of order $0.1c_A$, independent of S. Recently, Jafari et al. (2021) summarized this area as follows:

> At any scale ℓ in the turbulent inertial range, the magnetic field can be divided up into a large-scale component and a small-scale, high spatial frequency component which undergoes magnetic reversals. Such local reconnections, i.e., on any inertial scale ℓ, seem to be an inseparable part of MHD turbulence, whose collective outcome can lead to global reconnection with a rate independent of the small-scale physics dominant at dissipative scales. ... this picture, known as stochastic reconnection, is intimately related to nanoflare theory.

We are thus reminded of the conceptual picture represented by Eq. (2.25) of Sect. 1:

$$\mathbf{B}(\mathbf{r}, \mathbf{t}) = \mathbf{B}_0(\mathbf{r}) + \mathbf{b}(\mathbf{r}, t) \tag{C.16}$$

where $\mathbf{B}_0$ is fixed on scales above those smaller dynamical scales associated with a rapidly varying and reconnecting $\mathbf{b}(\mathbf{r}, t)$. This picture is quite different from the *in-situ* magnetic field measurements of large-scale "switchback" variations in the solar wind, obtained with the Parker Solar Probe (Bale et al., 2019), measured as low as $13R_\odot$ at the time of writing (May 2023, N. Raouafi, personal communication).

Broadly speaking, much of solar physics and space weather-related research concerns itself with observational changes in non-potential fields $\mathbf{B}_0$ which, before large-scale reconnection occurs, can be considered as constant as far as the dynamics of $\mathbf{b}(\mathbf{r}, t)$ is concerned. Again, because of scale invariance there is no difference between reconnection of $\mathbf{B}(\mathbf{r}, t)$, $\mathbf{B}_0$ and $\mathbf{b}(\mathbf{r}, t)$, it is all reconnection. However, the distinction helps the authors point to potential areas where coronal heating has been related to reconnection on observable scales of $\mathbf{B}_0$. This may be the case in some regions of activity through subsequent secondary processes of dissipation through shocks or collapse of smaller-scale current sheets such that the processes discussed above (Sect. 3.5). But we stress that the bulk of the volume of the quiet solar corona has no or little evidence that anything impulsive is involved, nor is there evidence of changes in morphology. Our point here is that the idea that coronal heating is a cascade of self-similar reconnection events extending downwards from observed scales of flares to nano-flares and below, is an *assumption* that is supported only by (sometimes discrepant) statistical analyses of large-scale observations. It begs the question of the final part of the journey of the large-scale injected energy to dissipation as heat. Further, reconnection is a phenomenon that is yet to be fully understood, and so it forms a poor basis on which to claim that related problems are essentially "understood". It is also not even necessary for coronal heating, even though it appears to be to explain energy storage and release as flares and CMEs. Indeed, there are ways to dissipate the free energy associated with $\mathbf{b}(\mathbf{r}, t)$ without it, such as wave damping resulting from large gradients in $\mathbf{u}$ and $\mathbf{b}$ arising through resonant absorption and phase mixing.

Returning to the ideas of ideal, aligned MHD structures modulated by plasmoid formation, we must note that three dimensional numerical work tends not to show the orderly procession of plasmoids expected above, but instead more like strips of vigorous MHD turbulence characterized by elongated structures parallel to $\mathbf{B}_0$, illustrated in Fig. C.2 (from Huang and Bhattacharjee, 2016). The idea

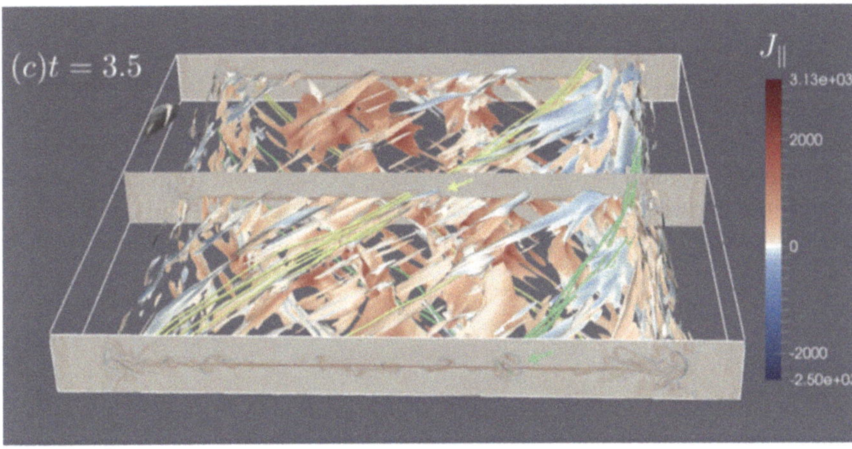

Fig. C.2 One time step is shown from a 3D numerical experiment in which plasmoid instabilities have occurred. The appearance is more like sheet-like MHD turbulence than processions of flux-rope like plasmoids. From Huang and Bhattacharjee (2016)

of "stochastic reconnection" has been developed parallel to the work mentioned above (Lazarian and Vishniac, 1999; Lazarian et al., 2020). Interestingly, the latter review points to different perceptions of reconnection between astrophysical, solar, magnetospheric, numerical and analytical MHD practitioners, and argues for a more coherent approach, based upon the inevitability of turbulent reconnection and flux-freezing violation (on scales ℓ with Rm $\sim$ 1) in turbulent fluids. The basic new idea is that reconnection is enhanced not through plasma effects (e.g. plasmoid dynamics), instead it is enhanced through the natural wandering of magnetic lines of force undergoing turbulence and heating. The concept of a fixed $\mathbf{B}_0$ is rejected as misleading. Instead the authors argue that the geometric and dynamically evolving wandering of field lines that results from continual turbulence-driven reconnections means that the concept of a fixed $\mathbf{B}_0(\mathbf{r})$ misses an effective way to transport, tangle and thereby reconnect magnetic fields perpendicular to an average $\mathbf{B}_0$ in a stochastic fashion. In this picture field lines wander in response to turbulent motions, they quickly reconnect, causing yet more turbulent motion, and so forth. The notion of field-line diffusion via kinetic diffusivities η ($\propto$ thermal speed times mean free path, of order unity in the solar corona) is conceptually replaced by a far larger "eddy diffusivity" familiar in hydrodynamics, of order $\Delta\delta_u$ where length scale Δ is of order the correlation length of field wandering across $\mathbf{B}_0$. Lazarian et al. (2020) show how Δ can be estimated, yielding $\Delta\delta_u \gg \eta$ (Table 2.1).

In geometrically opening up reconnection outflows in the perpendicular plane, reconnection is fast (independent of S), and an intrinsic part of the turbulent dynamics on all scales. This school of thought effectively yields the GS95 scalings, and has considerable support from numerical experiments.

In both pictures (ideal alignment and stochastic reconnection) the turbulent processes leading to irreversible heating all occur in the plane perpendicular to $\mathbf{B}_0$.

Again, the result is that ideal MHD sows the seeds of its own destruction, without having to deal with any physics at the diffusive scales.

Finally, it seems clear that at least one solar observable might reveal something about whether cascades terminated by plasmoid formation can be refuted. If plasmoids form they will present a very different signature of linearly polarized light emitted by coronal ions. If spatially unresolved, we should see almost no linear polarization because the plasma is threaded by closed field lines when oriented in the plane-of-the sky. Evidence in favor of this picture has already been found (French et al., 2019).

Appendix D
Viscous Dissipation in MHD

The viscosity of a magnetofluid is represented by a tensor possessing a cylindrical symmetry about the magnetic field $\mathbf{B}_0 = (0, 0, B_z)$ (Braginskii, 1965), dominated by differential ion motions. Diagonal tensor elements describe compressions ($\partial_i u_i$, $i = 0 \ldots 2$) with a scalar coefficient, dominated by protons, of $\eta_0 \approx n_p k T_p \tau_p$ (Table 1.2), a similar magnitude to the unmagnetized case. The off-diagonal components involve motion across the field $\mathbf{B}_0$ and scale as $\eta_0(\omega\tau)^{-2}$ and $\eta_0(\omega\tau)^{-1}$. The first case applies to fluid velocity gradients such as $\partial_y u_z$ which measure velocity variations across the $z-$directed magnetic field. Momentum transport in these directions is inhibited to an ion gyroradius r, and so momentum transport is inhibited by the factor $(\lambda/r = \omega\tau)^2$. The second involves changes in say u_x along y and are perpendicular to $\mathbf{u}$ so do not do any work (Braginskii, 1965). In the corona $\omega\tau \gg 1$ for electrons and ions.

The result is that viscosity can play one of two roles in dissipating gradients in the fluid velocity depending on the so-called rate-of-strain tensor,

$$W_{\alpha\beta} = \frac{\partial u_\alpha}{\partial x_\beta} + \frac{\partial u_\beta}{\partial x_\alpha} - \frac{2}{3}\,\delta_{\alpha\beta}\,\nabla\cdot\mathbf{u}, \tag{D.1}$$

an explicit function of the dynamics through $\mathbf{u}(\mathbf{r}, t)$. The viscous dissipation of the kinetic energy of fluid motions occurs at the rate (Braginskii, 1965)

$$Q_0 = \frac{3}{4}\eta_0 W_{zz}^2 = \frac{\eta_0}{3}\left(\frac{\partial u_x}{\partial x} + \frac{\partial u_y}{\partial y} - 2\frac{\partial u_z}{\partial z}\right)^2 \quad \mathrm{W\ m^{-3}} \tag{D.2}$$

which depends only on the diagonal components of W. When the motions are purely incompressible ($\nabla \cdot \mathbf{u} = 0$), then

$$Q_0 = 3\eta_0 \left(\frac{\partial u_z}{\partial z}\right)^2 \tag{D.3}$$

P. Judge, J. A. Ionson, *The Problem of Coronal Heating*, Astrophysics and Space Science Library 470, https://doi.org/10.1007/978-3-031-46273-3

If this term is zero, as might occur in the absence of slow or fast modes in strong magnetic fields, the next highest terms also involve off-diagonal components (shear motions) which are a factor of $(\omega\tau)^2$ slower (see Table 4.1):

$$Q_1 \approx \frac{1}{2}\frac{\eta_0}{\omega^2\tau^2}\left(W_{xx}^2 + W_{yy}^2 + 2W_{xy}^2 + 4W_{xz}^2 + 4W_{yz}^2\right), \tag{D.4}$$

These properties have been recognized in the heating problem, for example (Davila, 1987) showed that resonant absorption forms dissipative waves on scales determined by viscosity of a thickness 0.3 km for incompressible modes, to 250 km for compressible modes where $Q \neq 0$ in Eq. (D.2). Curiously, 250 km is the smallest thickness of bright "coronal strands" found in the highest resolution images of the corona (HiC data, Rachmeler et al., 2019). But these scales are perilously close to the angular resolution of the instrument. Dere (1989) found qualitatively similar results, from his analysis of spatial power spectra from transition region lines.

Davila's work is one example of an MHD process which generates dissipative scales, but as we have seen (Sects. 4.7 and 3.5) this is but one way in which small scales spontaneously arise from large scales in near-ideal MHD. So, *the nature of the motions induced by various energy transport mechanisms on MHD scales determines whether kinetic viscosity is sufficient to dissipate the energy, or whether we must examine the far more complex and less complete theories on scales of the ion inertial length.*

Two arguments might support a heating mechanism dominated by ion viscosity. Heating in the quiet Sun through the acceleration of energetic particles can be ruled out (Sect. 4.5, Paterson et al., 2023). A large viscosity would quench the electromagnetic free energy well before kinetic processes might accelerate particles. The second argument simply notes that in any models in which large-scale input energy is transported towards dissipative scales, be it via waves or current sheets or turbulence, compressible motions seem inevitable as a result of non-linear interactions and instabilities. Therefore it appears that fast viscous dissipation is also inevitable. Ideas to refute these arguments are discussed in Chap. 6.

If we pursue viscous dissipation a little further, we can compare heating rates required with some basic observational facts about the corona. Equation (D.2) can be written

$$Q = n_p k T_p \tau_p \frac{u^2}{\ell^2} \ \mathrm{W\,m^{-3}} \tag{D.5}$$

for characteristic speeds u and lengths ℓ. Anderson and Athay (1989a) showed that the quiet corona requires $Q_{QS} \sim 10^{-5}$ W m^{-3}. The rms fluctuation of non-thermal speeds ξ_0 in the red and green coronal forbidden lines is 20 km s^{-1} (Singh et al., 2006), then we find, with $Q = Q_{QS}$ and $u = \xi_0$,

$$\ell \approx \sqrt{\frac{n_p k T_p \tau_p}{Q}}\,\xi_0 \tag{D.6}$$

$$= 1.4\ 10^{-9}\, T_p^{5/4}\, Q^{-1/2}\, \xi_0 \tag{D.7}$$

$$\approx 100 - 300 \text{ km (Quiet and active Sun)} \tag{D.8}$$

using parameters from Table 1.2 and Sect. 1.8.2, and linewidths in active regions twice those of the quiet Sun. This implies that *fine scales beneath 100 km would be erased by viscosity*, something close to being testable with modern new instrumentation (see also Dere, 1989). In the cores of active regions, T_p sometimes can approach 10^7 K. If we adopt the scaling law of Eq. (2.24), with $Q = \mathcal{F}/L \propto T_e^{7/2} L^{5/2}$ we have for a loop length L, assuming first that $T_e \approx T_p \approx T$,

$$\ell \propto T^{-1/2} L^{-5/4} \xi_0, \tag{D.9}$$

implying that short loops would appear less structured than longer loops, but that loops at higher (ion) temperatures would appear sharper for the same length L. Hot plasma ($T_e > 5\ 10^6$ K) in the cores of active regions have broader lines ($\xi_0 \approx 60$ km s^{-1}, see, e.g., SUMER data presented by Parenti et al., 2017). Observed behavior (Sect. 5.3) shows that hotter loops have less structure than cool loops, without clear dependence on loop length L. However, electron-ion relaxation times vary from 80s at 10^6 K, varying as $T_p^{3/2}/n_p$ (Table 1.2). It appears premature to rule out viscous heating on this basis, since the relaxation times may exceed dynamical times, rendering Eq. (2.24) inapplicable. However, this discussion does indicate the kind of analysis which might serve to refute certain pictures.

References

Allen, C.W.: Astrophysical Quantities. University London, Athlone Press (1973)

Anderson, L.S., Athay, R.G.: Chromospheric and coronal heating. Astrophys. J. **336**, 1089–1091 (1989a)

Anderson, L.S., Athay, R.G.: Model solar chromosphere with prescribed heating. Astrophys. J. **346**, 1010 (1989b)

Athay, R.G.: The Solar Chromosphere and Corona: Quiet Sun. Reidel, Dordrecht (1976)

Ayres, T.R.: Chromospheric Scaling Laws, Width-Luminosity Variations and the Wilson-Bappu Effect. Astrophys. J. **228**, 509–520 (1979)

Bale, S.D., et al.: Highly structured slow solar wind emerging from an equatorial coronal hole. Nature **576**, 237–242 (2019). https://doi.org/10.1038/s41586-019-1818-7

Beresnyak, A.: Spectral slope and kolmogorov constant of MHD turbulence. Phys. Rev. Lett. **106**, 075001 (2011). https://doi.org/10.1103/PhysRevLett.106.075001

Biermann, L., Bruggencate, P.: ÜBer Die Ursachen Der Hohen Temperatur Der Sonnenkorona Nebst Einer Bemerkung ÜBer Das Nachthimmellicht. Veroeffentlichungen der Universitaets-Sternwarte zu Goettingen **5**, 223–228 (1947)

Boldyrev, S.: Spectrum of magnetohydrodynamic turbulence. Phys. Rev. Lett. **96**, 115002 (2006)

Braginskii, S.I.: Transport processes in a plasma. Rev. Plasma Phys. **1**, 205–311 (1965)

Burton, W.M., Ridgeley, A., Wilson, R.: The ultraviolet emission spectrum of the solar chromosphere and corona. Month. Not. R. Astron. Soc. **135**, 207–223 (1967)

Chandran, B.D.G., Schekochihin, A.A., Mallet, A.: Intermittency and alignment in strong RMHD turbulence. Astrophys. J. **807**, 39 (2015)

Chapman, G.A.: Active Regions from the photosphere to the chromosphere. In Orrall, F.Q. (ed.) Solar Active Regions: A Monograph from Skylab Solar Workshop III, pp. 43–82 (1981)

Coyner, A.J., Davila, J.M.: Determination of non-thermal velocity distributions from SERTS linewidth observations. Astrophys. J. **742**, 115 (2011). https://doi.org/10.1088/0004-637X/742/2/115

Culhane, J.L., et al.: The EUV imaging spectrometer for Hinode. Sol. Phys. **243**, 19–61 (2007). https://doi.org/10.1007/s01007-007-0293-1

Davila, J.M.: Heating of the solar corona by the resonant absorption of Alfvén waves. Astrophys. J. **317**, 514–521 (1987)

de Pontieu, B., et al.: Probing the physics of the solar atmosphere with the multi-slit solar explorer (MUSE). I. Coronal heating. Astrophys. J. **926**, 52 (2022). https://doi.org/10.3847/1538-4357/ac4222

Del Zanna, G., Mason, H.E.: Solar UV and X-ray spectral diagnostics. Living Rev. Solar Phys. **15**, 5 (2018). https://doi.org/10.1007/s41116-018-0015-3

Dere, K.P.: Turbulent power and dissipation in the solar transition zone. Astrophys. J. **340**, 599 (1989). https://doi.org/10.1086/167420

French, R.J., et al.: Spectropolarimetric insight into plasma sheet dynamics of a solar flare. Astrophys. J. Lett. **887**, L34, (2019). https://doi.org/10.3847/2041-8213/ab5d34

Goldreich, P., Sridhar, S.: Toward a theory of interstellar turbulence. II. Strong alfvenic turbulence. Astrophys. J. **438**, 763 (1995)

Hara, H., Ichimoto, K.: Astrophys. J. **513**, 969 (1999)

Huang, Y.M., Bhattacharjee, A.: Turbulent magnetohydrodynamic reconnection mediated by the plasmoid instability. Astrophys. J. **818**, 20 (2016). https://doi.org/10.3847/0004-637X/818/1/20

Iroshnikov, P.S.: Turbulence of a conducting fluid in a strong magnetic field. Sov. Astron. **7**, 566 (1964)

Jafari, A., Vishniac, E.T., Xu, S.: Nanoflare theory revisited. Astrophys. J. **906**, 109 (2021). https://doi.org/10.3847/1538-4357/abca47

Jordan, C.: The structure and energy balance of solar active regions. Philos. Trans. R. Soc. Lond. **A281** , 391 (1976)

Jordan, C.: Helium line emission - its relation to atmospheric structure. Philos. Trans. R. Soc. Lond. A **297**, 541–554 (1980)

Jordan, C.: Ultraviolet Fe II emission from cool star chromospheres. In: Viotti, A.V.R., Friedjung, M. (eds.) Physics of Formation of Fe II Lines Outside LTE, vol. 223. Dordrecht, Reidel (1988)

Jordan, C., Mendoza, B., Gill, R.S.: Observational constraints on heating processes, pp. 133–136. ESA-SP 220 (1984)

Judge, P., Casini, R., Paraschiv, A.R.: On single-point inversions of magnetic dipole lines in the corona. Astrophys. J. **912**, 1, 18 (2021). https://doi.org/10.3847/1538-4357/abebd8

Kolmogorov, A.: The local structure of turbulence in incompressible viscous fluid for very large Reynolds' numbers. Akad. Nauk SSSR Dokl. **30**, 301–305 (1941)

Kosugi, T., et al.: The Hinode (solar-B) mission: an overview. Sol. Phys. **243**, 3–17 (2007)

Kraichnan, R.H.: Inertial-Range spectrum of hydromagnetic turbulence. Phys. Fluids **8**, 1385–1387 (1965)

Krishna Prasad, S., Singh, J., Banerjee, D.: Variation of emission line width in mid- and high-latitude corona. Sol. Phys. **282**, 427–442 (2013). https://doi.org/10.1007/s11207-012-0160-010.48550/arXiv.1210.6434

Kulsrud, R.M.: Intuitive approach to magnetic reconnection. Phys. Plasmas **18**, 111201 (2011)

Lazarian, A., Vishniac, E.T.: Reconnection in a weakly stochastic field. Astrophys. J. **517**, 700–718 (1999). https://doi.org/10.1086/307233

Lazarian, A., et al.: 3D turbulent reconnection: theory, tests, and astrophysical implications. Phys. Plasmas **27**, 012305 (2020). https://doi.org/10.1063/1.5110603

Loureiro, N.F., Boldyrev, S.: Role of magnetic reconnection in magnetohydrodynamic turbulence. Phys. Rev. Lett. **118**, 245101 (2017). https://doi.org/10.1103/PhysRevLett.118.245101

Loureiro, N.F., Schekochihin, A.A., Cowley, S.C.: Instability of current sheets and formation of plasmoid chains. Phys. Plasmas **14**, 100703–100703 (2007). https://doi.org/10.1063/1.2783986

Mallet, A., Schekochihin, A.A.: A statistical model of three-dimensional anisotropy and intermittency in strong Alfvénic turbulence. Month. Not. R. Astron. Soc. **466**, 3918–3927 (2017). https://doi.org/10.1093/mnras/stw3251

Mallet, A., Schekochihin, A.A., Chandran, B.D.G.: Refined critical balance in strong Alfvénic turbulence. Month. Not. R. Astron. Soc. **449**, L77–L81 (2015)

Mallet, A., Schekochihin, A.A., Chandran, B.D.G.: Disruption of sheet-like structures in Alfvénic turbulence by magnetic reconnection. Month. Not. R. Astron. Soc. **468**, 4862–4871 (2017). https://doi.org/10.1093/mnras/stx670

Mariska, J.T.: The Solar Transition Region. Cambridge University Press, Cambridge (1992)

Osterbrock, D.E.: The heating of the solar chromosphere, plages, and corona by magnetohydrodynamic waves. Astrophys. J. **134**, 347 (1961)

Paraschiv, A.R., Judge, P.G.: Efficient and automated inversions of magnetically sensitive forbidden coronal lines: CLEDB - the coronal line emission database magnetic field inversion algorithm. Sol. Phys. **297**, 63 (2022). https://doi.org/10.1007/s11207-022-01996-5

Parenti, S., et al.: Spectroscopy of very hot plasma in non-flaring parts of a solar limb active region: spatial and temporal properties. Astrophys. J. **846**, 25 (2017). https://doi.org/10.3847/1538-4357/aa835f

Paterson, S., et al.: The first survey of quiet sun features observed in hard x-rays with NuSTAR. Sol. Phys. **298**, 3, 47 (2023). https://doi.org/10.1007/s11207-023-02135-4

Pietarila, A., Judge, P.G.: On the formation of the resonance lines of helium in the sun. Astrophys. J. **606**, 1239–1257 (2004)

Rachmeler, L.A., et al.: The high-resolution coronal imager, flight 2.1. Sol. Phys. **294**, 174 (2019)

Raghavendra Prasad, B. et al.: Visible emission line coronagraph on Aditya-L1. Curr. Sci. **113**(4), 613 (2017). https://doi.org/10.18520/cs/v113/i04/613-615

Raju, K.P., Sakurai, T., Ichimoto, K.: A spectroscopic study of the solar corona from norikura and SOHO data. In: Brekke, P., Fleck, B., Gurman, J.B. (eds.) Recent Insights into the Physics of the Sun and Heliosphere: Highlights from SOHO and other Space Missions, vol. 203, p. 488 (2001)

Schekochihin, A.A.: MHD turbulence: a biased review. J. Plasma Phys. **88**, 155880501 (2022). https://doi.org/10.1017/S0022377822000721

Singh, J., et al.: Spectroscopic studies of the solar corona II. Properties of green and red emission lines in open and closed coronal structures. Publ. Astron. Soc. Jpn. **54**, 793–806 (2002). https://doi.org/10.1093/pasj/54.5.793

Singh, J., Sakurai, T., Ichimoto, K.: Do the line widths of coronal emission lines increase with height above the limb? Astrophys. J. **639**, 475–483 (2006). https://doi.org/10.1086/499200

Singh, J., et al.: Exploring the outer emission corona spectroscopically by using visible emission line coronagraph (VELC) on board ADITYA-L1 mission. Adv. Space Res. **64**, 1455–1464 (2019). https://doi.org/10.1016/j.asr.2019.07.007

Strauss, H.R.: Nonlinear, three-dimensional magnetohydrodynamics of noncircular tokamaks. Phys. Fluids **19**, 134–140 (1976). https://doi.org/10.1063/1.861310

Thomas, R.J., Neupert, W.: Extreme ultraviolet spectrum of a solar active region from SERTS. Astrophys. J. Suppl. **91**, 461–482 (1994)

Uzdensky, D.A., Loureiro, N.F.: Magnetic reconnection onset via disruption of a forming current sheet by the tearing instability. Phys. Rev. Lett. **116**, 105003 (2016). https://doi.org/10.1103/PhysRevLett.116.105003

van Ballegooijen, A.A., Asgari-Targhi, M., Voss, A.: The heating of solar coronal loops by Alfvén wave turbulence. Astrophys. J. **849**, 46 (2017). https://doi.org/10.3847/1538-4357/aa9118

Vernazza, J.E., Avrett, E.H., Loeser, R.: Structure of the solar chromosphere. I. Basic computations and summary of the results. Astrophys. J. **184**, 605 (1973)

Vernazza, J.E., Avrett, E.H., Loeser, R.: Structure of the solar chromosphere. II. The photosphere and the temperature-minimum region. Astrophys. J. Suppl. Ser. **30**, 1 (1976)

Vernazza, J.E., Avrett, E.H., Loeser, R.: Structure of the chromosphere. III. Models of The EUV brightness components of the quiet sun. Astrophys. J. Suppl. Ser. **45**, 635–725 (1981)

Withbroe, G.L.: The temperature structure, mass and energy flow in the corona and inner solar wind. Astrophys. J. **325**, 442–467 (1988)

Withbroe, G.L., Noyes, R.W.: Mass and energy flow in the solar chromosphere and corona. Ann. Rev. Astron. Astrophys. **15**, 363–387 (1977)

Woolley, R.D.V.R., Allen, C.W.: The coronal emission spectrum. Month. Not. R. Astron. Soc. **108**, 292–305 (1948)

Index

P. Judge, J. A. Ionson, *The Problem of Coronal Heating*, Astrophysics and Space Science Library 470, https://doi.org/10.1007/978-3-031-46273-3